Lecture Notes in Applied and Computational Mechanics

Volume 26

Series Editors

Prof. Dr.-Ing. Friedrich Pfeiffer
Prof. Dr.-Ing. Peter Wriggers

Lecture Notes in Applied and Computational Mechanics

Edited by F. Pfeiffer and P. Wriggers

Vol. 26: Nowacki, J.P.
Static and Dynamic Coupled Fields in Bodies with Piezoeffects or Polarization Gradient
220 p. 2006 [3-540-31668-X]

Vol. 25: Chen C.-N.
Discrete Element Analysis Methods of Generic Differential Quadratures
294 p. 2006 [3-540-28947-X]

Vol. 24: Schenk C., Schuëller G. (Eds.)
Uncertainty Assessment of Large Finite Element Systems
400 p. 2005 [3-540-25343-2]

Vol. 23: Frémond M., Maceri F. (Eds.)
Mechanical Modelling and Computational Issues in Civil Engineering
400 p. 2005 [3-540-25567-3]

Vol. 22: Chang C.H.
Mechanics of Elastic Structures with Inclined Members: Analysis of Vibration, Buckling and Bending of X-Braced Frames and Conical Shells
190 p. 2004 [3-540-24384-4]

Vol. 21: Hinkelmann R.
Efficient Numerical Methods and Information-Processing Techniques for Modeling Hydroand Environmental Systems
305 p. 2005 [3-540-24146-9]

Vol. 20: Zohdi T.I., Wriggers P.
Introduction to Computational Micromechanics
196 p. 2005 [3-540-22820-9]

Vol. 19: McCallen R., Browand F., Ross J. (Eds.)
The Aerodynamics of Heavy Vehicles: Trucks, Buses, and Trains
567 p. 2004 [3-540-22088-7]

Vol. 18: Leine, R.I., Nijmeijer, H.
Dynamics and Bifurcations of Non-Smooth Mechanical Systems
236 p. 2004 [3-540-21987-0]

Vol. 17: Hurtado, J.E.
Structural Reliability: Statistical Learning Perspectives
257 p. 2004 [3-540-21963-3]

Vol. 16: Kienzler R., Altenbach H., Ott I. (Eds.)
Theories of Plates and Shells: Critical Review and New Applications
238 p. 2004 [3-540-20997-2]

Vol. 15: Dyszlewicz, J.
Micropolar Theory of Elasticity
356 p. 2004 [3-540-41835-0]

Vol. 14: Frémond M., Maceri F. (Eds.)
Novel Approaches in Civil Engineering
400 p. 2003 [3-540-41836-9]

Vol. 13: Kolymbas D. (Eds.)
Advanced Mathematical and Computational Geomechanics
315 p. 2003 [3-540-40547-X]

Vol. 12: Wendland W., Efendiev M. (Eds.)
Analysis and Simulation of Multifield Problems
381 p. 2003 [3-540-00696-6]

Vol. 11: Hutter K., Kirchner N. (Eds.)
Dynamic Response of Granular and Porous Materials under Large and Catastrophic Deformations
426 p. 2003 [3-540-00849-7]

Vol. 10: Hutter K., Baaser H. (Eds.)
Deformation and Failure in Metallic Materials
409 p. 2003 [3-540-00848-9]

Vol. 9: Skrzypek J., Ganczarski A.W. (Eds.)
Anisotropic Behaviour of Damaged Materials
366 p. 2003 [3-540-00437-8]

Vol. 8: Kowalski, S.J.
Thermomechanics of Drying Processes
365 p. 2003 [3-540-00412-2]

Vol. 7: Shlyannikov, V.N.
Elastic-Plastic Mixed-Mode Fracture Criteria and Parameters
246 p. 2002 [3-540-44316-9]

Vol. 6: Popp K., Schiehlen W. (Eds.)
System Dynamics and Long-Term Behaviour of Railway Vehicles, Track and Subgrade
488 p. 2002 [3-540-43892-0]

Vol. 5: Duddeck, F.M.E.
Fourier BEM: Generalization of Boundary Element Method by Fourier Transform
181 p. 2002 [3-540-43138-1]

Vol. 4: Yuan, H.
Numerical Assessments of Cracks in Elastic-Plastic Materials
311 p. 2002 [3-540-43336-8]

Vol. 3: Sextro, W.
Dynamical Contact Problems with Friction: Models, Experiments and Applications
159 p. 2002 [3-540-43023-7]

Static and Dynamic Coupled Fields in Bodies with Piezoeffects or Polarization Gradient

Jerzy Paweł Nowacki

Springer

Professor Jerzy Paweł Nowacki
Polish-Japanese Institute of Information
Technology
ul. Koszykowa 86
02-008 Warsaw, Poland
E-mail: jnowacki@pjwstk.edu.pl

With 24 Figures

ISSN print edition: 1613-7736
ISSN electronic edition: 1860-0816

ISBN 978-3-642-06855-3 e-ISBN 978-3-540-31670-1

Springer is a part of Springer Science+Business Media
springer.com

Printed in The Netherlands

Cover design: *design & production* GmbH, Heidelberg

Dedicated to Isabela

Contents

Introduction

The basic subject chosen for this book is directly associated with physics of piezoelectricity and some other electro-magneto-thermo-elastic coupling phenomena in solids. Fundamentals of physics of electromechanical interactions in dielectric bodies have been studied, both theoretically and experimentally, by a great number of researchers being reflected in multiple publications. It will be worthwhile to start our reference list from the old books by P. Curie [1] and W. Voigt [2]: the first discovered piezoelectricity experimentally (1880) and the second established its relation to the crystalline structure (1884). We shall not go to detailed history of this field of science. One can find it in many more recent monographs devoted, in part or totally, to linear piezoelectricity [3-9], to its nonlinear aspects [10-14] and to linear electromechanical coupling arising due to more delicate physical reasons and existing even in centrosymmetric media [15, 16].

The book is related to series of theoretical studies of electro-elastic fields in solids with piezoelectric coupling or polarization gradient. We shall consider three series of problems: i) some fundamental properties of line defects (dislocations and their four-dimensional extension) of arbitrary shape in infinite anisotropic media (for both types of coupling); ii) analytical derivation of electro-elastic fields of thermal inclusion in dielectrics with polarization gradient and of Geen's functions for a number of bounded piezoelectric anisotropic structures (such as an infinite strip or a layer on a substrate) with straight line sources in the bulk and/or at the surfaces; iii) theoretical analysis of some dynamic eigenwave fields in media with electro-elastic or magneto-elastic couplings. The motivation for these studies is related both to a fundamental character of the above problems and to their applicational aspect.

In recent years, due to technological applications of piezoelectric materials in multiple devices (resonators, filters, sensors, delay lines, etc.) the problem of electro-elastic fields produced in such materials by different sources have received considerable attention from researchers (see, for instance, [17-30]). The reason for such popularity is rather evident: piezoelectric devices of a new generation designed for functioning at very low signal levels call for materials and constructions of increasingly high quality and sensitivity. Such devices might be very dependent on the level of noise electric fields excited by various sources (e.g., by dislocations) in piezoelectric media due to electro-mechanical coupling. From this point of view, any progress in theoretical description of electro-elastic fields in piezoelectric structures, apart from its ordinary scientific aspect, might have a more wide address related to possible applications. And it is not accidental that most part of mentioned above publications is related to piezoelectric fields of line defects, in particular of dislocations, in different single-, bi- or multi-crystalline structures. Indeed, a dislocation is one of the most dangerous sources of piezoelectric fields, which dramatically increase their strength in its neighborhood.

It was Merten [31-34], who made the first theoretical approach to the problem of piezoelectric field generated by a straight dislocation in an infinite medium. Using the approximate theory (neglecting the piezoelectric stress and assuming the elastic isotropy), Merten has found the first analytical results for electro-elastic fields excited by a dislocation. It has been found that the distortion of elastic strain field caused by the inverse piezoelectric effect represents a correction of only a few percents to the primary elastic strain. However, in Merten's approximation the electric potential being sensitive to a direction was independent of the distance from the dislocation that looked to be unrealistic.

The next series of theoretical studies of the same problem has been accomplished by Kosevich, Pastur & Feldman [35], Saada [36] and Faivre & Saada [37]. In order to avoid Merten's approximations, they extended the analysis and generalized the exact plane-strain solution given by Eshelby, Read & Shockley [38] for a straight dislocation in anisotropic purely elastic medium, to include the complete piezoelectric coupling into the constitutive equations. The application of this exact method has shown that the electric potential due to the dislocation depends not only on a direction but also (logarithmically) on a distance. Following these fundamental considerations, the detailed calculations of the magnitudes and configurations of piezoelectric fields around rectilinear dislocations in crystals of the wuertzite structure have been made by Smirnova [39]. For example, it has been shown that a screw dislocation lying in the basal plane and dislocations oriented along the c axis are not piezoactive, i.e. do not excite any electric field. On the other hand, the electric field of an edge dislocation lying in the basal plane with a Burgers vector also belonging to the same plane has a considerable radial component. And this dislocation behaves like a positively charged filament. The further study of piezoelectric fields surrounding a static dislocation accomplished by Huston & Walker [40] was devoted to scrutinizing the previous results by means of computer calculations. However in this book we shall concentrate our attention basically on an analytical approach to the problem.

Specific character of theories related to piezoelectricity consists in principal impossibility of simplifications related to such a powerful approximation as assumption of isotropy of a medium. Indeed, piezoelectric effect is a purely anisotropic phenomenon, which cannot occur in media with a center of symmetry. Meanwhile a theoretical description of elastic fields caused by line defects (e.g., by dislocations) in anisotropic (even purely elastic) media looks to be extremely complicated, almost hopeless problem. Direct attempts to get explicit solutions lead to results only for very symmetric situations and even then final expressions turn out to be so cumbersome that their analytical analysis becomes practically impossible. The real progress came to this theory only after the recognition of advantages provided by an implicit approach, when solutions are expressed in terms of some parameters, which often cannot be found explicitly and need additional computing. This was the basic idea, which allowed solving many problems of anisotropic elasticity. The theory of dislocations in media of unrestricted anisotropy is almost completely based on an implicit description.

The principally important step here has been done by Stroh [41, 42] who developed his famous sextic formalism reducing the description of generalized plane strain in an arbitrary anisotropic medium to an eigenvalue problem, so that the 2D field of a straight dislocation was expressed in a very compact form in terms of eigenvalues and eigenvectors of the 6×6 Stroh matrix (see also [43]). Of course, these eigenvectors and eigenvalues are not known explicitly, but their computing is not a problem [44]. On the basis of the Stroh theory the so-called integral formalism was developed by Barnett and Lothe [45] (see also [46]), which in many cases has advantages over the eigenvector theory.

The other non-trivial implicit idea was related to using the field of a straight dislocation itself as a "brick" for building the more complex 3D dislocation fields. In particular, Indenbom & S.Orlov [47] have reduced the field of an arbitrary dislocation loop to a superposition of the fields of straight dislocations. And Lothe [48, 49] has proved that the self-action of a semi-infinite straight dislocation inclined to a planar free surface of an anisotropic half-space is directly determined by the orientational derivative of the line energy of a straight dislocation in the unbounded medium with the same elastic constants. These and many other beautiful theorems and a complete list of references can be found in the book edited and to large extent written by Indenbom and Lothe [50].

A separate story is associated with a description of the fields of straight dislocations in bounded anisotropic media. The Stroh approach has proved to be efficient also for 2D problems of this kind. In terms of the sextic formalism there were found the fields of straight dislocations in a half-space [51], in the structure of a layer on a substrate [52], and within an infinite strip [53, 54]. The mentioned results for anisotropic [53] (and possibly inhomogeneous [54]) strips became an extension of the previous theories for isotropic homogeneous strips [55-59].

As was shown by Jergan [60] (see also [61]), the Stroh theory can be modified for a description of 2D electro-elastic fields in piezoelectrics. In this case the formalism becomes eight-dimensional, so that the 6 (6 Stroh matrix is replaced by a similar 8×8 matrix. Later Alshits, Darinskii & Lothe [62] introduced a ten-dimensional modification of the Stroh method for media with coexisting piezoelectric, piezomagnetic and magneto-electric couplings. In both cases the extended Stroh formalism has proved to be very fruitful in the theory of surface acoustic waves [62-65] in such media.

Recently J.P.Nowacki, Alshits & Radowicz [27-30], combining the methods developed in [54, 60, 61], have solved a series of problems related to plain electro-elastic fields in bounded piezoelectric media with line sources in the bulk or at the surface. There have been found Green's functions describing 2D fields in piezoelectric plate [27, 28] and in piezoelectric layer-substrate structure [29, 30] of unrestricted anisotropy. Actually, these results represent an important extension of early papers [35-37], where the straight dislocation fields were found for an infinite piezoelectric body. The content of the series of publications [27-30] together with some other relevant results will form the Part II of the book.

In contrast to dislocations and other defects with their fields injurious for devices, electro-elastic and magneto-elastic waves in different structures represent a new more positive practical aspect of considered coupling effects. Such waves together with an ordinary mechanical field "slowly" (with the sound speed, instead of the light speed) carry quasistatic electric and/or magnetic signals, which are already widely used in various devices. It is sufficient to mention the specific for piezoelectrics Bleustein-Gulyaev surface acoustic wave [66, 67], which has turned out to be a convenient element for modern delay lines. Below in Ch. 10 of the book the new resonance methods of excitation of the magneto-elastic analog of the Bleustein-Gulyaev wave [68] and of the similar waveguide SH modes in a piezoelectric plate [69] will be considered.

Up to now we discussed either static piezoelectric fields created by line defects or dynamic eigenwave coupled fields in corresponding media. From this point of view, Minagawa [70] and Minagawa & Shintani [71, 72] have introduced a combined approach. They developed the theory of moving dislocations and disclinations in thermo-piezoelectric continua. This approach was based on the modeling of defects by the initial plastic distortion. The final expressions for thermo-piezoelectric fields are obtained in terms of the defect density and the defect-flux tensor by the aid of Green's functions and dynamic Green's potential. Later Minagawa, Hsieh, J.P. Nowacki & Shintani [73] have studied the similar problem of dynamic thermo-mechanical fields accompanied moving dislocations in a coupled-thermoelastic solid with thermal conductivity. The results of this study will be included into Part I, Ch. 4 of the book.

The considered above classical piezoeffects in their continual description do not provide a complete list of all possible electro-elastic couplings. It is evident that on this traditional way we must inevitably miss any effects arising from the discreteness of a crystal lattice. In particular, it is well known that the continual theory of elasticity cannot describe degrees of freedom related to intermolecular motions of the unit cell and therefore in this framework there are no optical phonons. There were introduced quite a lot of different approaches to a continuous description of the discreteness aspect. In fact, most of them represent different modifications of the continuum theory by E. and F. Cosserat [74], the so-called "micropolar elasticity". The contributions with various versions of micropolar theories have been done by Aero & Kuvshinskii [75-77], Eringen [78, 79], Green & Rivlin [80], Mindlin [81, 82], Neuber [83], Sandru [84], Schaefer [85], Cowin [86, 87], W. Nowacki [88, 89], J.P. Nowacki & W. Nowacki [90], J.P. Nowacki [91-93], J.P. Nowacki & P.G. Glockner [94], J.P. Nowacki & Shindo [95]. The corresponding modifications in the dislocation and disclination theories have been pointed out by de Wit [96], Antony [97], Claus & Eringen [98], J.P.Nowacki [99-102]. For details we refer readers to the review article by Mindlin [103] and to original papers.

Below we shall limit ourselves to a consideration of only such intermolecular motions, which lead to electro-elastic coupling. For example, in ionic crystals like NaCl, where the unit cell represents a dipole (Na^{+}-Cl^{-}), the intermolecular motions must be accompanied by a change of the corresponding electric polarization. It was Mindlin [103, 104], who found the way, how to describe continuously the above very specific coupling effects. He has extended the classical Voigt [2] theory of piezoelectricity in Toupin's formulation [105] by supposing that the stored energy function depends not only on the strain tensor and polarization vector but also on the polarization gradient tensor. As a result, he came to a new type of electro-elastic coupling, which may exist even in centrosymmetric crystals. The non-linear version of this theory was developed by Suhubi [106]. Of course, such sort of coupling can manifest itself in rather specific phenomena related either to very high frequencies (optical phonons) or to a space scale comparable with a lattice parameter. And indeed, the Mindlin's theory of elastic dielectrics with polarization gradient accommodates the observed and experimentally measured phenomena [107-111], such as electromechanical interaction in centrosymmetric materials, capacitance of thin dielectric films, surface energy of polarization, deformation and optical activity in quartz.

This new electro-elastic coupling also contributes to the electro-elastic fields of defects and some their basic properties. The appropriate version of the dislocation theory has been developed in a series of papers by J.P.Nowacki & Hsieh [112] and J.P. Nowacki [113-116]. The distortion and electro-elastic fields due to a dislocation have been calculated and the relation between the "linear" and "surface" formulation of the theory has been discussed. This theory will be presented below in the Chapter 5.

A linear theory of thermoelastic media with polarization gradient was developed by Chowdhury & Glockner [117] while a complete nonlinear theory for such materials was given by Chowdhury, Epstein & Glockner [118]. The related constitutive equations were derived by Chowdhury & Glockner [117] and by J.P. Nowacki & Glockner [119]. In [119] there was also solved the particular problem of electro-thermo-elastic fields excited in the layer by the heat sources. The analogous fields induced in the infinite medium by a cylindrical thermal inclusion have been found by J.P. Nowacki & Trimarco [120]. A series of similar problems for plane inclusions in infinite, simi-infinite or finite media was solved recently by J.P. Nowacki [121]. The results of [119 - 121] will be included in Chs. 2 and 6.

Dynamical problems for such elastic dielectric continua including the wave propagation in media with polarization gradient were studied in the series of publications [122-129], including the papers by J.P. Nowacki & Glockner [123, 124], J.P. Nowacki [126, 127, 129] and J.P. Nowacki & Trimarco [128].

In its structure the book is divided into two parts. Part I (Chs 1-5) is started from the two introductory chapters where the basic concepts and equations are introduced for the theories of electro- and magneto-mechanical couplings due to piezoeffects (Ch. 1) and polarization gradient (Ch. 2). Then on this basis we shall extend the existing dislocation theory for the case of generalized 4D dislocations in arbitrary piezoelectric bodies of unrestricted anisotropy (Ch. 3). Each generalized dislocation represents a combination of an ordinary dislocation of an arbitrary form with its electrostatic analog along the same line. In the next Ch. 4 we shall study the reversible thermo-electro-elastic fields accompanied plastic deformation and, in particular, the temperature distribution around a moving edge straight dislocation in an infinite medium of cubic or isotropic symmetry. The last Ch. 5 of the first part contains some necessary extensions of the dislocation theory for materials with polarization gradient.

In the Part II we consider static and dynamic thermo-electro-elastic fields in media with surfaces and interfaces. Ch. 6 is devoted to the analysis of 1D thermo-electro-elastic fields excited by a plane or a cylindrical thermal inclusion in Mindlin's media (infinite or bounded) with polarization gradient. In particular, in Ch. 6 we study a distribution of mentioned coupled fields in a plate with a plane inclusion at the surface or in the bulk. In the next two chapters (Chs. 7 and 8) we consider 2D electroelastic fields in an arbitrary anisotropic piezoelectric plate with line sources in the bulk or at the surfaces. In fact, here the Green's functions are derived which describe coupled fields produced by the arbitrary 2D distributions of dislocations, mechanical forces, electric potentials and electric charges. In Ch. 9 the same problem is solved for the piezoelectric layer-substrate structure of unrestricted anisotropy. And in the last chapter of the book (Ch. 10) the new resonance methods of excitation of the magneto-elastic analog of the SH surface Bleustein-Gulyaev wave and/or of the waveguide SH modes in a piezoelectric plate will be considered.

Part I

Two Types of Electro-Elastic Coupling and Their Role in Properties of Line Defects in Unbounded Anisotropic Media

1

Piezoelectricity and Piezomagnetism: Basic Concepts and Equations

1.1 Thermodynamic potentials

(a) Some preliminary remarks

Below we shall consider media with piezoeffects (i.e., electromechanical or magnetomechanical coupling), where the elastic distortions create electric or magnetic fields and, vise versa, electric or magnetic polarizations course in the same points the proportional elastic deformations. The basic properties of such media and their thermodynamics are described in many monograph publications (see, e.g., [3-16]). In this chapter we shall limit ourselves only to a very brief survey of basic features of linear electro-elastic and magneto-elastic coupling effects that will be necessary for derivations in the other parts of the book.

The notations for electro-magnetic vector characteristics will be customary: **E** for electric field, **D** for electric displacement, **H** for magnetic field and **B** for magnetic induction. They are connected by Maxwell's equations:

$$\nabla \times \mathbf{E} = -\frac{1}{c}\frac{\partial \mathbf{B}}{\partial t}, \qquad \nabla \times \mathbf{H} = \frac{1}{c}\frac{\partial \mathbf{D}}{\partial t} \tag{1.1}$$

For a convenience of estimations, eqs. (1.1) are deliberately presented in the CGSE system. That is why they explicitly contain the light velocity c. In further considerations we are going to deal either with static fields of line sources or with steady-state dynamic problems (e.g., with moving dislocations or acoustic waves). This allows putting into Maxwell's equations (1.1) the functions

$$\mathbf{E}(\mathbf{r},t) = \mathbf{E}(\mathbf{r}-\mathbf{v}t), \qquad \mathbf{D}(\mathbf{r},t) = \mathbf{D}(\mathbf{r}-\mathbf{v}t), \tag{1.2}$$

$$\mathbf{H}(\mathbf{r},t) = \mathbf{H}(\mathbf{r}-\mathbf{v}t), \qquad \mathbf{B}(\mathbf{r},t) = \mathbf{B}(\mathbf{r}-\mathbf{v}t) \tag{1.2'}$$

where **v** is the velocity not exceeding the sound speed. In pure piezoelectrics, where the magnetic field **H** is not coupled with mechanical distortions, it does not vanish only due to the right-hand-side in $(1.1)_2$ and in view of $(1.2)_2$ can be estimated as a small parameter $\mathbf{H} \sim (v/c)\mathbf{D}$. In contrast to $\mathbf{H}$, the electric field $\mathbf{E}$ is determined by the electro-elastic coupling and generally is not at all as small as $\mathbf{H}$. However, rot $\mathbf{E}$ in $(1.1)_1$ must be extremely small because in accordance with the above consideration it is proportional to $(v/c)^2$.

Thus, for pure piezoelectrics we have come to the following conclusions:

i) up to terms $\sim v/c$ one may neglect by magnetic fields;

ii) up to terms $\sim (v/c)^2$ electric fields are potential, i.e. one may introduce the electric potential φ^E :

$$\mathbf{E} = -\nabla\varphi^E. \tag{1.3}$$

For pure piezomagnetics, obviously, everything is the other way around: one may neglect by electric field and introduce the magnetic potential φ^H :

$$\mathbf{H} = -\nabla\varphi^H. \tag{1.4}$$

This is the so-called quasi-static approximation in the theory of piezoelectricity and piezomagnetism. In the same approximation the electromagnetic energy flux $\mathbf{J}^{em}$ (the Pointing vector) is equal, for piezoelectrics,

$$\mathbf{J}^{em} = \varphi^E \frac{\partial \mathbf{D}}{\partial t} \tag{1.5}$$

and for piezomagnetics

$$\mathbf{J}^{em} = \varphi^H \frac{\partial \mathbf{B}}{\partial t} \tag{1.6}$$

And the rate of change of the electromagnetic energy density is given by

$$\dot{U}^{em} = \mathbf{E}\cdot\dot{\mathbf{D}} + \mathbf{H}\cdot\dot{\mathbf{B}} \tag{1.7}$$

In (1.5)-(1.7) and in further equations we use the more convenient MKS units system.

(b) Thermodynamic potentials and constitutive equations

Let us introduce the basic quantities characterizing elasticity of solids: the displacement vector $\mathbf{u}(\mathbf{r},t)$, the distortion tensor $\boldsymbol{\beta}(\mathbf{r},t)$, the strain tensor $\varepsilon^{str}(\mathbf{r},t)$,

$$\varepsilon_{ij}^{str} = \frac{1}{2}(\beta_{ij} + \beta_{ji}) \tag{1.8}$$

and the stress tensor $\boldsymbol{\sigma}(\mathbf{r},t), \sigma_{ij} = \sigma_{ji}$. Note that only for media, where there are no sources of incompatibility (e.g., dislocations), one may write the identity:

$$\beta_{ji} = u_{i,j} \equiv \frac{\partial u_i}{\partial x_j} \tag{1.9}$$

In the terms introduced above the rate of change of the stored elastic energy is equal

$$\dot{U}^{str} = \sigma_{ij}\dot{\varepsilon}_{ij}^{str} = \sigma_{ij}\dot{\beta}_{ij} = tr(\boldsymbol{\sigma}\dot{\boldsymbol{\beta}}). \tag{1.10}$$

Adding to (1.7) and (1.10) the rate of change of the thermal energy density

$$\dot{U}^{th} = T\dot{S}, \tag{1.11}$$

(T is a temperature, S is an entropy) we come to the energy potential $U(\boldsymbol{\beta}, \mathbf{D}, \mathbf{B}, S)$ characterized by the rate of change

$$\dot{U} = tr(\boldsymbol{\sigma}\dot{\boldsymbol{\beta}}) + \mathbf{E}\cdot\dot{\mathbf{D}} + \mathbf{H}\cdot\dot{\mathbf{B}} + T\dot{S}. \tag{1.12}$$

Basing on this potential we can construct a series of other thermodynamic functions. For instance, the free energy potential is defined by $F(\boldsymbol{\beta}, \mathbf{D}, \mathbf{B}, T) = U - TS$.

Let us return to a pure piezoelectric medium where the magnetic field is supposed to be of negligible magnitude. It will be useful for the forthcoming considerations to form the two new thermodynamic potentials: the electric enthalpy $H(\boldsymbol{\beta}, \mathbf{E}, S)$,

$$H(\boldsymbol{\beta}, \mathbf{E}, S) = U(\boldsymbol{\beta}, \mathbf{D}, \mathbf{0}, S) - \mathbf{E}\cdot\mathbf{D}, \tag{1.13}$$

$$\dot{H} = tr(\boldsymbol{\sigma}\dot{\boldsymbol{\beta}}) - \mathbf{D}\cdot\dot{\mathbf{E}} + T\dot{S}, \tag{1.14}$$

and the Gibbs electric enthalpy $H^G(\boldsymbol{\beta}, \mathbf{E}, T)$,

$$H^G(\boldsymbol{\beta}, \mathbf{E}, T) = F(\boldsymbol{\beta}, \mathbf{D}, \mathbf{0}, T) - \mathbf{E}\cdot\mathbf{D} = H(\boldsymbol{\beta}, \mathbf{E}, S) - TS, \tag{1.15}$$

$$\dot{H}^G = tr(\boldsymbol{\sigma}\dot{\boldsymbol{\beta}}) - \mathbf{D}\cdot\dot{\mathbf{E}} - S\dot{T}. \tag{1.16}$$

The first potential H is convenient for the description of adiabatic motions, and the second potential H^G - for isothermal processes. Eqs. (1.14) and (1.16) provide the following constitutive relations:

$$\boldsymbol{\sigma} = \left(\frac{\partial H}{\partial \boldsymbol{\beta}}\right)_{\mathbf{E},S}, \quad \mathbf{D} = \left(\frac{\partial H}{\partial \mathbf{E}}\right)_{\boldsymbol{\beta},S}, \quad T = \left(\frac{\partial H}{\partial S}\right)_{\boldsymbol{\beta},\mathbf{E}}; \tag{1.17}$$

$$\boldsymbol{\sigma} = \left(\frac{\partial H^G}{\partial \boldsymbol{\beta}}\right)_{\mathbf{E},T}, \quad \mathbf{D} = \left(\frac{\partial H^G}{\partial \mathbf{E}}\right)_{\boldsymbol{\beta},T}, \quad S = -\left(\frac{\partial H^G}{\partial T}\right)_{\boldsymbol{\beta},\mathbf{E}}. \tag{1.18}$$

Below the thermodynamic functions H and H^G will be expanded with respect to a referential state free of deformations or biasing electromagnetic fields. This means that in the ground state

$$\begin{aligned} &\left(\frac{\partial H}{\partial \boldsymbol{\beta}}\right)_0 = \left(\frac{\partial H}{\partial \mathbf{E}}\right)_0 = \mathbf{0}, \quad \left(\frac{\partial H}{\partial S}\right)_0 = \bar{T}, \\ &\left(\frac{\partial H^G}{\partial \boldsymbol{\beta}}\right)_0 = \left(\frac{\partial H^G}{\partial \mathbf{E}}\right)_0 = \mathbf{0}, \quad \left(\frac{\partial H^G}{\partial T}\right)_0 = -\bar{S}, \end{aligned} \tag{1.19}$$

where $\bar{T}$ and $\bar{S}$ are the uniform values of temperature and entropy in the medium. The electric and magnetic bias fields introduced at the end of the chapter are considered as excitations of the ground state (1.19).

1.2 Linear piezoelectricity

(a) Isothermal and adiabatic constitutive equations

Thus, we are starting from expanding the potential $H^G(\boldsymbol{\beta}, \mathbf{E}, T)$ near the ground state in small perturbations $\boldsymbol{\beta}, \mathbf{E}$ and $\theta = T - \bar{T}$, up to second order terms:

$$\begin{aligned} H^G(\boldsymbol{\beta}, \mathbf{E}, T) &= H^G(\mathbf{0}, \mathbf{0}, \bar{T}) - \bar{S}\theta + \frac{1}{2}\tilde{c}_{ijkl}\beta_{ij}\beta_{kl} - \frac{1}{2}\tilde{\varepsilon}_{kl}E_k E_l \\ &- \tilde{e}_{kij}E_k\beta_{ij} - \gamma_{ij}\beta_{ij}\theta - g_k E_k\theta - \frac{1}{2}\frac{c_v}{\bar{T}}\theta^2. \end{aligned} \tag{1.20}$$

Hear $\tilde{c}_{ijkl}$ are the isothermal moduli of elasticity, $\tilde{\varepsilon}_{kl}$ is the isothermal tensor of electric permittivity, $\tilde{e}_{kij}$ are the isothermal piezoelectric moduli, γ_{ij} are the temperature coefficients of stress, g_k are piroelectric coefficients, and c_v is the specific heat.

Combining expansion (1.20) with eqs. (1.18) one easily obtains the explicit form of the constitutive equations of linear piezoelectricity:

$$\sigma_{ij} = \tilde{c}_{ijkl}\beta_{kl} - \tilde{e}_{kij}E_k - \gamma_{ij}\theta, \tag{1.21}$$

$$D_i = \tilde{e}_{ikl}\beta_{kl} + \tilde{\varepsilon}_{ik}E_k + g_i\theta, \tag{1.22}$$

$$S - \bar{S} = \gamma_{kl}\beta_{kl} + g_k E_k + \frac{c_v}{\bar{T}}\theta. \tag{1.23}$$

For isothermal conditions, when $T = \bar{T}$ and $\theta = 0$, eqs. (1.21) and (1.22) lead to the very well known system of constitutive equations for piezoelectric coupling:

$$\sigma_{ij} = \tilde{c}_{ijkl}\beta_{kl} - \tilde{e}_{kij}E_k, \tag{1.24}$$

$$D_i = \tilde{e}_{ikl}\beta_{kl} + \tilde{e}_{ik}E_k. \tag{1.25}$$

By their derivation, these equations correctly describe static deformations or sufficiently slow dynamic processes (like relatively slow motions of dislocations), until they may be considered as quasi-isothermal. However, for such dynamic processes as acoustic wave propagation the adiabatic condition $S = \bar{S}$ is a much better approximation than the isothermal condition $T = \bar{T}$. So, let us also derive the adiabatic variant of the constitutive relations of piezoelectricity and establish the relations between the corresponding material tensors.

This time one should operate with the electric enthalpy $H(\boldsymbol{\beta}, \mathbf{E}, S)$ (1.13). Its expansion in the small variables $\boldsymbol{\beta}, \mathbf{E}$ and $S - \bar{S}$ is very similar to (1.20):

$$\begin{aligned} H(\boldsymbol{\beta}, \mathbf{E}, S) &= H(\mathbf{0}, \mathbf{0}, \bar{S}) + \bar{T}(S - \bar{S}) + \frac{1}{2}\hat{c}_{ijkl}\beta_{ij}\beta_{kl} - \frac{1}{2}\hat{\varepsilon}_{kl}E_k E_l - \\ &- \hat{e}_{kij}E_k\beta_{ij} - \chi_{ij}\beta_{ij}(S - \bar{S}) - r_k E_k(S - \bar{S}) + \frac{1}{2}\frac{\bar{T}}{c_v}k(S - \bar{S})^2. \end{aligned} \tag{1.26}$$

Here the hat $\wedge$ over tensors indicates the adiabatic material constants. In this case the explicit form of the constitutive equations, by (1.19), is given by the system

$$\sigma_{ij} = \hat{c}_{ijkl}\beta_{kl} - \hat{e}_{kij}E_k - \chi_{ij}(S - \bar{S}), \tag{1.27}$$

$$D_i = \hat{e}_{ikl}\beta_{kl} + \hat{\varepsilon}_{ik}E_k + r_i(S - \bar{S}), \tag{1.28}$$

$$\theta = -\chi_{kl}\beta_{kl} - r_k E_k + \frac{\bar{T}}{c_v}(S - \bar{S}) \;. \tag{1.29}$$

For $S = \bar{S}$ the system (1.27), (1.28) takes the same form as (1.24), (1.25) with isothermal moduli replaced by the adiabatic ones. Of course, both systems of constitutive equations, (1.21)-(1.23) and (1.27)-(1.29), must be identical at $S = \bar{S}$. This enables establishing the relations between two sets of material constants [15]. Combining (1.23) with (1.29) at $S = \bar{S}$ one obtains

$$\theta = -\chi_{kl}\beta_{kl} - r_k E_k = -\frac{\bar{T}}{c_v}(\gamma_{kl}\beta_{kl} + g_k E_k), \tag{1.30}$$

which provides the two identities

$$\chi_{kl} = \frac{\bar{T}}{c_v}\gamma_{kl}, \quad r_k = \frac{\bar{T}}{c_v}g_k. \tag{1.31}$$

On the other hand, substituting $(1.30)_2$ into eqs. (1.21), (1.22) and comparing the results with eqs. (1.27), (1.28) taken for $S = \bar{S}$ we find

$$\hat{c}_{ijkl} = \tilde{c}_{ijkl} + \frac{\bar{T}}{c_v}\gamma_{ij}\gamma_{kl}, \tag{1.32}$$

$$\hat{e}_{kij} = \tilde{e}_{kij} - \frac{\bar{T}}{c_v}g_k\chi_{ij}, \tag{1.33}$$

$$\hat{\varepsilon}_{ik} = \tilde{\varepsilon}_{ik} - \frac{\bar{T}}{c_v}g_i g_k. \tag{1.34}$$

In the forthcoming we shall for brevity omit the hats $\wedge$ and $\sim$ over the material constants when the type of moduli used is clear from the context. Accordingly, the constitutive equations of linear piezoelectricity both for isothermal and for adiabatic conditions will simply read

$$\sigma_{ij} = c_{ijkl}\beta_{kl} - e_{kij}E_k, \tag{1.35}$$

$$D_i = e_{ikl}\beta_{kl} + \varepsilon_{ik}E_k. \tag{1.36}$$

Now let us shortly stand on the symmetry properties of the introduced material tensors. It is easily proved that

$$c_{ijkl} = c_{klij} = c_{jikl} = c_{ijlk}, \tag{1.37}$$

$$e_{kij} = e_{kji}, \tag{1.38}$$

$$\varepsilon_{ik} = \varepsilon_{ki}. \tag{1.39}$$

The explicit structure of these tensors for crystals of different symmetry can be found in the monographs [130, 131].

(b) Equations of motion and boundary conditions

Consider an arbitrary piezoelectric medium with some distributions of bulk forces $\mathbf{f}(\mathrm{r})$ and charges $q(\mathrm{r})$. The motions of elements of such a continuum are described by a known system of equations, which is nonspecific for piezoelectrics:

$$\rho\ddot{\mathbf{u}} = \operatorname{div}\boldsymbol{\sigma} + \mathbf{f}\ , \tag{1.40}$$

$$\operatorname{div}\mathbf{D} = q. \tag{1.41}$$

When we are interested in thermoelastic effects, the standard equation of heat conductivity [15] must be added to (1.40), (1.41),

$$\bar{T}\dot{S} = \kappa_{ij}\theta_{,ji} + W \tag{1.42}$$

where κ_{ij} is the tensor of thermal conductivity and W is the power of the heat supply. The situation becomes more definite after the substitution into this system the constitutive relations (1.21)-(1.23) (or (1.35), (1.36), when we deal with isothermal or adiabatic processes. However, even after this substitution the above system, eqs. (1.40)-(1.42), remains unclosed until we introduce a clear relation between the distortion tensor $\boldsymbol{\beta}$ and the displacement vector $\mathbf{u}$ in the left-hand-side of eq. (1.40). The problem is that the distortion, which is present in all constitutive equations, is purely elastic, but the displacement determining the inertial term in (1.40) is total, i.e. it might include some plastic contribution. Below we shall consider three different cases.

i) The medium has no topological sources of internal electro-elastic fields (like dislocations) and thermoelasticity can be ignored.

In this case one can use relation (1.9), which makes the system of eqs. (1.40), (1.41) and (1.35), (1.36) to be closed. The result reduces to the equations

$$c_{ijkl}u_{k,lj} - e_{lij}E_{l,j} = \rho\ddot{u}_i - f_i, \tag{1.43}$$

$$e_{jlk}u_{k,lj} + \varepsilon_{jl}E_{l,j} = q. \tag{1.44}$$

In particular, this situation is realized in the theory of plane acoustic waves adiabatically propagating in piezoelectric media (see Ch. 10). In the latter case we shall use the above system in the simplified version putting there $f_i = 0,\ q = 0$.

ii) The medium contains some topological sources continuously described by the tensor of plastic distortion $\boldsymbol{\beta}^0$, and thermoelasticity again does not contribute into considered processes.

In this case, as was shown by Kröner [132] (see also the beginning of Chapter 3), one should write, instead of (1.9),

$$\beta_{ji} + \beta^0_{ji} = u_{i,j}. \tag{1.45}$$

Accordingly, the equations of motion for such piezoelectric medium are given by

$$c_{ijkl}u_{k,lj} - e_{lij}E_{l,j} = \rho\ddot{u}_i - f_i + c_{ijkl}\beta^0_{kl,j}\ , \tag{1.46}$$

$$e_{jlk}u_{k,lj} + \varepsilon_{jl}E_{l,j} = q + e_{jlk}\beta^0_{kl,j}\ . \tag{1.47}$$

This system describes both statics and dynamics of dislocations in piezoelectric solids. At $\boldsymbol{\beta}^0 = 0$ it is naturally transformed into (1.43), (1.44).

iii) The most general case of thermo-piezoelectric media with dislocations or other sources of plastic distortion β^0. The corresponding system is formed by three equations:

$$c_{ijlk}u_{k,lj} - e_{lij}E_{l,j} - \gamma_{ij}\theta_{,j} = \rho\ddot{u}_i - f_i + c_{ijkl}\beta^0_{kl,j}, \tag{1.48}$$

$$e_{jlk}u_{k,lj} + \varepsilon_{jl}E_{l,j} + g_j\theta_{,j} = q + e_{jlk}\beta^0_{kl,j}, \tag{1.49}$$

$$\bar{T}\gamma_{lk}\dot{u}_{k,l} + \bar{T}g_k\dot{E}_k + c_v\dot{\theta} - \kappa_{ij}\theta_{,ji} = W + \bar{T}\gamma_{lk}\dot{\beta}^0_{kl}. \tag{1.50}$$

This form of equations of motion will be used in Ch. 4.

Sometimes the electric field $\mathbf{E}^0$ of some sources is very well known in vacuum. In such cases it can be convenient to count off the unknown total electric field $\mathbf{E}$ in the medium from $\mathbf{E}^0$:

$$\mathbf{E} = \mathbf{E}^0 + \mathbf{E}', \tag{1.51}$$

where $\mathbf{E}'$ is the addition arising due to the presence of the dielectric medium with its boundaries and possibly with some additional sources. Then the fields $\boldsymbol{\sigma}$ and $\mathbf{D}$ should have similar forms:

$$\boldsymbol{\sigma} = \boldsymbol{\sigma}^0 + \boldsymbol{\sigma}', \quad \mathbf{D} = \mathbf{D}^0 + \mathbf{D}', \tag{1.52}$$

where the terms $\boldsymbol{\sigma}^0$ and $\mathbf{D}^0$ should be proportional $\mathbf{E}^0$. Accordingly, the potentials $H^G(\boldsymbol{\beta}, \mathbf{E}, T)$ and $H(\boldsymbol{\beta}, \mathbf{E}, S)$ should be expanded close to the ground state characterized by the electric field $\mathbf{E}^0$. As a result, in each of expansions (1.20) and (1.26) one should add the term $\sigma^0_{ij}\beta_{ij} - D^0_i E_i$ and replace all E_i by E'_i; in all constitutive equations, e.g., in (1.35), (1.36), the quantities σ_{ij}, D_i and E_i must be respectively replaced by σ'_{ij}, D'_i and E'_i. Then, comparing the resulting eq. (1.52)

$$\sigma_{ij} = \sigma^0_{ij} + c_{ijkl}\beta_{kl} - e_{lij}E'_l, \quad D_j = D^0_j + e_{jlk}\beta_{kl} + \varepsilon_{jl}E'_l \tag{1.53}$$

with (1.35), (1.36) after the substitution there (1.51), one can easily find: $\sigma^0_{ij} = -e_{lij}E^0_l$, $D^0_j = \varepsilon_{jl}E^0_l$. Let us denote

$$\mathrm{div}\boldsymbol{\sigma}^0 = -\mathrm{div}(\hat{e}\mathbf{E}^0) = -\mathbf{f}^0, \quad \mathrm{div}\mathbf{D}^0 = \mathrm{div}(\hat{\varepsilon}\mathbf{E}^0) = q^0.$$

In this terms eqs. (1.40), (1.41) take the form:

$$\begin{aligned} \rho\ddot{\mathbf{u}} - \mathrm{div}\boldsymbol{\sigma}' &= \mathbf{f} - \mathbf{f}^0 \equiv \mathbf{f}', \\ \mathrm{div}\mathbf{D}' &= q - q^0 \equiv q'. \end{aligned} \tag{1.54}$$

Let us introduce the potential $\varphi, E_i' = -\varphi_{,i}$, and present the appropriate new form of eqs. (1.46) and (1.47):

$$\begin{aligned} \rho\ddot{u}_i - c_{ijkl}u_{k,lj} - e_{lij}\varphi_{,lj} &= f_i - c_{ijkl}\beta^0_{kl,j} - e_{lij}E^0_{l,j}, \\ e_{jlk}u_{k,lj} - \varepsilon_{jl}\varphi_{,lj} &= q + e_{jlk}\beta^0_{kl,j} - \varepsilon_{jl}E^0_{l,j}. \end{aligned} \tag{1.55}$$

The above form of the equations of motion will be used in Ch. 3 and in Part 2 of this book.

Now let us discuss boundary conditions for different problems related to piezoelectric media. Mechanically, we shall distinguish a loaded surface,

$$\boldsymbol{\sigma}\mathbf{n} = \mathbf{f}^s \tag{1.56}$$

($\mathbf{n}$ is the unit normal to the surface, $\mathbf{f}^s$ is the force density at the surface), and a prescribed displacements $\mathbf{u}^s$ at the surface,

$$\mathbf{u} = \mathbf{u}^s. \tag{1.57}$$

In particular, a free surface is described by the condition $\boldsymbol{\sigma}\mathbf{n} = \mathbf{0}$, and a clamped surface by $\mathbf{u} = \mathbf{0}$.

Electrically, we shall consider three types of boundary conditions: the prescribed charge density q^s at the surface,

$$\mathbf{D} \cdot \mathbf{n} = -q^s, \tag{1.58}$$

the prescribed potential φ^s at the surface,

$$\varphi^E = \varphi^s, \tag{1.59}$$

and the contact with the adjoined dielectric medium described by the continuity of the electric potential φ^E and the projection $\mathbf{D}{\cdot}\mathbf{n}$ at the interface,

$$\mathbf{D} \cdot \mathbf{n} = \mathbf{D}_{adj} \cdot \mathbf{n}. \tag{1.60}$$

In particular, by (1.59), the metallized surface is described, as usual, by the condition

$$\varphi^E = 0. \tag{1.61}$$

(c) The reciprocity theorem

The physical coupled fields under discussion can be excited by various combinations of different sources, which may reveal themselves both in equations and in boundary conditions. In the most general case of thermo-electro-elastoplasticity described by system (1.48)-(1.50) the following bulk sources can be listed: forces $\mathbf{f}(\mathbf{r},t)$, charges $q(\mathbf{r},t)$, plastic distortions $\beta^0(\mathbf{r},t)$ and a heat supply $W(\mathbf{r},t)$. They can act independently in any combinations. Surface sources mathematically involved through boundary conditions, as we have seen, exist in the following alternatives: prescribed tractions ($\boldsymbol{\sigma}\mathbf{n} = \mathbf{f}^s$) or displacements ($\mathbf{u} = \mathbf{u}^s$), prescribed surface charge density ($\mathbf{D} \cdot \mathbf{n} = -q^s$) or electro-potential ($\varphi^E = \varphi^s$), prescribed normal component of the heat flux ($h_n = \mathbf{h} \cdot \mathbf{n} = \kappa_{ij} n_i \theta_{,j}$) or temperature ($\theta = \theta^s$). In addition, initial conditions also can serve as sources.

Consider in the given body the two independent sets of sources and corresponding excited physical fields. Say, let the sources $\mathbf{f}, q, \beta^0, W$, etc produce the fields $\mathbf{u}, \mathbf{E}, \theta$, etc and the sources $\mathbf{f}', q', \beta^{0'}, W'$, etc excite the fields $\mathbf{u}', \mathbf{E}', \theta'$, etc. In this subsection we establish a general relation between these two sets which is known as the reciprocity theorem. We start from the first equation of motion, eq. (1.40):

$$\sigma_{ji,j} + f_i = \rho \ddot{u}_i, \qquad \sigma'_{ji,j} + f'_i = \rho \ddot{u}'_i. \tag{1.62}$$

For brevity of final expressions let us choose the simplest initial conditions

$$u_i(\mathbf{r}, 0) = \dot{u}_i(\mathbf{r}, 0) = 0, \qquad u'_i(\mathbf{r}, 0) = \dot{u}'_i(\mathbf{r}, 0) = 0. \tag{1.63}$$

After the Laplace transformation

$$L[\sigma_{ij}(\mathbf{r}, t)] = \bar{\sigma}_{ij}(\mathbf{r}, p) = \int_0^\infty \sigma_{ij}(\mathbf{r}, t) e^{-pt} dt, \quad etc \tag{1.64}$$

eqs. (1.62) take the form

$$\bar{\sigma}_{ji,j} + \bar{f}_i = \rho p^2 \bar{u}_i, \tag{1.65}$$

$$\bar{\sigma}'_{ji,j} + \bar{f}'_i = \rho p^2 \bar{u}'_i. \tag{1.66}$$

Let us scalarly multiply (1.65) by $\bar{u}'_i$ and (1.66) by $\bar{u}_i$, subtract the results and integrate the difference over the bulk of the body:

$$\int_V \left[\bar{\sigma}_{ji,j} \bar{u}'_i - \bar{\sigma}'_{ji,j} \bar{u}_i\right] dV + \int_V \left[\bar{f}_i \bar{u}'_i - \bar{f}'_i \bar{u}_i\right] dV = 0. \tag{1.67}$$

Making use of the identities

$$\bar{\sigma}_{ji,j} \bar{u}'_i = (\bar{\sigma}_{ji} \bar{u}'_i)_{,j} - \bar{\sigma}_{ji} \bar{u}'_{i,j}. \quad \bar{\sigma}'_{ji,j} \bar{u}_i = (\bar{\sigma}'_{ji} \bar{u}_i)_{,j} - \bar{\sigma}'_{ji} \bar{u}_{i,j} \tag{1.68}$$

and of Gauss-Ostrogradski theorem, one can transform (1.67) into

$$\begin{aligned} \int_V \left[\bar{f}_i \bar{u}'_i - \bar{f}'_i \bar{u}_i\right] dV + \int_S n_i \left(\bar{\sigma}_{ji} \bar{u}'_i - \bar{\sigma}'_{ji} \bar{u}_i\right) dS = \\ = \int_V \left[\bar{\sigma}_{ji} \bar{u}'_{i,j} - \bar{\sigma}'_{ji} \bar{u}_{i,j}\right] dV, \end{aligned} \tag{1.69}$$

Substituting to the right-hand-side of (1.69) the Laplace transformations of $u_{i,j}$ and $u'_{i,j}$, eq. (1.45), and of the corresponding constitutive relations for σ_{ij} and σ'_{ij} (see (1.21)) we have

$$\int_V (\bar{f}_i \bar{u'}_i - \bar{f'}_i \bar{u}_i) dV + \int_V (\bar{\beta}^0_{ij} \bar{\sigma'}_{ji} - \bar{\beta}^{0'}_{ij} \bar{\sigma}_{ji}) dV + \int_S n_i \left(\bar{\sigma}_{ji} \bar{u'}_i - \bar{\sigma'}_{ji} \bar{u}_i \right) dS =$$
$$= \int_V \left[-e_{kij} (\bar{E}_k \bar{\beta}'_{ij} - \bar{E}'_k \bar{\beta}_{ij}) + \gamma_{ij} (\bar{\beta}_{ij} \bar{\theta}' - \bar{\beta}'_{ij} \bar{\theta}) \right] dV. \tag{1.70}$$

Quite similarly combining equations

$$\bar{D}_{k,k} = \bar{q}, \qquad \bar{D'}_{k,k} = \bar{q}' \tag{1.71}$$

multiplied by $\bar{\varphi}'_i$ and $\bar{\varphi}_i$, respectively, with due regards for (1.22) one obtains

$$- \int_V (\bar{q} \bar{\varphi}' - \bar{q}' \bar{\varphi}) dV + \int_S n_i \left(\bar{D}_i \bar{\varphi}' - \bar{D'}_i \bar{\varphi} \right) dS =$$
$$= \int_V \left[e_{kij} (\bar{E}_k \bar{\beta}'_{ij} - \bar{E}'_k \bar{\beta}_{ij}) + g_k (\bar{E}_k \bar{\theta}' - \bar{E}'_k \bar{\theta}) \right] dV. \tag{1.72}$$

The last equation of motion (1.50) after analogous manipulations leads to the relation

$$\int_V (\bar{W} \bar{\theta}' - \bar{W}' \bar{\theta}) dV + \kappa_{ij} \int_S n_i \left(\bar{\theta}_{,j} \bar{\theta}' - \bar{\theta}'_{,j} \bar{\theta} \right) dS =$$
$$= p\bar{T} \int_V \left[\gamma_{ij} (\bar{\beta}_{ij} \bar{\theta}' - \bar{\beta}'_{ij} \bar{\theta}) + g_k (\bar{E}_k \bar{\theta}' - \bar{E}'_k \bar{\theta}) \right] dV, \tag{1.73}$$

where $\bar{T}$ as before is the ground level of temperature and should not be mixed with a Laplace image of T (the latter presents in (1.73) by means of $\bar{\theta}$).

Now let us sum eqs. (1.70) and (1.72), and subtract the result multiplied by $p\bar{T}$ from (1.73). This provides an exclusion of the right-hand-sides of all three equations. After some rearrangements of the retained terms we obtain

$$\int_V \left\{ \bar{T} p \left[(\bar{f}_i \bar{u'}_i - \bar{f'}_i \bar{u}_i) + (\bar{\beta}^0_{ij} \bar{\sigma'}_{ji} - \bar{\beta}^{0'}_{ij} \bar{\sigma}_{ji}) - (\bar{q} \bar{\varphi}' - \bar{q}' \bar{\varphi}) \right] - (\bar{W} \bar{\theta}' - \bar{W}' \bar{\theta}) \right\} dV$$
$$+ \int_S n_i \left\{ \bar{T} p \left[(\bar{\sigma}_{ij} \bar{u'}_j \bar{\sigma'}_{ij} \bar{u}_j) + (\bar{D}_i \bar{\varphi}' - \bar{D'}_i \bar{\varphi}) \right] - \kappa_{ij} (\bar{\theta}_{,j} \bar{\theta}' - \bar{\theta}'_{,j} \bar{\theta}) \right\} dS = 0. \tag{1.74}$$

And now the back Laplace transformation of (1.74) gives the final equation representing the reciprocity theorem

$$\int_V \left\{ \bar{T} \left[(f_i \circ u'_i - f'_i \circ u_i) + (\beta^0_{ij} \circ \sigma'_{ji} - \beta^{0'}_{ij} \circ \sigma_{ji}) - (q \circ \varphi' - q' \circ \varphi) \right] \right.$$
$$\left. -(W * \theta' - W' * \theta) \right\} dV + \int_S n_i \left\{ \bar{T} \left[(\sigma_{ij} \circ u'_j - \sigma'_{ij} \circ u_j) \right. \right.$$
$$\left. \left. +(D_i \circ \varphi' - D'_i \circ \varphi) \right] - \kappa_{ij} (\theta_{,j} * \theta' - \theta'_{,j} * \theta) \right\} dS = 0 \tag{1.75}$$

Here the notation is introduced

$$L^{-1}\left[p\bar{f}_i\bar{u'}_i\right] = f_i \circ u'_i = \int_0^t f_i(\mathbf{r}, t-\tau)\frac{\partial u'_i(\mathbf{r},\tau)}{\partial\tau}d\tau, \quad etc, \tag{1.76}$$

$$L^{-1}\left[\bar{W}\bar{\theta}'\right] = W * \theta' = \int_0^t W(\mathbf{r}, t-\tau)\theta'(\mathbf{r},\tau)d\tau, \qquad etc \tag{1.77}$$

The obtained form of the reciprocity theorem given by eq. (1.75) provides some extension of the initial formulation [133] for accounting a couple of new sources: bulk charges q and plastic distortions $\boldsymbol{\beta}^0$.

As an example, let us present the alternative appearance of this theorem for a particular choice of surface sources. Let at the surface there are prescribed the tractions $\boldsymbol{\sigma}\mathbf{n} = \mathbf{f}^s$, the charge distribution $-\mathbf{D}\cdot\mathbf{n} = q^s$ and the normal projection of the heat flux $h_n = \mathbf{h}\cdot\mathbf{n} = \kappa_{ij}n_i\theta_{,j}$. In this case eq. (1.75) takes the form

$$\begin{aligned}
&\int_V \{\bar{T}\left(f_i \circ u'_i + \beta^0_{ij} \circ \sigma'_{ji} - q \circ \varphi'\right) - W * \theta'\}dV + \\
&\qquad \int_S \{\bar{T}\left(f^s_i \circ u'_i - q^s \circ \varphi'\right) - h_n * \theta'\}\, dS \\
&= \int_V \left\{\bar{T}\left(f'_i \circ u_i + \beta^{0'}_{ij} \circ \sigma_{ji} - q' \circ \varphi\right) - W' * \theta\right\}dV + \\
&\qquad \int_S \left\{\bar{T}\left(f^{s'}_i \circ u_i - q^{s'} \circ \varphi\right) - h'_n * \theta\right\} dS.
\end{aligned} \tag{1.78}$$

In conclusion, consider the particular case, when all sources and field variables are harmonic in time:

$$u_i(\mathbf{r},t) = u^*_i(\mathbf{r})e^{-i\omega t}, \qquad f_i(\mathbf{r},t) = f^*_i(\mathbf{r})e^{-i\omega t} \quad etc. \tag{1.79}$$

It is clear that now there is no need in Laplace transformations and the result obtained by a straightforward repetition of all above manipulations will take the form very close to (1.74):

$$\begin{aligned}
&\int_V \left\{i\omega\bar{T}[(f^*_i u'^*_i - f'^*_i u^*_i) + (\beta^{0^*}_{ij}\sigma'^*_{ji} - \beta^{0'^*}_{ij}\sigma^*_{ji}) - (q^*\varphi'^* - q'^*\varphi^*)]\right. \\
&\qquad \left. +(W^*\theta'^* - W'^*\theta^*)\right\} dV + \\
&+ \int_S n_i \left\{i\omega\bar{T}[(\sigma^*_{ij}u'^*_j - \sigma'^*_{ij}u^*_j) + (D^*_i\varphi'^* - D'^*_i\varphi^*)]\right. \\
&\qquad \left. +\kappa_{ij}(\theta^*_{,j}\theta'^* - \theta'^*_{,j}\theta^*)\right\} dS = 0.
\end{aligned} \tag{1.80}$$

Again this can be rearranged like (1.78).

Finally, let us consider the even more simplified (but practically important) particular case of a static problem formulated for the infinite isothermal piezoelectric medium with physical fields vanishing at infinities. In this case again there is no need in Laplace transformation, and in addition now we can also put everywhere $\theta = \theta' = 0$ and exclude all surface integrals. So, instead of (1.80) one can write

$$\int_V [(f_i u'_i - f'_i u_i) + (\beta^0_{ij}\sigma'_{ji} - \beta^{0'}_{ij}\sigma_{ji}) - (q\varphi' - q'\varphi)]dV = 0 \tag{1.81}$$

It is convenient to rewrite here

$$\begin{aligned}\beta^0_{ij}\sigma'_{ji} - \beta^{0'}_{ij}\sigma_{ji} &= \beta^0_{ij}[c_{jikl}(u'_{l,k} - \beta^{0'}_{kl}) + e_{kij}\varphi^{E'}_{,k}] \\ &\quad -\beta^{0'}_{ij}[c_{jikl}(u_{l,k} - \beta^0_{kl}) + e_{kij}\varphi^E_{,k}] \\ &= [\beta^0_{ij}(c_{jikl}u'_l + e_{kij}\varphi^{E'}) - \beta^{0'}_{ij}(c_{jikl}u_l + e_{kij}\varphi^E)]_{,k} \\ &\quad +(f^{\mathrm{pl}}_l u'_l - f^{\mathrm{pl}'}_l u_l) - (q^{\mathrm{pl}}\varphi^{E'} - q^{\mathrm{pl}'}\varphi^E),\end{aligned} \tag{1.82}$$

where the notation is introduced

$$f^{\mathrm{pl}}_i = -c_{ijkl}\beta^0_{kl,j}, \quad q^{\mathrm{pl}} = e_{jkl}\beta^0_{kl,j}. \tag{1.83}$$

After the substitution (1.82) into the integral (1.81) the first term is immediately transformed to the surface integral, which is supposed to be zero. Thus, the new appearance of the reciprocity theorem in the given case takes the form

$$\int_V [(f_i + f^{\mathrm{pl}}_i)u'_i - (f'_i + f^{\mathrm{pl}'}_i)u_i - (q + q^{\mathrm{pl}})\varphi' + (q' + q^{\mathrm{pl}'})\varphi]dV = 0 \tag{1.84}$$

which might be more convenient in some applications.

1.3 Piezomagnetic effects of linear and linearized origins

(a) Constitutive relations and equations of motion for purely piezomagnetic media

Linear piezomagnetism is a phenomenon absolutely similar to the considered above piezoelectricity. Just in this case elastic distortions turn out to be coupled with magnetic, instead of electric field. Accordingly, all the theory can be almost identically repeated with some necessary replacements: **E** by **H**, **D** by **B**, electric permittivity ε_{kl} by magnetic permibility μ_{kl}, etc. Instead of the electric enthalpy $H(\beta, \mathbf{E}, S)$ we introduce the magnetic enthalpy $H^m(\beta, \mathbf{H}, S)$,

$$H^m(\boldsymbol{\beta}, \mathbf{H}, S) = V(\boldsymbol{\beta}, \mathbf{0}, \mathbf{B}, S) - \mathbf{H} \cdot \mathbf{B}, \tag{1.85}$$

$$\dot{H}^m = tr(\boldsymbol{\sigma}\dot{\boldsymbol{\beta}}) - \mathbf{B} \cdot \dot{\mathbf{H}} + T\dot{S}, \tag{1.86}$$

with the corresponding constitutive equations

$$\boldsymbol{\sigma} = \left(\frac{\partial H^m}{\partial \boldsymbol{\beta}}\right)_{\mathbf{H},S}, \quad \mathbf{B} = -\left(\frac{\partial H^m}{\partial \mathbf{H}}\right)_{\beta,S}, \quad T = \left(\frac{\partial H^m}{\partial S}\right)_{\beta,\mathbf{H}}. \tag{1.87}$$

This thermodynamic potential again can be expanded close to the ground state:

$$\begin{aligned} H^m(\boldsymbol{\beta}, \mathbf{H}, S) = H^m(\mathbf{0}, \mathbf{0}, \bar{S}) + \bar{T}(S - \bar{S})\frac{1}{2}c_{ijkl}\beta_{ij}\beta_{kl} - \frac{1}{2}\mu_{kl}H_k H_l - \\ -m_{kij}H_k\beta_{ij} - \gamma_{ij}\beta_{ij}(S - \bar{S}) - s_k H_k(S - \bar{S}) + \frac{1}{2}\frac{\bar{T}}{c_v}(S - \bar{S})^2. \end{aligned} \tag{1.88}$$

Combining (1.87) and (1.88) one obtains the explicit constitutive equations

$$\sigma_{ij} = c_{ijkl}\beta_{kl} - m_{kij}H_k - \gamma_{ij}(S - \bar{S}), \tag{1.89}$$

$$B_i = m_{ikl}\beta_{kl} + \mu_{ik}H_k + s_i(S - \bar{S}), \tag{1.90}$$

$$\theta = -\gamma_{kl}\beta_{kl} - s_k H_k + \frac{\bar{T}}{c_v}(S - \bar{S}). \tag{1.91}$$

The adiabatic limit $S = \bar{S}$ of the first pair of this system is

$$\sigma_{ij} = c_{ijkl}\beta_{kl} - m_{kij}H_k, \tag{1.92}$$

$$B_i = m_{ikl}\beta_{kl} + \mu_{ik}H_k \,'. \tag{1.93}$$

Of course, for the isothermal conditions we would obtain again the same system (1.92), (1.93), where all material moduli must be considered as isothermal.

In this book we shall meet with piezomagnetic effect only in Ch. 10 related to plane acoustic waves. Therefore for our aims it is sufficient to present the equation of motion in the form of the system analogous to (1.43), (1.44) taken at $f_i = q = 0$:

$$c_{ijkl}u_{k,lj} - m_{lij}H_{l,j} = \rho\ddot{u}_i, \tag{1.94}$$

$$m_{jlk}u_{k,lj} + \mu_{jl}H_{l,j} = 0. \tag{1.95}$$

(b) Linearized piezomagnetism under the bias magnetic field

The considered effects of linear coupling of elastic distortions with electric or magnetic fields may occur only in anisotropic materials without such an element of symmetry as a center of inversion. There are many known crystals with strong piezoelectric effect which have multiple applications. However, as noted by Auld [4], the linear piezomagnetism is of no practical significance because of the lack of available piezomagnetic crystals with a reasonable level of coupling. On the other hand, there are nonlinear coupling effects, such as electro- and magnetostriction, which are present in media of any symmetry including isotropic bodies. Here the situation is opposite: the electrostriction cannot compete in practice with the piezoelectricity, but the biased magnetostriction has proved to be a strong effect in certain materials and has many technical applications [4, 134-136]. So, here we limit ourselves by consideration of only biased magnetostriction.

In order to include the magnetostriction into our above formalism one should just add to the potential $H^m(\boldsymbol{\beta}, \mathbf{H}, S)$ (1.88) the term

$$-\frac{1}{2} q_{klij} H_k H_l \beta_{ij}, \tag{1.96}$$

where q_{klij} are the moduli of magnetostriction, which have the symmetry

$$q_{klij} = q_{lkij} = q_{klji}. \tag{1.97}$$

Let us consider a small dynamic field $\mathbf{H}$ coupled with small dynamic distortions β on the background of high enough external bias magnetic field $\mathbf{H}^0$. After the linearization of the combined square terms

$$(H_k^0 + H_k)(H_l^0 + H_l) \tag{1.98}$$

over small $\mathbf{H}$, eq. (1.96) is transformed to

$$-q_{klij} H_l^0 H_k \beta_{ij}. \tag{1.99}$$

Thus, one can see that the addition of the term (1.96) to the expansion (1.88) in a considered situation is equivalent to some renormalization of the piezomoduli m_{kij} :

$$m_{kij} \to m_{kij} + m'_{kij}, \tag{1.100}$$

where the bias component of the new piezomoduli is

$$m'_{kij} = H_l^0 q_{lkij}. \tag{1.101}$$

The important feature of this modified material tensor is that its magnitude can be regulated by a choice of the amplitude of the bias field $\mathbf{H}^0$.

The formal theory of biased piezoelectricity is practically completely similar to the above consideration. The result is expressed by the renormalization of the type (1.100): $e_{kij} \to e_{kij} + e'_{kij}$, where the biased term is given by (see also [13]):

$$e'_{kij} = E_l^0 q_{lkij}^E. \tag{1.102}$$

Here $\mathbf{E}^0$ is the electric bias field and q_{lkij}^E is the tensor of electrostriction having the same symmetry as its magnetostrictional analog, eq. (1.97).

2

Mindlin's Electro-Elastic Coupling due to Polarization Gradient

2.1 Toupin's alternative formulation of the theory of piezoelectricity

(a) On the physical meaning of the polarization vector

As it turned out, the electrostriction is not at all the only alternative to the piezoelectric coupling. It was already mentioned in the Introduction about the much less traditional and really fruitful idea of the polarization gradient [103 - 104], which provides a new linear mechanism of electro-elastic coupling. However, in order to start with this new ideology we have to revise the above classical theory of linear piezoelectricity and to introduce some formalism, which allows explicit involving of the vector of polarization $\mathbf{P}$, instead of the electric displacement $\mathbf{D}$, in constitutive relations and equations of motion.

A dielectric body placed into an external electric field $\mathbf{E}_0$ reacts on the latter by a polarization of the molecules, which is macroscopically characterized by the distributed dipole moment $\mathbf{P}$ per unit volume. The vector $\mathbf{P}$ is customary named the polarization vector. This polarization of the medium creates the so-called depolarization field $\mathbf{E}^d$, which weakens the initial electric field $\mathbf{E}^0$, so that the resulting electric field

$$\mathbf{E} = \mathbf{E}^0 + \mathbf{E}^d \tag{2.1}$$

is always less than $\mathbf{E}^0$. Of course, if the dielectric body came to its place together with its own sources (e.g., charges in the bulk or at the surface) creating the additional field $\mathbf{E}^s$, the resulting electric field will be

$$\mathbf{E} = \mathbf{E}^0 + \mathbf{E}', \tag{2.2}$$

where $\mathbf{E}' = \mathbf{E}^d + \mathbf{E}^s$ (compare with (1.51)). It is essential that the insertion of the dielectric body into the field $\mathbf{E}^0$ of external sources perturbs this field not only in the bulk of the body but, as a rule, also outside of it. That is why the electrodynamics of condensed media ordinarily deals with a resulting field $\mathbf{E}$, rather than with its constitutive components (2.2). Instead of decomposing electric field (2.2) it has turned out to be much more convenient to combine two physically meaningful quantities, the electric field $\mathbf{E}$ and the polarization vector $\mathbf{P}$, composing a formal but very convenient vector-function, the so-called electric displacement

$$\mathbf{D} = \varepsilon_0 \mathbf{E} + \mathbf{P}, \tag{2.3}$$

where ε_0 is the electric permittivity of the vacuum. As is well known, in terms of the pair, $\mathbf{E}$ and $\mathbf{D}$, together with analogous magnetic pair, $\mathbf{B}$ and $\mathbf{H}$ (in this case $\mathbf{B}$ is a physical quantity and $\mathbf{H}$ is a formal function), Maxwell's equations and boundary conditions look very compact and convenient (see, e.g., (1.1), (1.41), (1.58), (1.60)).

(b) Toupin's potential U^L of polarization energy

Paradoxically following to the opposite tendency of dividing of the electrical characteristics into vacuum and polarization contributions, Toupin [105] suggested counting off the energy U of the dielectric body from the energy of the resulting electric field in the vacuum $U_0 = \frac{1}{2}\varepsilon_0 \mathbf{E}^2$, so that

$$U = \frac{1}{2}\varepsilon_0 \mathbf{E}^2 + U^L(\boldsymbol{\beta}, \mathbf{P}). \tag{2.4}$$

For simplicity we omit here a thermal variable. In accordance with eq. (1.12) under the adiabatic conditions and in the absence of the magnetic field the rate of change of the energy density $U(\boldsymbol{\beta}, \mathbf{D})$ is determined by

$$\dot{U} = \sigma_{ij}\dot{\beta}_{ij} + E_i \dot{D}_i \,. \tag{2.5}$$

This gives

$$\sigma_{ij} = \left(\frac{\partial U}{\partial \beta_{ij}}\right)_{\mathbf{D}}, \quad E_i = \left(\frac{\partial U}{\partial D_i}\right)_{\boldsymbol{\beta}}. \tag{2.6}$$

Now let us express the potential $U^L(\boldsymbol{\beta}, \mathbf{P})$ using eqs. (2.4) and (2.3):

$$U^L(\boldsymbol{\beta}, \mathbf{P}) = U[\boldsymbol{\beta}, \varepsilon_0 \mathbf{E}(\boldsymbol{\beta}, \mathbf{P}) + \mathbf{P}] - \frac{1}{2}\varepsilon_0 [\mathbf{E}(\boldsymbol{\beta}, \mathbf{P})]^2, \tag{2.7}$$

and calculate the following derivatives

$$\left(\frac{\partial U^L}{\partial \beta_{ij}}\right)_{\mathbf{P}} = \left(\frac{\partial U}{\partial \beta_{ij}}\right)_{\mathbf{D}} + \varepsilon_0 \left(\frac{\partial U}{\partial D_l}\right)_{\boldsymbol{\beta}} \left(\frac{\partial E_l}{\partial \beta_{ij}}\right)_{\mathbf{P}} - \varepsilon_0 E_l \left(\frac{\partial E_l}{\partial \beta_{ij}}\right)_{\mathbf{P}}, \tag{2.8}$$

$$\left(\frac{\partial U^L}{\partial P_i}\right)_{\boldsymbol{\beta}} = \left(\frac{\partial U}{\partial D_l}\right)_{\boldsymbol{\beta}} \left[\varepsilon_0 \left(\frac{\partial E_l}{\partial P_i}\right)_{\boldsymbol{\beta}} + \delta_{li}\right] - \varepsilon_0 E_l \left(\frac{\partial E_l}{\partial P_i}\right)_{\boldsymbol{\beta}}. \tag{2.9}$$

In (2.9) δ_{li} is the Kronecker delta. In view of eqs. (2.6), the latter derivatives are reduced to the constitutive relations

$$\sigma_{ij} = \left(\frac{\partial U^L}{\partial \beta_{ij}}\right)_{\mathbf{P}}, \quad E_i = \left(\frac{\partial U^L}{\partial P_i}\right)_{\boldsymbol{\beta}}. \tag{2.10}$$

On the other hand, with (2.10) the rate of change of the potential $U^L(\boldsymbol{\beta}, \mathbf{P})$ is equal

$$\dot{U}^L = \sigma_{ij}\dot{\beta}_{ij} + E_i\dot{P}_i. \tag{2.11}$$

(c) New forms for constitutive relations and equations of motion

An expansion of the function $U^L(\boldsymbol{\beta}, \mathbf{P})$ up to the second order terms has the form

$$U^L(\boldsymbol{\beta}, \mathbf{P}) = \frac{1}{2}c^P_{ijkl}\beta_{ij}\beta_{kl} + \frac{1}{2}a_{ij}P_iP_j + f_{kij}\beta_{ij}P_k, \tag{2.12}$$

where c^P_{ijkl} are the adiabatic moduli, however different from those in eq. (1.26): in contrast to the latter, they are supposed to be measured under the constant $\mathbf{P}$, instead of constant $\mathbf{E}$. Substituting the above expansion into eqs. (2.10) we find the explicit form of the constitutive equations replacing (1.35), (1.36):

$$\sigma_{ij} = c^P_{ijkl}\beta_{kl} + f_{kij}P_k, \tag{2.13}$$

$$E_i = f_{ikl}\beta_{kl} + a_{ik}P_k. \tag{2.14}$$

The symmetry properties of the introduced material constants are very similar to the analogous relations (1.37)-(1.39):

$$c^P_{ijkl} = c^P_{klij} = c^P_{jikl} = c^P_{ijlk}, \qquad f_{kij} = f_{kji}, \qquad a_{ik} = a_{ki}. \tag{2.15}$$

In fact, combining eqs. (2.13), (2.14) with (1.35), (1.36) and (2.3) it is possible to express the above constants c^P_{ijkl}, f_{kij} and a_{ik} in terms of material tensors of classical theory of piezoelectricity, c_{ijkl}, e_{kij} and ε_{ik}. Omitting bulky however rather trivial calculations we present the results [103]:

$$c^P_{ijkl} = c_{ijkl} - e_{sij}[(\varepsilon_0\mathbf{I} - \boldsymbol{\varepsilon})^{-1}]_{st}e_{tkl}, \tag{2.16}$$

$$f_{ikl} = [(\varepsilon_0\mathbf{I} - \boldsymbol{\varepsilon})^{-1}]_{ij}e_{jkl}, \tag{2.17}$$

$$a_{ik} = -[(\varepsilon_0\mathbf{I} - \boldsymbol{\varepsilon})^{-1}]_{ik}, \tag{2.18}$$

where $\mathbf{I}$ is the unit tensor and $\boldsymbol{\varepsilon} = \{\varepsilon_{ij}\}$ is the tensor of electric permittivity of the medium.

Equations of motion in this formalism are given by a substitution of constitutive equations (2.13), (2.14) into the system (1.40), (1.41) and (2.3). In the general case admitting possible presence of plastic deformation, eq. (1.45), one has

$$c_{ijkl}u_{k,lj} + f_{lij}P_{l,j} = \rho\ddot{u}_i - f_i + c_{ijkl}\beta^0_{kl,j}, \tag{2.19}$$

$$\varepsilon_0 f_{jlk}u_{k,lj} + (\delta_{jl} + \varepsilon_0 a_{jl})P_{l,j} = q + \varepsilon_0 f_{jlk}\beta^0_{kl,j}. \tag{2.20}$$

For boundary conditions one can repeat eqs. (1.56)-(1.61) with $\mathbf{D}$ replaced by $\varepsilon_0\mathbf{E} + \mathbf{P}$ (see eq. (2.3)).

As before, the above formalism can be slightly rearranged basing on the decomposition (1.51), (1.52). This would allow excluding electric sources from the piezoelectric equations of motion. Such approach was used in [15, 105]. We stress that eqs. (2.19), (2.20) represent just an alternative form to earlier obtained eqs. (1.46), (1.47) being absolutely equivalent to the latter. The only advantage of the last presentation consists in its convenience for the Mindlin's extension, which will be developed in the next subsection.

2.2 Mindlin's extension of the classical theory of piezoelectricity

(a) Mindlin's generalization of Toupin's potential U^L and some thermodynamical consequences

As was already explained in Introduction, the aim of Mindlin [104] was to find a continual description of phenomena and properties of a crystal lattice arising due to its discreteness. We mean such features as a dispersion of short-wavelength waves, an existence of additional wave modes at very high frequencies for any wavelength (optical phonon branches), and a possibility of electro-elastic coupling even in centrosymmetric crystals. The last property is naturally expected in ionic crystals (like NaCl) with elementary cells, which are electric dipoles. Electric polarization coupled with deformation may arise in such a medium only as a reaction on very inhomogeneous distortions, for instance, close to the dislocation core, or in the elastic field of very short waves (e.g., phonons).

Mindlin's idea of extension of the classical theory for a model continuous description of the above physical features of real materials was technically reduced to an addition of a new independent thermodynamic variable into a standard theory of Toupin's type. In order to describe the dispersion phenomena this variable should contain at least one derivative. And for including into the consideration the new electro-elastic coupling it should be some electrical characteristic. So, Mindlin [103, 104] has chosen the polarization gradient $\nabla \otimes \mathbf{P} \equiv \boldsymbol{\pi}$, $(P_{j,i} \equiv \pi_{ij})$. He supposed that the potential U^L (2.11), (2.12) depends not only on $\boldsymbol{\beta}$ and $\mathbf{P}$, but also on $\nabla \otimes \mathbf{P}$:

$$U^L = U^L(\boldsymbol{\beta}, \mathbf{P}, \boldsymbol{\pi}). \tag{2.21}$$

This time, instead of (2.7), one should write

$$U^L = (\boldsymbol{\beta}, \mathbf{P}, \boldsymbol{\pi}) = U(\boldsymbol{\beta}, \mathbf{D}, \mathbf{X}) - \frac{1}{2}\varepsilon_0[\mathbf{E}(\boldsymbol{\beta}, \mathbf{P}, \boldsymbol{\pi})]^2. \tag{2.22}$$

Here $\mathbf{X} = \{X_{ij}\}$ is the appropriate canonical thermodynamic variable describing the sensitivity of the energy U to the polarization gradient. Let us denote

$$Y_{kl} = \left(\frac{\partial U}{\partial X_{lk}}\right)_{\boldsymbol{\beta},\mathbf{D}}. \tag{2.23}$$

By the sense of Mindlin's model, the gradient effects should provide some addition to the classical energy $U(\boldsymbol{\beta}, \mathbf{D})$. In other words, the extension for eq. (2.5) must read

$$\dot{U}(\boldsymbol{\beta}, \mathbf{D}, \mathbf{X}) = \sigma_{ij}\dot{\beta}_{ij} + E_i\dot{D}_i + Y_{ij}\dot{X}_{ij}. \tag{2.24}$$

Making time derivative in (2.22) with due regards for (2.24) and (2.3) one obtains

$$\dot{U}^L(\boldsymbol{\beta}, \mathbf{P}, \boldsymbol{\pi}) = \sigma_{ij}\dot{\beta}_{ij} - E_i^L\dot{P}_i + Q_{ij}\dot{\pi}_{ij}, \tag{2.25}$$

where the new physical quantities known as the local electric force vector $\mathbf{E}^L$ and the electric tensor $\mathbf{Q}$ are, respectively, defined by

$$\begin{aligned} -E_i^L &= E_i + Y_{lk}\left(\frac{\partial X_{kl}}{\partial P_i}\right)_{\boldsymbol{\pi}} = \left(\frac{\partial U^L}{\partial P_i}\right)_{\boldsymbol{\beta},\boldsymbol{\pi}}, \\ Q_{ij} &= Y_{lk}\left(\frac{\partial X_{kl}}{\partial \pi_{ij}}\right)_{\mathbf{P}} = \left(\frac{\partial U^L}{\partial \pi_{ij}}\right)_{\boldsymbol{\beta},\mathbf{P}}. \end{aligned} \tag{2.26}$$

Here X is considered as a function $X(\mathbf{P}, \boldsymbol{\pi})$ independent of $\boldsymbol{\beta}$. This is a rather delicate place in Mindlin's model. One can see that at $\pi_{ij} = 0$ eq. (2.25) does not coincide with its classical analog (2.11) until $-\mathbf{E}^L \neq \mathbf{E}$. Such inequality is not so trivial, because both E_i and $-E_i^L$ by eqs. $(2.10)_2$ and $(2.26)_1$ are derivatives of U^L over P_i. In fact, we deal here with functional gradients of U^L. However, the first gradient is calculated for any variations of "trajectories" $\delta P_i(\mathbf{r}, t)$ while the second one - for only such "trajectories", along which $P_{i,j}(\mathbf{r}, t) = 0$. It is rather evident that those two derivatives generally must differ from each other and one should expect $-\mathbf{E}_i^L \neq \mathbf{E}_i$. That is why the derivative $(\partial X_{kl}/\partial P_i)_{\boldsymbol{\pi}}$ should not vanish, i.e. X_{kl} must depend not only on the polarization gradient $P_{i,j}$, but also on the polarization P_i itself: $X_{kl} = X_{kl}(P_i, \pi_{ij})$. On the other hand, with such a structure of X_{kl} its variation must be determined by two independent variations

$$\delta X_{kl} = \left(\frac{\partial X_{kl}}{\partial P_i}\right)_{\boldsymbol{\pi}} \delta P_i + \left(\frac{\partial X_{kl}}{\partial \pi_{ij}}\right)_{\mathbf{P}} \delta\pi_{ij}. \tag{2.27}$$

This means that at $\pi_{ij} = 0$ and $\delta P_i \neq 0$ the time derivative $\dot{X}_{kl} = \delta X_{kl}/\delta t$ does not vanish. But then the rate of change of energy $\dot{U}$ (2.24) at $\pi_{ij} = 0$ also (similarly to $\dot{U}^L$) does not coincide with its classical variant (2.5). In fact, this is just the other face of the same medal.

(b) Constitutive relations and equations of motion

Now let us return to the potential $U^L(\boldsymbol{\beta}, \mathbf{P}, \boldsymbol{\pi})$. Its expansion close to the ground state is now lengthier than (2.12):

$$
\begin{aligned}
U^L(\boldsymbol{\beta},\mathbf{P},\boldsymbol{\pi}) = b^0_{ij}\pi_{ij} + \frac{1}{2}a^{\pi}_{ij}P_iP_j + \frac{1}{2}b_{ijkl}\pi_{ij}\pi_{kl} + \frac{1}{2}c^{P,\pi}_{ijkl}\beta_{ij}\beta_{kl} + \\
+ d^P_{ijkl}\pi_{ij}\beta_{kl} + f^{\pi}_{kij}P_k\beta_{ij} + j_{ikl}P_i\pi_{kl}.
\end{aligned} \tag{2.28}
$$

Taking into account that the gradient corrections are expected to be rather small it seems to be quite reasonable to put into (2.26) for $c^{P,\pi}_{ijkl}$, f^{π}_{kij} and a^{π}_{ij} the material parameters c^P_{ijkl}, f_{kij} and a_{ij} given by eqs. (2.16)-(2.18). With regards to the new introduced tensors, we can only state that generally they are less symmetrical as compared with classical material tensors of the same rank:

$$
b^0_{ij} \neq b^0_{ji}, \quad j_{ikl} \neq j_{ilk}, \quad d^P_{ijkl} = d^P_{ijlk}, \quad b_{ijkl} = b_{klij}. \tag{2.29}
$$

In addition, they should be small in a sense that three of them (b^0_{ij}, j_{ikl} and d^P_{ijkl}) must be linear in some length parameter, which is expected to be of order of the lattice discreteness measure, and the fourth tensor (b_{ijkl}) - even quadratic in this parameter. Below for brevity we shall omit the superscripts π and P from all moduli notations. Eq. (2.28) provides the explicit form for appropriate constitutive equations:

$$
\sigma_{ij} = c_{ijkl}\beta_{kl} + f_{kij}P_k + d_{klij}\pi_{kl}, \tag{2.30}
$$

$$
-E^L_i = f_{ikl}\beta_{kl} + a_{ik}P_k + j_{ikl}\pi_{kl}, \tag{2.31}
$$

$$
Q_{ij} = d_{ijkl}\beta_{kl} + j_{kij}P_k + b_{ijkl}\pi_{kl} + b^0_{kl}. \tag{2.32}
$$

One can see from eq. (2.30) that even when there is no piezoelectric effect, $f_{kij} = 0$, the electro-elastic coupling retains due to the gradient term $d_{klij}P_{l,k}$. We must stress that the linear first term in (2.28) and the respective constant last (but not least) constant term in (2.32) were retained not at all accidentally. Their significance and quite nontrivial physical meaning will be separately discussed in the next subsection.

Let us now derive the new form for equations of motion. As before, one can operate by the two classical equations (1.40), (1.41) combined with (1.44) and (2.3). However, in the considered extended variant of theory we have introduced two new unknown values, E^L_i and Q_{ij}, and have got only one additional equation - the third constitutional equation (2.32). So, in order to close the system we need to establish the relation between E_i, E^L_i and Q_{ij} and to find the boundary condition for Q_{ij}. Following to Mindlin [104], this can be done basing on standard Hamilton's principle reducing in our case to a minimization of the functional:

$$
\delta \int_{t_1}^{t_2} dt \left\{ \int_{V+V'} dV(K-H) + \int_V dV(f_iu_i - q\varphi) + \int_S dS f^s_i u_i \right\} = 0\,. \tag{2.33}
$$

Here the first volume integral is taken over all the space, the bulk of the body V and the surrounding vacuum V'. The external forces in the bulk f_i and at the surface f_i^s and the charge q are supposed to be constant as well as the displacements at the moments t_1 and t_2,

$$\delta \mathbf{u}(\mathbf{r}, t_1) = 0, \quad \delta \mathbf{u}(\mathbf{r}, t_2) = 0. \tag{2.34}$$

K and H are, respectively, the density of the kinetic energy,

$$K = \frac{1}{2}\rho \dot{u}_i^2, \tag{2.35}$$

and the electric enthalpy,

$$H = U - \mathbf{E}\cdot\mathbf{D} = U^L + \frac{1}{2}\varepsilon_0 \mathbf{E}^2 - \mathbf{E}\cdot(\mathbf{P} + \varepsilon_0\mathbf{E}) = U^L - \frac{1}{2}\varepsilon_0\mathbf{E}^2 - \mathbf{E}\cdot\mathbf{P}. \tag{2.36}$$

The variation of the kinetic energy is

$$\delta \int_{t_1}^{t_2} dt \int_{V+V'} dV K = \int_{t_1}^{t_2} dt \int_V dV \rho \dot{u}_i \delta \dot{u}_i = -\int_{t_1}^{t_2} dt \int_V dV \rho \ddot{u}_i \delta u_i, \tag{2.37}$$

where we have made use of the identity

$$\dot{u}_i \delta \dot{u}_i = \frac{\partial}{\partial t}(\dot{u}_i \delta u_i) - \ddot{u}_i \delta u_i. \tag{2.38}$$

In view of (2.25), the variation of the electric enthalpy can be presented in the form

$$\delta H = \sigma_{ji}\delta\beta_{ij} - (E_i + E_i^L)\delta P_i + Q_{ji}\delta P_{i,j} - \varepsilon_0 E_i \delta E_i - P_i \delta E_i. \tag{2.39}$$

Substituting here

$$\sigma_{ji}\delta\beta_{ij} = \sigma_{ji}\delta(u_{i,j} - \beta_{ij}^0) = \sigma_{ji}\delta u_{i,j} = (\sigma_{ji}\delta u_i)_{,j} - \sigma_{ji,j}\delta u_i \ , \tag{2.40}$$

$$Q_{ji}\delta P_{i,j} = (Q_{ji}\delta P_i)_{,j} - Q_{ji,j}\delta P_i, \tag{2.41}$$

$$(\varepsilon_0 E_i + P_i)\delta E_i = -[(\varepsilon_0 E_i + P_i)\delta\varphi]_{,i} + (\varepsilon_0 E_i + P_i)_{,i}\delta\varphi \ , \tag{2.42}$$

and combining the result with (2.33) and (2.36) one obtains

$$\begin{aligned} &\int_{t_1}^{t_2} dt \left\{ \int_V dV[(\sigma_{ji,j} + f_i - \rho\ddot{u}_i)\sigma u_i + (Q_{ji,j} + E_i - E_i^L)\delta P_i + \right. \\ &\qquad +[(\varepsilon_0 E_i + P_i)_{,i} - q]\delta\varphi] - \int_{V'} dV \varepsilon_0 \varphi'_{,ii}\delta\phi' - \\ &\left. \int_S dS[(n_i\sigma_{ij} - f_j^s)\delta u_j + n_i Q_{ij}\delta P_j + n_i(P_i - \varepsilon_0[\varphi]_{,i})\delta\varphi] \right\} = 0. \end{aligned} \tag{2.43}$$

Here φ' is the electric potential field in the surrounding vacuum and $[\varphi]$ is the difference $\varphi - \varphi'$ at the surface.

Thus, from (2.43) we have the equations of motion

$$\mathrm{div}\sigma + \mathbf{f} = \rho\ddot{\mathbf{u}}, \tag{2.44}$$

$$\mathrm{div}(\varepsilon_0\mathbf{E} + \mathbf{P}) = q, \tag{2.45}$$

$$\mathrm{div}\mathbf{Q} + \mathbf{E} = -\mathbf{E}^L. \tag{2.46}$$

As was expected, the first couple, eqs. (2.44) and (2.45), are just a repetition of classical equations (1.40) and (1.41) with (2.3). However, the third one, eq. (2.46), provides just the necessary additional relation between the unknown quantities $\mathbf{Q}$, $\mathbf{E}^L$ and $\mathbf{E}$. Now one can see, why the inequality $\mathbf{E} \neq -\mathbf{E}^L$ is so essential for the considered theory (see the discussion in the end of Sec. 2 (a)). Indeed, for $\mathbf{E} = -\mathbf{E}^L$, which would occur at $\partial X_{kl}/\partial P_i = 0$, the one of the unknown quantities disappears and the above system completely changes its appearance.

It is worthy to note that system (2.44)-(2.46) is very convenient for counting the unknown electric field $\mathbf{E}$ from the background field $\mathbf{E}^0$ in the vacuum produced by some sources, for instance, by the charge density $q_0 = q - q_c$. Indeed, after the substitution $\mathbf{E} = \mathbf{E}' + \mathbf{E}^0$ into eqs. (2.45) and (2.46), with making use of the identity

$$\mathrm{div}\varepsilon_0\mathbf{E}^0 = q_0, \tag{2.47}$$

one obtains instead of these equations

$$\mathrm{div}(\varepsilon_0\mathbf{E}' + \mathbf{P}) = q_c, \tag{2.48}$$

$$\mathrm{div}\mathbf{Q} + \mathbf{E}' + \mathbf{E}^0 = -\mathbf{E}^L. \tag{2.49}$$

This is the form of equations used in [15].

Let us now discuss the boundary conditions following from (2.42). The mechanical part of these conditions is described by the equation

$$(\mathbf{n}\boldsymbol{\sigma} - \mathbf{f}^s)\cdot\delta\mathbf{u} = 0, \tag{2.50}$$

which is decomposed on the two different boundary conditions related either to a prescribed force density at the surface:

$$\mathbf{n}\boldsymbol{\sigma} = \mathbf{f}^s, \tag{2.51}$$

or to a prescribed displacement distribution along the surface

$$\mathbf{u} = \mathbf{u}^s. \tag{2.52}$$

We already met the both above conditions, see eqs. (1.56) and (1.57). The electrical boundary conditions following from (2.43) are determined by

$$[\mathbf{n}\cdot(\varepsilon_0\mathbf{E} + \mathbf{P}) - \mathbf{n}\cdot\mathbf{D}_{adj}]\delta\varphi = 0, \tag{2.53}$$

where $\mathbf{D}_{adj} = -\varepsilon_0 \nabla \varphi'$ is the electric induction on the vacuum side of the boundary. Quite similarly eq. (2.53) can be decomposed on the known electrical conditions at the surface, see eqs. (1.59) and (1.60). In fact, by a slight change of the above Hamilton's procedure we would easily come to an alternative condition of a prescribed charge density at the surface, eq. (1.58). The only new boundary condition provided by (2.42) is related to a new tensor quantity Q_{ij} :

$$\mathbf{nQ} = \mathbf{0}. \tag{2.54}$$

Thus, eqs. (2.46) and (2.54) close the system extending the classical theory. However, one should admit that, in contrast to the latter, this new system looks a bit too cumbersome, at least in its electrical part. Indeed, combining equations of motion (2.44)-(2.46) with constitutive equations (2.30)-(2.32) we come to the following extension of the classical equations (2.19), (2.20):

$$c_{ijkl}u_{l,kj} + f_{lij}P_{l,j} + d_{klij}P_{l,kj} = \rho\ddot{u}_i - f_i + c_{ijkl}\beta^0_{kl,j}, \tag{2.55}$$

$$\begin{aligned} &\varepsilon_0 f_{jkl}u_{l,kj} - d_{ijkl}u_{l,kij} + (\delta_{jl}\varepsilon_0^{-1} + a_{jl})P_{l,j} \\ &+(j_{jkl} - j_{lkj})P_{l,kj} - b_{ijkl}P_{l,kij} \\ &= q\varepsilon_0^{-1} + f_{jkl}\beta^0_{kl,j} - d_{ijkl}\beta^0_{kl,ij}. \end{aligned} \tag{2.56}$$

It is worthwhile to notice that the material tensor b^0_{ij} is not involved in these equations. However, as we shall see, it describes surface effects and principally acts through boundary condition (2.54).

(c) Surface tension in Mindlin's formulation

One of the important dispersion effects, existing due to a discreetness of the lattice and not describ by the classical theory of elasticity, is a so-called surface tension. The atoms situated in a body close to a surface have a less number of neighbors from one side than from the other side. So the conditions of equilibrium for such atoms must be different than for atoms in the bulk. This is a physical reason for a known dispersion of a lattice parameter in the vicinity of surfaces. On the other hand, the corresponding elastic displacements in the pre-surface regions must create there some additional elastic energy known as a surface tension. It is a remarkable feature of Mindlin's theory that, in contrast to the classical theory, it has an instrument, the material tensor b^0_{ij}, for a description of the surface tension effect. Let us demonstrate this directly.

Combining constitutive relations (2.30)-(2.32) with expansion (2.28) of the energy density U^L, one can express the latter in the form

$$U^L = \frac{1}{2}b^0_{ij}P_{j,i} + \frac{1}{2}\sigma_{ij}\beta_{ij} - \frac{1}{2}E^L_i P_i + \frac{1}{2}Q_{ij}P_{j,i}. \tag{2.57}$$

Then, in view of (2.4), the total energy of the bulk V bounded by the surface S is given by

$$\int\limits_V UdV = \frac{1}{2}\int\limits_V [(b^0_{ij}+Q_{ij})P_{j,i}+\sigma_{ij}(u_{i,j}-\beta^0_{ij})-(E^L_i+E_i)P_i+E_iD_i]dV, \quad (2.58)$$

where eq. (1.45) is used. Substituting here the identities

$$(b^0_{ij}+Q_{ij})P_{j,i} = \left[(b^0_{ij}+Q_{ij})P_j\right]_{,i} - Q_{ij,i}P_j, \quad (2.59)$$

$$\sigma_{ij}u_{i,j} = (\sigma_{ij}u_i)_{,j} - \sigma_{ij,j}u_i, \quad (2.60)$$

$$E_iD_i = -\varphi_{,i}D_i = -(\varphi D_i)_{,i} + \varphi D_{i,i}, \quad (2.61)$$

and making use of the divergence theorem we obtain

$$\begin{aligned}\int\limits_V UdV = \frac{1}{2}\int\limits_S n_i\left[(b^0_{ij}+Q_{ij})P_j + \sigma_{ij}u_j - \varphi D_i\right]dS-\\ -\int\limits_V \left[(Q_{ij,i}+E_j+E^L_j)P_j + \sigma_{ij,i}u_j + \sigma_{ij}\beta^0_{ij} - \varphi D_{i,i}\right]dV.\end{aligned} \quad (2.62)$$

Substituting here eqs. (2.44)-(2.46) for a static equilibrium together with boundary conditions (1.56), (1.58) and (2.54) one has

$$\int\limits_V UdV = \frac{1}{2}\int\limits_S (n_ib^0_{ij}P_j + f^s_iu_i + q^s\varphi)dS + \frac{1}{2}\int\limits_V (f_iu_i - \sigma_{ij}\beta^0_{ij} + q\varphi)dV. \quad (2.63)$$

In the absence of any external forces, electric charges or plastic distortions it gives

$$\int\limits_V UdV = \frac{1}{2}\int\limits_S n_ib^0_{ij}P_jdS. \quad (2.64)$$

Thus even without external sources the Mindlin formalism gives a non-vanishing electro-elastic energy, which is expressed by a surface integral having a clear meaning of a surface energy. The integrand representing a surface energy per unit area (the surface tension),

$$U^{surf} = \frac{1}{2}[n_ib^0_{ij}P_j]_S, \quad (2.65)$$

is completely determined by the material constant b^0_{ij}, which serves as a sources for pre-surface physical fields. Below we shall demonstrate how it works.

2.3 Mindlin's equations for centrosymmetric and isotropic media

(a) Equations for centrosymmetric media and the example of their solution

As was already mentioned, the discussed gradient effects should manifest themselves in very specific conditions, when the involved sources of fields are of microscopic scale (dislocation lines, phonon waves etc.). It is clear that electro-elastic phenomena of this kind could hardly be extracted on the background of strong classical piezoelectric coupling. On the other hand, in a medium with the center of inversion where classical piezoeffects are forbidden by symmetry the above polarization gradient influence would be the only source for linear electro-elastic coupling. Thus, the favorable conditions for studies of the new mechanism of electromechanical coupling are provided just in centrosymmetric media. As is well known [130, 131], in such media all material tensors of odd rank must vanish. In our case this means that

$$f_{ikl} = 0, \quad j_{ikl} = 0. \tag{2.66}$$

Accordingly, all equations are simplified. The new form for constitutive equations (2.30)-(2.32) looks much more compact

$$\sigma_{ij} = c_{ijkl}\beta_{kl} + d_{klij}P_{l,k} \ , \tag{2.67}$$

$$-E_i^L = a_{ik}P_k \ , \tag{2.68}$$

$$Q_{ij} = d_{ijkl}\beta_{kl} + b_{ijkl}P_{l,k} + b^0_{ij} \ . \tag{2.69}$$

Equations of motion also become shorter

$$c_{ijkl}u_{l,kj} + d_{klij}P_{l,kj} = \rho\ddot{u}_i - f_i + c_{ijkl}\beta^0_{kl,j}, \tag{2.70}$$

$$d_{ijkl}u_{l,ki} + b_{ijkl}P_{l,ki} - a_{jl}P_l + E_j = d_{ijkl}\beta^0_{kl,ij}, \tag{2.71}$$

$$\varepsilon_0 E_{i,i} + P_{i,i} = q. \tag{2.72}$$

This system allows even a direct solving some physical problems. Below on two examples we shall demonstrate how it works in practice.

1^0. Longitudinal bulk wave in a cubic crystal

Let us consider a plane longitudinal acoustic wave propagating along the symmetry axis of an even order directed along the x-axis in a crystal where

$$c_{1112} = c_{1113} = b_{1112} = b_{1113} = d_{1112} = d_{1113} = 0. \tag{2.73}$$

In particular, it might be a cubic crystal along the direction [100]. The electro-elastic field in such wave is described by the combination

$$\begin{gathered}\mathbf{u}(x,t) = \mathbf{A}\exp[i(kx-\omega t)],\\ \mathbf{P}(x,t) = \mathbf{B}\exp[i(kx-\omega t)], \quad \mathbf{E}(x,t) = \mathbf{C}\exp[i(kx-\omega t)],\end{gathered} \tag{2.74}$$

where

$$\mathbf{A} = [A,0,0], \quad \mathbf{B} = [B,0,0], \quad \mathbf{C} = [C,0,0]. \tag{2.75}$$

Substitution (2.74) with (2.75) into eqs. (2.70)-(2.72) taken at $\mathbf{f} = \mathbf{0}$, $q = 0$ and $\boldsymbol{\beta}^0 = \mathbf{0}$ leads to a linear algebraic system of homogeneous equations with respect to the unknown amplitudes A, B and C. After a trivial exclusion of $C = \varepsilon_0^{-1} B$ it reduces to

$$(c_{11}k^2 - \rho\omega^2)A + d_{11}k^2 B = 0, \tag{2.76}$$

$$d_{11}k^2 A + (\varepsilon_0^{-1} + a + b_{11}k^2)B = 0, \tag{2.77}$$

where the notation is introduced

$$c_{11} = c_{1111}, \quad d_{11} = d_{1111}, \quad b_{11} = b_{1111}, \quad a_{ij} = a\delta_{ij}. \tag{2.78}$$

The condition for an existence of a nontrivial solution for the above homogeneous system is reduced, as usually, to the vanishing determinant

$$\begin{vmatrix} c_{11}k^2 - \rho\omega^2 & d_{11}k^2 \\ d_{11}k^2 & \varepsilon_0^{-1} + a + b_{11}k^2 \end{vmatrix} = 0. \tag{2.79}$$

Solving this equation one readily obtains the following dispersion dependence $\omega(k)$

$$\omega = c_l k \left[1 - \frac{(d_{11}^2/c_{11})k^2}{\varepsilon_0^{-1} + a + b_{11}k^2}\right]^{\frac{1}{2}}, \tag{2.80}$$

where c_l is the longitudinal sound speed,

$$c_l = \sqrt{c_{11}/\rho}. \tag{2.81}$$

The reality of the expression (2.80) is guaranteed by the inequalities

$$a > 0, \quad b_{11}c_{11} - d_{11}^2 > 0, \tag{2.82}$$

which follow from the positive definiteness of the energy.

The dependence (2.80) looks to be qualitatively very reasonable. We recall that the material parameters d_{11} and b_{11} are small as compared with classical moduli being proportional to the lattice parameter (d_{11}) or to its square (b_{11}). This means that $d_{11}k \propto k/k_D$ and $b_{11}k^2 \propto (k/k_D)^2$, where k_D is the Debye limit in the phonon spectrum. In the long-wavelength limit at $k \ll k_D$ we have from (2.80) the classical dispersionless dependence $\omega = c_l k$ known for ultrasound. However, at $k \sim k_D$ the found dependence $\omega(k)$ acquires the typical features for phonons with a noticable decrease of the group velocity $\partial\omega/\partial k$ upon growing k.

The relation between the amplitudes A and B of mechanical and electrical components of the wave,

$$B = \left(\frac{(d_{11}k^2}{\varepsilon_0^{-1} + a + b_{11}k^2}\right) A, \tag{2.83}$$

manifests the same reasonable quality. At small $k \ll k_D$ there is practically no electromechanical coupling, as it should be in a considered centrosymmetric crystal. However at $k \sim k_D$ the coefficient between A and B in (2.83) has no small parameters. Thus, we can state that Mindlin's theory works in accordance with our expectations.

2^0. Electro-elastic fields excited by a surface tension in a cubic crystal

Consider a semi-infinite medium ($x \geq 0$) of the same cubic symmetry with a free surface belonging to the system (100). So, again all three unknown vectors **u**, **P** and **E** must be parallel to the symmetry axis x being dependent on only x :

$$\mathbf{u} = [u(x), 0, 0], \quad \mathbf{P} = [P(x), 0, 0], \quad \mathbf{E} = [E(x), 0, 0]. \tag{2.84}$$

This time the problem is static, however with an existing surface the boundary conditions must be involved. Accordingly, system (2.70)-(2.72) reduce to the form

$$c_{11}u_{,xx} + d_{11}P_{,xx} = 0, \tag{2.85}$$

$$d_{11}u_{,xx} + b_{11}P_{,xx} - aP + E = 0, \tag{2.86}$$

$$\varepsilon_0 E_{,x} + P_{,x} = 0 \tag{2.87}$$

and the boundary conditions at $x = 0$ become

$$c_{11}u_{,x} + d_{11}P_{,x} = 0, \tag{2.88}$$

$$d_{11}u_{,x} + b_{11}P_{,x} = b_0, \tag{2.89}$$

$$\varepsilon_0 E + P = 0. \tag{2.90}$$

We shall look for solutions in the form

$$u(x) = A\exp(-\lambda x), \quad P(x) = B\exp(-\lambda x), \quad E(x) = C\exp(-\lambda x). \tag{2.91}$$

The substitution of (2.91) into the equilibrium equations (2.85)-(2.87) immediately provides the relations

$$C = -B/\varepsilon_0 = c_{11}A/\varepsilon_0 d_{11}, \tag{2.92}$$

$$\lambda = \sqrt{\frac{c_{11}(a + \varepsilon_0^{-1})}{c_{11}b_{11} - d_{11}^2}}. \tag{2.93}$$

The reality of λ is provided by the condition (2.82). With (2.92) the boundary conditions (2.88)-(2.90) give finally

$$A = -\frac{\lambda b_0 d_{11}}{c_{11}(a + \varepsilon_0^{-1})}, \qquad B = \frac{\lambda b_0}{a + \varepsilon_0^{-1}}, \qquad C = -\frac{\lambda b_0}{a\varepsilon_0 + 1}. \tag{2.94}$$

As one could expect, the depth of "penetration" λ^{-1} proves to be of order of the lattice parameter being independent of the modulus b_0. On the other hand, the all three amplitudes A, B and C are linear in b_0. The surface tension (2.65) with the field $P(x)$ found, eqs. (2.91) and (2.94), is

$$U^{surf} = -\frac{\lambda b_0^2}{2(a + \varepsilon_0^{-1})}. \tag{2.95}$$

Thus, in Mindlin's theory indeed the modulus b_0 proves to be a source of pre-surface physical fields.

(b) Equations for isotropic media

W. Nowacki [15] has derived a very convenient isotropic representation for Mindlin's equations and solved on this basis a series of physical problems, both static and dynamic. Having in mind some further applications of this approach to the dislocation theory, let us extend Nowacki's equations and add there the source terms related to plastic distortions β^0. We start from the constitutive relations (2.67)-(2.69). An isotropic form for the corresponding material tensors is given by

$$a_{ij} = a\delta_{ij}, \qquad b^0_{ij} = b_0\delta_{ij} \ , \tag{2.96}$$

$$c_{ijkl} = c_{12}\delta_{ij}\delta_{kl} + c_{44}(\delta_{ik}\delta_{jl} + \delta_{il}\delta_{jk}), \tag{2.97}$$

$$d_{ijkl} = d_{12}\delta_{ij}\delta_{kl} + d_{44}(\delta_{ik}\delta_{jl} + \delta_{il}\delta_{jk}), \tag{2.98}$$

$$b_{ijkl} = b_{12}\delta_{ij}\delta_{kl} + b_{44}(\delta_{ik}\delta_{jl} + \delta_{il}\delta_{jk}) + b_{77}(\delta_{ik}\delta_{jl} - \delta_{il}\delta_{jk}), \tag{2.99}$$

where the coefficients are isotropic moduli. We recall that in terms of the above moduli the conditions for isotropy read

$$c_{11} - c_{12} - 2c_{44} = 0 \ , \quad d_{11} - d_{12} - 2d_{44} = 0, \quad b_{11} - b_{12} - 2b_{44} = 0, \tag{2.100}$$

with c_{11}, d_{11} and b_{11} defined by eq. (2.78. With representations (2.96)-(2.99) one obtains instead of (2.67)-(2.69):

$$\sigma_{ij} = c_{12}\beta_{kk}\delta_{ij} + c_{44}(\beta_{ij} + \beta_{ji}) + d_{12}P_{k,k}\delta_{ij} + d_{44}(P_{i,j} + P_{j,i}), \tag{2.101}$$

$$E^L_i = aP_i \ , \tag{2.102}$$

$$\begin{aligned} Q_{ij} = d_{12}\beta_{kk}\delta_{ij} + d_{44}(\beta_{ij} + \beta_{ji}) + b_{12}P_{k,k}\delta_{ij} + b_{44}(P_{i,j} + P_{j,i}) + \\ + b_{77}(P_{j,i} - P_{i,j}) + b_0\delta_{ij}. \end{aligned} \tag{2.103}$$

Substitution of these constitution equations together with the Kröner relation

$$\beta_{ij} = u_{i,j} - \beta^0_{ij} \tag{2.104}$$

into system (2.44)-(2.46) gives the final form for isotropic equations of motion

$$c_{44}\nabla^2\mathbf{u} + (c_{12}+c_{44})\nabla\nabla\cdot\mathbf{u} + d_{44}\nabla^2\mathbf{P} + (d_{12}+d_{44})\nabla\nabla\cdot\mathbf{P} + \mathbf{f} + \mathbf{f}^{pl} = \rho\ddot{u}, \tag{2.105}$$

$$\begin{aligned} d_{44}\nabla^2\mathbf{u} + (d_{12} + d_{44})\nabla\nabla\cdot\mathbf{u} + (b_{44} + b_{77})\nabla^2\mathbf{P} + (b_{12} + \\ + b_{44} - b_{77})\nabla\nabla\cdot\mathbf{P} - a\mathbf{P} + \mathbf{E} = \mathbf{0}, \end{aligned} \tag{2.106}$$

$$\nabla\cdot(\mathbf{P} + \varepsilon_0\mathbf{E}) = q. \tag{2.107}$$

In eq. (2.105) we have introduced the effective force $\mathbf{f}^{pl}$ induced by the plastic distortion:

$$\mathbf{f}^{pl} = -c_{12}\text{grad}\,\text{Sp}\,\varepsilon^0 - 2c_{44}\text{div}\varepsilon^0, \tag{2.108}$$

where ε^0 is the tensor of plastic strain, i.e. the symmetric part of the distortion β^0 :

$$\varepsilon^0_{ij} = \frac{1}{2}(\beta^0_{ij} + \beta^0_{ji}). \tag{2.109}$$

Below we shall also use the more compact symbolic Navier's form of the above system of equations of motion. Let us introduce three differential operators

$$C_{ji} = c_{jpis}\nabla_p\nabla_s = [c_{12}\delta_{jp}\delta_{is} + c_{44}(\delta_{ps}\delta_{ij} + \delta_{js}\delta_{pi})]\nabla_p\nabla_s, \tag{2.110}$$

$$D_{ji} = d_{jpis}\nabla_p\nabla_s = [d_{12}\delta_{jp}\delta_{is} + d_{44}(\delta_{ps}\delta_{ij} + \delta_{js}\delta_{pi})]\nabla_p\nabla_s, \tag{2.111}$$

$$\begin{aligned} B_{ji} = b_{jpis}\nabla_p\nabla_s = [b_{12}\delta_{jp}\delta_{is} + b_{44}(\delta_{ps}\delta_{ij} + \delta_{js}\delta_{pi}) \\ +b_{77}(\delta_{ps}\delta_{ij} - \delta_{js}\delta_{pi})]\nabla_p\nabla_s - a\delta_{ij}\ . \end{aligned} \tag{2.112}$$

In these terms the equations of motion read [112-116]

$$C_{ji}u_i + D_{ji}P_i + f_j + f_j^{pl} = \rho\ddot{u}_j, \tag{2.113}$$

$$D_{ij}u_i + B_{ij}P_i + E_j = 0, \tag{2.114}$$

$$\varepsilon_0 E_{i,i} + P_{i,i} = q. \tag{2.115}$$

The remarkable similarity of the first pair of equations with respect to differential operations acting on the unknown vectors **u** and **P** in the left hand side will simplify some of our further calculations. However, this formal similarity should not mislead us to wrong physical conclusions. One should not forget that the displacement vector **u** is a potential function defined only to the accuracy of any constant vector and providing physically measurable quantities only after its differentiation. And the polarization **P** itself is a physical value manifesting itself in experiments.

2.4 Thermoelasticity of dielectrics with polarization gradient

(a) Extension of the formalism for thermal effects

The above description was given for simplified situations when one can ignore the processes of heat conduction. The extension of Mindlin's theory for thermoelastic dielectrics has been developed by Chowdhury & Glockner [117]. Later J.P. Nowacki & Glockner [119] have introduced a new approach to the derivation of the same equations obtained in [117]. This approach is reproduced in the book [15]. Below we shall not repeat rather cumbersome developments of [117, 119]. In our case it is more convenient to choose an alternative more compact procedure based on the ready elements prepared in above considerations. Let us introduce the potential U^L (2.21), (2.22) not omitting in that an appropriate thermal variable - the entropy S :

$$U^L(\boldsymbol{\beta}, \mathbf{P}, \boldsymbol{\pi}, S) = U(\boldsymbol{\beta}, \mathbf{D}, \mathbf{X}, S) - \frac{1}{2}\varepsilon_0[\mathbf{E}(\boldsymbol{\beta}, \mathbf{P}, \boldsymbol{\pi}, S)]^2. \tag{2.116}$$

Bearing in mind the corresponding extension of (2.5), (2.24) based on (1.11), (1.12)

$$\dot{U}(\boldsymbol{\beta}, \mathbf{D}, \mathbf{X}, S) = \sigma_{ij}\dot{\beta}_{ij} + E_i\dot{D}_i + Y_{ji}\dot{X}_{ij} + T\dot{S}, \tag{2.117}$$

one can easily find the coefficients in the corresponding decomposition for $\dot{U}^L$:

$$\begin{aligned} \delta\dot{U}^L(\boldsymbol{\beta}, \mathbf{P}, \boldsymbol{\pi}, S) = &\left(\frac{\partial U^L}{\partial \beta_{ij}}\right)_{\mathbf{P},\boldsymbol{\pi},S} \delta\dot{\beta}_{ij} + \left(\frac{\partial U^L}{\partial P_i}\right)_{\boldsymbol{\beta},\boldsymbol{\pi},S} \delta\dot{P}_i + \\ &+ \left(\frac{\partial U^L}{\partial \pi_{ij}}\right)_{\boldsymbol{\beta},\mathbf{P},S} \delta\dot{\pi}_{ij} + \left(\frac{\partial U^L}{\partial S}\right)_{\boldsymbol{\beta},\mathbf{P},\boldsymbol{\pi}} \delta\dot{S}. \end{aligned} \tag{2.118}$$

Upon calculating derivatives of the potential U^L (2.116) the function $U(\boldsymbol{\beta}, \mathbf{D}, \mathbf{X}, S)$ should be expressed in terms of the arguments $\boldsymbol{\beta}, \mathbf{P}, \boldsymbol{\pi}, S$:

$$U(\boldsymbol{\beta}, \mathbf{D}, \mathbf{X}, S) = U[\boldsymbol{\beta}; \varepsilon_0\mathbf{E}(\boldsymbol{\beta}, \mathbf{P}, \boldsymbol{\pi}, S) + \mathbf{P}; \mathbf{X}(\mathbf{P}, \boldsymbol{\pi}); S]. \tag{2.119}$$

The procedure is completely similar to that in eqs. (2.7)-(2.10). We shall present only results:

$$\begin{aligned} &\left(\frac{\partial U^L}{\partial \beta_{ij}}\right)_{\mathbf{P},\boldsymbol{\pi},S} = \sigma_{ij}, \quad \left(\frac{\partial U^L}{\partial P_i}\right)_{\boldsymbol{\beta},\boldsymbol{\pi},S} = -E_i^L, \\ &\left(\frac{\partial U^L}{\partial \pi_{ij}}\right)_{\boldsymbol{\beta},\mathbf{P},S} = Q_{ij}, \quad \left(\frac{\partial U^L}{\partial S}\right)_{\boldsymbol{\beta},\mathbf{P},\boldsymbol{\pi}} = T. \end{aligned} \tag{2.120}$$

Here one should consider σ_{ij} and T as physical quantities, mechanical stress tensor and temperature, identically equal to those in eq. (2.117). On the other hand, the local electric force E_i^L and the electric tensor Q_{ij} in (2.120) are clearly the same as in (2.26).

With (2.120) eq. (2.118) takes the form

$$\delta\dot{U}^L(\boldsymbol{\beta}, \mathbf{P}, \boldsymbol{\pi}, S) = \sigma_{ij}\delta\dot{\beta}_{ij} - E_i^L\delta\dot{P}_i + Q_{ij}\delta\dot{\pi}_{ij} + T\delta\dot{S}. \tag{2.121}$$

For further development it is convenient to introduce the free energy $F^L = U^L - TS$, which is a function of the arguments $\boldsymbol{\beta}, \mathbf{P}, \boldsymbol{\pi}, T$:

$$\delta\dot{F}^L(\boldsymbol{\beta}, \mathbf{P}, \boldsymbol{\pi}, T) = \sigma_{ij}\delta\dot{\beta}_{ij} - E_i^L\delta\dot{P}_i + Q_{ij}\delta\dot{\pi}_{ij} - S\delta\dot{T}, \tag{2.122}$$

$$\left(\frac{\partial F^L}{\partial \beta_{ij}}\right)_{\mathbf{P},\boldsymbol{\pi},T} = \sigma_{ij}, \quad \left(\frac{\partial F^L}{\partial P_i}\right)_{\boldsymbol{\beta},\boldsymbol{\pi},T} = -E_i^L,$$
$$\left(\frac{\partial F^L}{\partial \pi_{ij}}\right)_{\boldsymbol{\beta},\mathbf{P},T} = Q_{ij}, \quad \left(\frac{\partial F^L}{\partial T}\right)_{\boldsymbol{\beta},\mathbf{P},\boldsymbol{\pi}} = -S. \tag{2.123}$$

Now let us expand $F^L((\boldsymbol{\beta},\mathbf{P},\boldsymbol{\pi},T)$ close to the ground state similarly to (2.28)

$$F^L(\boldsymbol{\beta},\mathbf{P},\boldsymbol{\pi},T) = b^0_{ij}\pi_{ij} + \frac{1}{2}a_{ij}P_iP_j + \frac{1}{2}b_{ijkl}\pi_{ij}\pi_{kl} +$$
$$+\frac{1}{2}c_{ijkl}\beta_{ij}\beta_{kl} - \frac{c_v}{2\bar{T}}\theta^2 + d_{ijkl}\pi_{ij}\beta_{kl} + f_{kij}P_k\beta_{ij} + \tag{2.124}$$
$$+j_{ikl}P_i\pi_{kl} - \gamma_{ij}\beta_{ij}\theta - p_iP_i\theta - \eta_{ij}\pi_{ij}\theta.$$

Combining (2.123) with (2.124) one obtains the extended constitutive relations:

$$\sigma_{ij} = c_{ijkl}\beta_{kl} + f_{kij}P_k + d_{klij}\pi_{kl} - \gamma_{ij}\theta, \tag{2.125}$$

$$-E_i^L = f_{ikl}\beta_{kl} + a_{ik}P_k + j_{ikl}\pi_{kl} - p_i\theta, \tag{2.126}$$

$$Q_{ij} = d_{ijkl}\beta_{kl} + j_{kij}P_k + b_{ijkl}\pi_{kl} - \eta_{ij}\theta + b^0_{ij}, \tag{2.127}$$

$$S = \gamma_{kl}\beta_{kl} + p_kP_k + \eta_{kl}\pi_{kl} + \frac{c_v}{\bar{T}}\theta. \tag{2.128}$$

Substitution of the constitutive relations (2.125)-(2.128) into the equations of motion, given by eqs. (2.44)-(2.46), (1.42), provides a final general form of system describing thermoelasticity of the considered dielectric medium. Here we shall limit ourselves by presenting only its particular form related to an isotropy. In this case we again have a simplified representation of material tensors (2.66), (2.96)-(2.99), which should be added by the relations

$$\gamma_{ij} = \gamma\delta_{ij}, \quad \eta_{ij} = \eta\delta_{ij}, \quad \kappa_{ij} = \kappa\delta_{ij}, \quad p_i = 0. \tag{2.129}$$

As a result, the constitution relations (2.125)-(2.128) reduce to

$$\sigma_{ij} = c_{12}\beta_{kk}\delta_{ij} + c_{44}(\beta_{ij}+\beta_{ji}) + d_{12}P_{k,k}\delta_{ij} + d_{44}(P_{i,j}+P_{j,i}) - \gamma\delta_{ij}\theta, \tag{2.130}$$

$$E_i^L = aP_i, \tag{2.131}$$

$$Q_{ij} = d_{12}\beta_{kk}\delta_{ij} + d_{44}(\beta_{ij}+\beta_{ji}) + b_{12}P_{k,k}\delta_{ij} + b_{44}(P_{i,j}+P_{j,i}) +$$
$$+b_{77}(P_{j,i} - P_{i,j}) - \eta\delta_{ij}\theta + b_0\delta_{ij}, \tag{2.132}$$

$$S = \gamma\beta_{kk} + \eta P_{k,k} + \frac{c_v}{\bar{T}}\theta. \tag{2.133}$$

In its turn the system (2.105)-(2.107) is extended to four equations

$$c_{44}\nabla^2\mathbf{u} + (c_{12}+c_{44})\nabla\nabla\cdot\mathbf{u} + d_{44}\nabla^2\mathbf{P} +$$
$$+(d_{12}+d_{44})\nabla\nabla\cdot\mathbf{P} + \mathbf{f} + \mathbf{f}^{pl} = \rho\ddot{u} + \gamma\nabla\theta\ , \tag{2.134}$$

$$\begin{aligned} d_{44}\nabla^2\mathbf{u} + (d_{12}+d_{44})\nabla\nabla\cdot\mathbf{u} + (b_{44}+b_{77})\nabla^2\mathbf{P} + \\ +(b_{12}+b_{44}-b_{77})\nabla\nabla\cdot\mathbf{P} - a\mathbf{P} + \mathbf{E} = \eta\nabla\theta, \end{aligned} \tag{2.135}$$

$$\nabla\cdot(\mathbf{P}+\varepsilon_0\mathbf{E}) = q, \tag{2.136}$$

$$\bar{T}(\gamma\nabla\cdot\dot{\mathbf{u}} + \eta\nabla\cdot\dot{\mathbf{P}}) = W + \kappa\nabla^2\theta - c_v\dot{\theta}. \tag{2.137}$$

Boundary conditions have the same standard appearance, which was found above, eqs. (2.50)-(2.54), with some natural additional conditions for θ.

(b) The reciprocity theorem for dynamic problems in thermoelastic dielectrics

For the case of thermoelastic dielectrics with polarization gradient this theorem was first proved by J.P. Nowacki & Glockner [123]. Below we shall extend this proof additionally taking into consideration a plastic distortion as a source of thermo-electro-elastic fields. The derivation will be very similar to that accomplished in Ch.1 (sec. 1.2(c)). We shall just add to the equations of motion (1.40)-(1.42) the additional equation (2.46) (at that certainly eq. (1.41) will be written in the form (2.45)). And upon manipulating with all these equations the more complete system of constitutive relations, eqs. (2.125)-(2.128), will be used.

Thus we consider the two identical thermoelastic dielectric bodies with polarization gradient excited by the two series of different sources of the same kinds acting in the bulk and at the surfaces. They can be distributed in the bulk: forces f, charges q, plastic distortions β^0 and heat supply W, in line with prescribed at the surface: tractions $\sigma_{ij}n_j = f_i^s$ or displacements u_i^s, charge density q^s or electric potential φ^s, a normal component of the heat flux $\kappa_{ij}\eta_i\theta_{,j}$ or temperature θ. As we have seen, the only additional source specific for dielectrics with polarization gradient is related to a surface tension determined by the modulus b_{ij}^0. The reciprocity theorem expresses the cross-relation between two sets of sources $(\mathbf{f}, q, \beta^0, W, etc$ and $\mathbf{f}', q', \beta^{0'}, W', etc)$ and the excited by them physical fields $(\mathbf{u}, \mathbf{E}, \mathbf{P}, \theta, etc$ and $\mathbf{u}', \mathbf{E}', \mathbf{P}', \theta', etc)$.

Repeating speculations of Sec. 1.2 (c) with indicated above extensions one can find that in the considered case eqs. (1.70), (1.72) and (1.73), respectively, reduce to

$$\begin{aligned} &\int_V (\bar{f}_i\bar{u'}_i - \bar{f'}_i\bar{u}_i)dV + \\ &+\int_V (\bar{\beta}^0_{ij}\bar{\sigma'}_{ji} - \bar{\beta}^{0'}_{ij}\bar{\sigma}_{ji})dV + \int_S n_i\left(\bar{\sigma}_{ji}\bar{u'}_i - \bar{\sigma'}_{ji}\bar{u}_i\right)dS = \\ &= \int_V \left[f_{kij}(\bar{P}_k\bar{\beta}'_{ij} - \bar{P'}_k\bar{\beta}_{ij}) + \right. \\ &\left. + d_{ijkl}(\bar{\pi}_{ij}\bar{\beta}'_{kl} - \bar{\pi}'_{ij}\bar{\beta}_{kl}) + \gamma_{ij}(\bar{\beta}_{ij}\bar{\theta}' - \bar{\beta}'_{ij}\bar{\theta})\right]dV, \end{aligned} \tag{2.138}$$

$$
\begin{aligned}
-\int_V (\bar{q}\bar{\varphi}' - \bar{q}'\bar{\varphi})dV + \int_S n_i \left(\bar{D}_i\bar{\varphi}' - \bar{D}'_i\bar{\varphi}\right) dS = \\
= \int_V (\bar{E}_k\bar{P}'_k - \bar{E}'_k\bar{P}_k)dV,
\end{aligned}
\tag{2.139}
$$

$$
\begin{aligned}
\int_V (\bar{W}\bar{\theta}' - \bar{W}'\bar{\theta})dV + \kappa_{ij}\int_S n_i \left(\bar{\theta}_{,j}\bar{\theta}' - \bar{\theta}'_{,j}\bar{\theta}\right) dS = \\
= p\bar{T}\int_V \left[\gamma_{ij}(\bar{\beta}_{ij}\bar{\theta}' - \bar{\beta}'_{ij}\bar{\theta}) + p_k(\bar{P}_k\bar{\theta}' - \bar{P}'_k\bar{\theta}) + \eta_{kl}(\bar{\pi}_{kl}\bar{\theta}' - \bar{\pi}'_{kl}\bar{\theta})\right] dV.
\end{aligned}
\tag{2.140}
$$

After quite analogous operations with eq. (2.46) and the constitutive relations (2.126) and (2.127) we obtain the forth necessary equation

$$
\begin{aligned}
\int_S n_i \left[(\bar{Q}_{ij} - b^0_{ij})\bar{P}'_j - (\bar{q}'_{ij} - b^0_{ij})\bar{P}_j\right] dS = \\
= \int_V \left[\eta_{kl}(\bar{\pi}_{kl}\bar{\theta}' - \bar{\pi}'_{kl}\bar{\theta}) - d_{ijkl}(\bar{\pi}_{ij}\bar{\beta}'_{kl} - \bar{\pi}'_{ij}\bar{\beta}_{kl}) + \right. \\
\left. + p_k(\bar{P}_k\bar{\theta}' - \bar{P}'_k\bar{\theta}) - f_{kij}(\bar{P}_k\bar{\beta}'_{ij} - \bar{P}'_k\bar{\beta}_{ij}) - (\bar{E}_k\bar{P}'_k - \bar{E}'_k\bar{P}_k)\right] dV.
\end{aligned}
\tag{2.141}
$$

Now let us form the linear combination of the above four equations which can be symbolically described as

$$
\{p\bar{T}[(2.134) + (2.135) + (2.137)] - (2.136)\}. \tag{2.142}
$$

It is easily checked that in the resulting equation the right-hand side is zero. In addition, due to the boundary condition (2.54), the left-hand side of eq. (2.141) reduces to

$$
\int_S n_i b^0_{ij}(\bar{P}_j - \bar{P}'_j)dS = 2\int_S (\bar{U}^{surf} - \bar{U}^{surf'})dS, \tag{2.143}
$$

which is the double difference of the surface energies (2.64) related to the two considered states. On the other hand, the left-hand sides of the other three equations respectively coincide with those in eqs. (1.70), (1.72) and (1.73). Therefore the final equation, related to the symbolic operation (2.142) and expressing the reciprocity theorem in the Laplace space, will differ from the corresponding eq. (1.74) by only one additional term

$$
p\bar{T}\int_S n_i b^0_{ij}(\bar{P}_j - \bar{P}'_j)dS \tag{2.144}
$$

in the left-hand side. And the formulation of the reciprocity theorem in a real space is obtained from eq. (1.75) by the addition to its left-hand side of the term

$$
p\bar{T}\int_S n_i b^0_{ij}(P_j - P'_j)dS. \tag{2.145}
$$

We shall not present again these lengthy equations. Actually the final result is very natural. Indeed, as we have seen, the Mindlin theory brings the only new source - the modulus b^0_{ij}, determining a surface tension (2.65). So the only expected addition to the balance equation (1.75) should be the term of the type (2.145).

Thus, the particular static case for an infinite isothermal Mindlin's medium will be described by the same eq. (1.81). However, its appearance (1.84) must be changed because of the different form of constitutive equations for a body with polarization gradient. Let us choose their simplified variant (2.67)-(2.69) for centrosymmetric crystals. In this case the transformation (1.82) should be replaced by

$$\begin{aligned} &\beta^0_{ij}\sigma'_{ji} - \beta^{0'}_{ij}\sigma_{ji} = \\ &= \beta^0_{ij}[c_{jikl}(u'_{l,k} - \beta^{0'}_{kl}) + d_{klij}P'_{l,k}] - \beta^{0'}_{ij}[c_{jikl}(u_{l,k} - \beta^0_{kl}) + d_{klij}P_{l,k}] \\ &= [\beta^0_{ij}(c_{jikl}u'_l + d_{klij}P'_l) - \beta^{0'}_{ij}(c_{jikl}u_l + d_{klij}P_l)]_{,k} + \\ &+ (f^{\mathrm{pl}}_l u'_l - f^{\mathrm{pl}\prime}_l u_l) + (E^{\mathrm{pl}}_l P'_l - E^{\mathrm{pl}\prime}_l P_l). \end{aligned} \tag{2.146}$$

Here f^{pl}_l is defined by eq. (1.83) and

$$E^{\mathrm{pl}}_l = -d_{klij}\beta^0_{ji,k}. \tag{2.147}$$

Substituting (2.147) into eq. (1.81) one can easily obtain instead of (1.84)

$$\int_V [(f_i + f^{\mathrm{pl}}_i)u'_i - (f'_i + f^{\mathrm{pl}\prime}_i)u_i + (E^{\mathrm{pl}}_l P'_l - E^{\mathrm{pl}\prime}_l P_l) - (q\varphi' - q'\varphi)]dV = 0. \tag{2.148}$$

This form of the reciprocity theorem will be used below in Ch. 5.

3

General Electro-Elastic Line Defect in Unbounded Piezoelectric Body

3.1 Dislocation and its electrostatic analog as a combined line defect

(a) Three definitions of a dislocation

As a purely topological line defect, a dislocation in a piezoelectric medium is defined in a completely the same way as that in any other type of solids. There are at least three independent but completely equivalent definitions of a dislocation. The idea of the first definition [137] was based on the Burgers circuit in a crystal lattice. One can form a circuit in a crystal making equal numbers of steps from atom to atom on the opposite sides of the path (say, n steps down, m steps to the right, n steps up and m steps to the left). In a perfect crystal the last step would bring to the initial point. However, if the crystal contains a dislocation and the circuit links the latter, the circuit will fail to close. The vector going from the start point of the circuit to its end is defined as the Burgers vector $\mathbf{b}$, which is a basic characteristic of a dislocation.

The second definition [138] is just a continual variant of the first one. The line integral of the elastic displacements over the closed contour γ, which links a dislocation, must fail to vanish, defining the value of the same Burgers vector,

$$b_i = \oint_{\gamma} du_i \tag{3.1}$$

In eq. (3.1) a direction of integration is supposed to be related to a chosen direction of the dislocation by the rule of a right-hand screw. Thus, a change of the dislocation direction leads to a change of the sign of the Burgers vector. This reflects the physical difference between dislocations of different signs (e.g., right-hand and left-hand screw dislocations or two edge dislocations with extra-planes situated on the opposite sides from a dislocation line).

A rather paradoxical appearance of eq. (3.1) just indicates that the vector field of displacements $\mathbf{u}(\mathbf{r})$ in a crystal with a dislocation must be discontinuous or multivalued. This reflects the fundamental new quality, which a crystal acquires after a topological transformation creating a dislocation. The atoms of such crystal cannot be brought into one-to-one correspondence with the atoms in the initial perfect crystal. Accordingly, it is principally impossible to decompose the displacement field $\mathbf{u}(\mathbf{r})$ on elastic and inelastic parts. However, it can be done for a total distortion $u_{i,j}$ [132, 139]:

$$u_{i,j} = \beta_{ji} + \beta^0_{ji}. \tag{3.2}$$

Here the first term describes the elastic distortion determined by the local change in interplane distances and lattice rotations, and the second term, known as the plastic distortion, is a characteristic of the transformation created the dislocation. In these terms eq. (3.1) is reduced to [139, 140]

$$b_i = \oint_\gamma \beta_{ji} dx_j, \tag{3.3}$$

or by (3.2),

$$b_i = -\oint_\gamma \beta^0_{ji} dx_j. \tag{3.3a}$$

The latter equation directly leads to the third definition of a dislocation. This definition describes specific transformation procedure necessary for creating a dislocation and simultaneously removing a multivalued character of the displacement field $\mathbf{u}(\mathbf{r})$. Let us make the cut over a smooth, however arbitrary, surface S bounded by the dislocation line C with a requirement of a jump of the displacement vector $\mathbf{u}$ at this surface by the value of the Burgers vector $\mathbf{b}$ (see Fig. 3.1):

$$\mathbf{u}^{S^-} - \mathbf{u}^{S^+} = \mathbf{b}. \tag{3.4}$$

As a convenient alternative to the cut along S, one can introduce the singular plastic distortions at S (however retaining a multivaluedness of $\mathbf{u}$ $(\mathbf{r})$ field)

$$\beta^0_{ji}(\mathbf{r}) = -b_i n^S_j \delta[(\mathbf{r} - \mathbf{r}^S) \cdot \mathbf{n}^S], \tag{3.5}$$

where $\mathbf{r}^S$ indicates points of the surface S and $\mathbf{n}^S$ is the unit normal to the surface at the point $\mathbf{r}^S$. The positive direction of $\mathbf{n}^S$ is related to the chosen positive direction of the dislocation line C by the right-hand screw rule. Indeed, one can easily check the validity of (3.3a) with (3.5) taking into account that close to the surface S

$$(\mathbf{r} - \mathbf{r}^S) \cdot \mathbf{n}^S = s(\mathbf{t}^\gamma \cdot \mathbf{n}^S), \quad dx_j = t^\gamma_j ds, \tag{3.6}$$

where $\mathbf{t}^\gamma$ is the unit vector tangent to the contour γ at the point s (Fig. 3.1). So that

$$-\oint_{\gamma} dx_j \beta^0_{ji} = b_i \int_{-\varepsilon}^{\varepsilon} ds(\mathbf{n}^S \cdot \mathbf{t}^{\gamma})\delta[(\mathbf{n}^S \cdot \mathbf{t}^{\gamma})s] = b_i \tag{3.7}$$

We stress that the above exercise is absolutely insensitive to the choice of the surface S. In other words, the described surface transformation related to the third definition of a dislocation indeed creates the same line defect as that defined by condition (3.1). On the other hand, this means that the plastic deformation is not a function of the body state, but depends on particular ways of formation or motion of dislocations [141].

We shall not go further in this brief introduction to the theory of dislocations. One can find details in many available books (see, e.g., [142-146]).

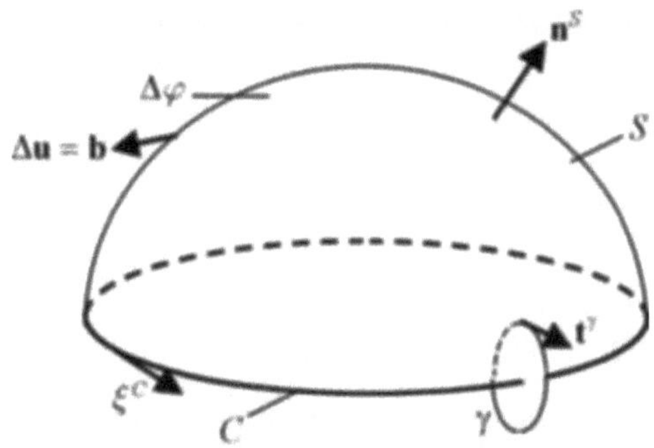

Fig. 3.1. The dislocation line C, the surface S bounded by C and the contour γ linking C

(b) Electrostatic analog of a dislocation

It is customary to make use of the analogy between dislocation stress fields in elastostatics and magnetic induction fields created by line electric currents in magnetostatics [140]. Indeed, these line sources and their fields are mathematically very similar. However, for dielectric media considered in this work it seems to be more relevant to look for electrostatic analogies with dislocations. The more so that an electromechanical coupling occurring in piezoelectrics creates there a very specific situation, when elastic and electric fields do not exist without each other.

On the other hand, as we shall see, it is really possible to introduce a very close electrostatic analog of a dislocation. The idea is based on the Dirichlet problem: to find the electrostatic field distribution in a dielectric medium excited by a given constant jump of potential $\Delta\varphi$ prescribed at some internal non-closed surface S, finite or semi-infinite. The latter requirement is very essential, being equivalent to a necessary existence of a line bounding the surface. Indeed, the electric potential φ is defined to the accuracy of an arbitrary constant. So, if in some closed domain of the body the potential φ differs from that in the surrounding medium by any constant $\Delta\varphi$, this means just the only thing: those potentials are counted from different reference points. Such jumps of a potential certainly cannot excite any electric fields. The situation, however, is radically changed when the surface S is not closed but is bounded by some line C (Fig. 3.1). It is rather evident that in this case we acquire an electrical singularity along the line C. And in complete analogy with a dislocation the excited electric field in the medium will be sensitive only to the position of the line C, but not to the form of the surface S. Indeed, let us add to the surface S the any other surface S' bounded by the same line C. Consider the closed domain between S and S'. It is clear that by changing its potential by the constant $\Delta\varphi$ we do not change any physical fields, however simultaneously the Dirichlet jump $\Delta\varphi$ is removed from the surface S to S'. Thus, the introduced electrostatic source is a line defect. And its similarity to a dislocation is even closer than one could expect.

This is just the case when it is very convenient to look for the electric field $\mathbf{E}$ counting it off from the "external" source field $\mathbf{E}^0$ produced by the same jump of potential $\Delta\varphi$ at the surface S in vacuum:

$$\mathbf{E} = \mathbf{E}^0 + \mathbf{E}'. \tag{3.8}$$

The form of the field $\mathbf{E}^0$ can be found immediately,

$$\mathbf{E}^0(\mathbf{r}) = -\Delta\varphi \mathbf{n}^S \delta[(\mathbf{r} - \mathbf{r}^S) \cdot \mathbf{n}^S] \tag{3.9}$$

which is very similar to the source term (3.5) in the theory of dislocations. The further steps are described in Ch. 1, eqs. (1.51)-(1.55). In particular, for the static case with no other sources, apart from the two considered line defects, final equation (1.55), takes the form

$$c_{ijkl}u_{k,lj} + e_{lij}\varphi_{,lj} = c_{ijkl}\beta^0_{lk,j} + e_{lij}E^0_{l,j}, \tag{3.10}$$

$$e_{jlk}u_{k,lj} - \varepsilon_{jl}\varphi_{,lj} = e_{jlk}\beta^0_{lk,j} - \varepsilon_{jl}E^0_{l,j}, \tag{3.11}$$

with

$$-\varphi_{,i} = E'_i = E_i - E^0_i, \quad u_{k,l} = \beta_{lk} + \beta^0_{lk}. \tag{3.12}$$

As follows from eqs. (3.5) and (3.9), beyond the arbitrary cut surface S those pairs of electric fields and distortions coalesce into one field $\mathbf{E} = \mathbf{E}'$ and one distortion $\nabla \otimes \mathbf{u} = \boldsymbol{\beta}$.

(c) Combined electro-elastic line defect - a 4D dislocation

Thus with a double jump at the same surface S of both a displacement, $\Delta\mathbf{u} = \mathbf{b}$ eq. (3.4), and an electric potential, $\Delta\varphi$, we obtain along the line C bounding S the combined electro-elastic line defect, which can be conveniently mathematically described as a 4D dislocation characterized by the "Burgers" 4-vector

$$\mathbf{b}^{(4)} = \begin{pmatrix} \mathbf{b} \\ \Delta\varphi \end{pmatrix}. \tag{3.13}$$

Let us formally introduce the corresponding four-dimensional "displacements"

$$U_K = \begin{cases} u_k, & K = k = 1,2,3; \\ \varphi, & K = 4; \end{cases} \tag{3.14}$$

"elastic" and "plastic distortions", respectively,

$$U_{lK} = \begin{cases} \beta_{lk}, & K = k = 1,2,3; \\ -E_l, & K = 4; \end{cases} \qquad U^0_{lK} = \begin{cases} \beta^0_{lk}, & K = k = 1,2,3; \\ E^0_l, & K = 4; \end{cases} \tag{3.15}$$

"stresses"

$$\Sigma_{iJ} = \begin{cases} \sigma_{ij}, & J = j = 1,2,3; \\ D_i, & J = 4 \end{cases} \tag{3.16}$$

and "elastic moduli"

$$\begin{aligned} C_{iJKl} &= c_{ijkl} \text{ for } J,K = j,k = 1,2,3, \\ &= e_{lij} \text{ for } J = j = 1,2,3, K = 4, \\ &= e_{ikl} \text{ for } J = 4, K = k = 1,2,3, \\ &= -\varepsilon_{il} \text{ for } J = 4, K = 4. \end{aligned} \tag{3.17}$$

It is easily checked that this tensor retains some symmetry

$$C_{iJKl} = C_{lKJi}. \tag{3.17'}$$

With the above notation the standard constitutive equations

$$\sigma_{ij} = c_{ijkl}\beta_{kl} - e_{kij}E_k, \tag{3.18}$$

$$D_i = e_{ikl}\beta_{lk} + \varepsilon_{ik}E_k, \tag{3.19}$$

take an appearance of the modified Hooke's law

$$\Sigma_{iJ} = C_{iJKl}U_{lK}. \tag{3.20}$$

And the equations of equilibrium,

$$\mathrm{div}\boldsymbol{\sigma} = 0, \qquad \mathrm{div}\mathbf{D} = 0 \tag{3.21}$$

are combined into

$$\Sigma_{iJ,i} = 0. \tag{3.22}$$

Substituting here (3.20) with

$$U_{lK} = U_{K,l} - U^0_{lK}, \tag{3.23}$$

one finally obtains

$$C_{iJKl}U_{K,li} = C_{iJKl}U^0_{lK,i}, \tag{3.24}$$

which certainly is just a short representation of system (3.10), (3.11).

Thus, we have come to rather compact equation (3.24) describing electro-elastic fields in an arbitrary piezoelectric medium excited by the combined defect, which will be called below a generalized line defect or a 4D dislocation. We recall that this defect is formed by the transformation, which is described by the generalized plastic distortion $\mathbf{U}^0 = \{U^0_{lK}\}$ defined by

$$\mathbf{U}^0(\mathbf{r}) = -\mathbf{n}^S \otimes \mathbf{b}^{(4)}\delta[(\mathbf{r} - \mathbf{r}^S)\cdot \mathbf{n}^S] \tag{3.25}$$

which differs from corresponding eq. (3.5) for an ordinary dislocation only by the replacement of the Burgers vector $\mathbf{b}$ by the 4-vector $\mathbf{b}^{(4)}$, eq. (3.13). Here we must stress that even at $\Delta\varphi = 0$, when our line defect is reduced to an ordinary dislocation, $\mathbf{b}^{(4)} = (\mathbf{b}, 0)$, the considered formalism remains four dimensional due to a 4D character of coupled electro-elastic fields in piezoelectrics.

3.2 Fundamental properties of 4D dislocations of arbitrary shape

(a) Generalized Green's tensor for the considered problem

Eq. (3.24) has the form very similar to a standard equation of elastostatics,

$$C_{iJKl}U_{K,li} + F_J = 0 \tag{3.26}$$

where F_J is the four dimensional "force" ($J = 1, 2, 3, 4$)

$$F_J = -C_{iJKl}U^0_{lK,i}. \tag{3.27}$$

Accordingly, it is natural to construct for its solution a Green's function analogous to that used in the theory of elasticity. The corresponding Green tensor, $G_{KM}(\mathbf{R})$ ($K, M = 1, 2, 3, 4$), describing electro-elastic fields in piezoelectrics must satisfy the equation

$$C_{iJKl}G_{KM,li} + \delta_{MJ}\delta(\mathbf{R}) = 0. \tag{3.28}$$

After a Fourier transformation of this equation with

$$G_{KM}(\mathbf{R}) = \frac{1}{(2\pi)^3}\int d^3\mathbf{k}G_{KM}(\mathbf{k})e^{i\mathbf{k}\cdot\mathbf{R}} \tag{3.29}$$

and

$$\delta(\mathbf{R}) = \frac{1}{(2\pi)^3}\int d^3\mathbf{k}e^{i\mathbf{k}\cdot\mathbf{R}} \tag{3.30}$$

one obtains

$$-(k_i C_{iJKl} k_l) G_{KM}(\mathbf{k}) + \delta_{MJ} = 0. \tag{3.31}$$

Let us introduce the unit vector

$$\boldsymbol{\kappa} = \mathbf{k}/k \tag{3.32}$$

and the 4×4 matrix $(\kappa\kappa)$ with the components

$$(\kappa\kappa)_{JK} = \kappa_i C_{iJKl} \kappa_l. \tag{3.33}$$

In these terms the Fourier amplitude $G_{KM}(\mathbf{k})$ given by (3.31) reads

$$G_{KM}(\mathbf{k}) = \frac{1}{k^2}[(\kappa\kappa)^{-1}]_{KM}. \tag{3.34}$$

This is a homogeneous function of $\mathbf{k}$ of degree -2. For any such function the Fourier integral (3.29) is identically reduced to a line integral along the unit circle in the plane orthogonal to $\mathbf{R}$ (see Fig. 3.2):

$$\int d^3k G_{KM}(\mathbf{k}) e^{i\mathbf{k}\cdot\mathbf{R}} = \frac{\pi}{R} \int_0^{2\pi} d\varphi G_{KM}(\mathbf{n}_\varphi), \tag{3.35}$$

where $\mathbf{n}_\varphi$ is a unit vector scanning the circle of integration and remaining normal to $\mathbf{R}$. Thus, combining (3.29) with (3.34) and (3.35) we finally obtain the extended form for Green's tensor $\mathbf{G}(\mathbf{R})$ describing electro-elastic fields in piezoelectrics

$$\mathbf{G}(\mathbf{R}) = \frac{1}{8\pi^2 R} \int_0^{2\pi} d\varphi (n_\varphi n_\varphi)^{-1}. \tag{3.36}$$

Here the 4×4 symmetric matrix $(n_\varphi n_\varphi)$ is defined in the same way as (3.33). Formally expression (3.36) has precisely the same appearance, which is known for Green's tensor in the theory of elasticity [147, 148]. However, one should remember that in pure elastic case both $\mathbf{G}(\mathbf{R})$ and $(n_\varphi n_\varphi)$ are 3×3 matrices. And our extension of the formalism is reduced to an increase of their dimension by means of replacement of the elastic stiffness tensor c_{ijkl} by the combined tensor C_{iJKl} (3.17) including all material constants of a piezoelectric body.

With a known Green's tensor a solution of initial eq. (3.26) for an arbitrary source vector $\mathbf{F}(\mathbf{r})$ is found immediately

$$\begin{aligned} U_K(\mathbf{r}) &= \int d^3\mathbf{r}' G_{KM}(\mathbf{r} - \mathbf{r}') F_M(\mathbf{r}') = \\ &= \frac{1}{8\pi^2} \int d^3\mathbf{r}' \frac{F_M(\mathbf{r}')}{|\mathbf{r} - \mathbf{r}'|} \int_0^{2\pi} d\varphi [(n_\varphi n_\varphi)^{-1}]_{MK}, \end{aligned} \tag{3.37}$$

where the unit vector $\mathbf{n}_\varphi$ belongs to the plane orthogonal to the vector $\mathbf{R} = \mathbf{r} - \mathbf{r}'$. In particular, eq. (3.37) gives the general solution for electro-elastic

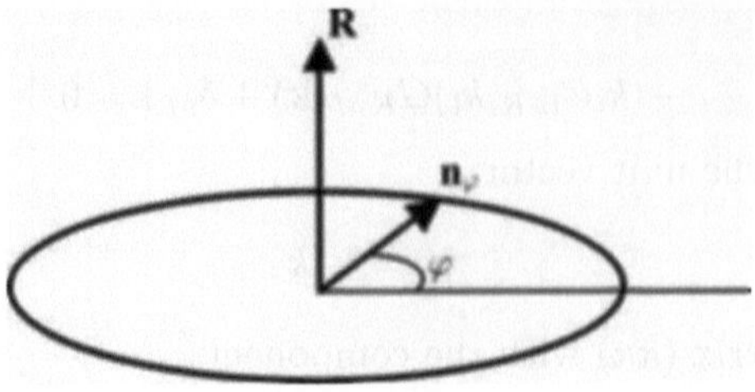

Fig. 3.2. The contour of integration $|\mathbf{n}_\varphi| = 1$ in the plane $\mathbf{n}_\varphi \cdot \mathbf{R} = 0$.

"displacement" fields $\mathbf{U} = (\mathbf{u}, \varphi)^t$ excited by arbitrary distributions of the force $\mathbf{f}(\mathbf{r})$ and charge $q(\mathbf{r})$ densities $\mathbf{F} = \mathbf{f}^{(4)} = (\mathbf{f}, -q)^t$ in a piezoelectric medium of unrestricted anisotropy.

It will be useful also for further applications to present here an explicit form of the ordinary Green function for isotropic elasticity [46]:

$$G_{km}(\mathbf{R}) = \frac{\tau_k \tau_m + (3 - 4\nu)\delta_{km}}{16\pi\mu(1-\nu)R} \tag{3.38}$$

where μ and ν are the shear modulus and the Poisson ratio, respectively, and $\boldsymbol{\tau}$ is the unit vector along $\mathbf{R}$:

$$\boldsymbol{\tau} = \mathbf{R}/R. \tag{3.39}$$

(b) "Displacement" and "distortion" fields of a general line defect

Now we are ready to derive the "displacement" field excited by our 4D dislocation line C of arbitrary shape situated in the infinite anisotropic piezoelectric medium free of body forces or charges. One should just substitute into (3.37) the corresponding source "force" given by (3.27). The result is

$$U_K(\mathbf{r}) = -\int d^3\mathbf{r}' G_{KM}(\mathbf{r} - \mathbf{r}') C_{iMNl} U^0_{lN,i}(\mathbf{r}') \tag{3.40}$$

where $U^0_{lN}(\mathbf{r}')$ is defined by eq. (3.25). Let us make use here of the identity

$$\begin{aligned} &G_{KM}(\mathbf{r} - \mathbf{r}') U^0_{lN,i}(\mathbf{r}') = \\ &= \frac{\partial}{\partial x'_i}[G_{KM}(\mathbf{r} - \mathbf{r}') U^0_{lN}(\mathbf{r}')] + G_{KM,i}(\mathbf{r} - \mathbf{r}') U^0_{lN}(\mathbf{r}'), \end{aligned} \tag{3.41}$$

where it was taken into account that

$$\frac{\partial}{\partial x'_i} G_{KM}(\mathbf{r} - \mathbf{r}') = -\frac{\partial}{\partial x_i} G_{KM}(\mathbf{r} - \mathbf{r}'). \tag{3.42}$$

The volume integral in (3.40) with the first term in (3.41) by means of the Gauss theorem is immediately reduced to the integral over the infinite surface of our unbounded body, where both G_{KM} and U^0_{lN} must vanish. Thus, eq. (3.40) takes the form

$$U_K(\mathbf{r}) = -\int d^3\mathbf{r}' G_{KM,i}(\mathbf{r}-\mathbf{r}')C_{iMNl}U^0_{lN}(\mathbf{r}'), \tag{3.43}$$

which with (3.25) is automatically transformed into the integral over the surface S resting on the dislocation line (Fig. 3.1),

$$U_K(\mathbf{r}) = C_{iMNl}b^{(4)}_N \int_S dS_l G_{KM,i}(\mathbf{r}-\mathbf{r}^S), \tag{3.44}$$

where by definition

$$dS_l = dS n^S_l. \tag{3.45}$$

Basing on eq. (3.44) it is possible also to find a representation of the vector field U($\mathbf{r}$) in the form of a line integral along the contour C coinciding with the dislocation line (Fig. 3.1). The corresponding result for an elastically isotropic medium is known as the Burgers solution [137]

$$\mathbf{u}(\mathbf{r}) = -\frac{\mathbf{b}\Omega}{4\pi} - \frac{1}{4\pi}\oint_C dl \frac{\mathbf{b}\times\boldsymbol{\xi}^C}{R} - \frac{1}{8\pi(1-\nu)}\nabla\oint_C dl \frac{(\mathbf{b}\times\mathbf{R})\cdot\boldsymbol{\xi}^C}{R}, \tag{3.46}$$

where $\boldsymbol{\xi}^C$ is the unit tangent vector to the line C (Fig. 3.1) at the point $\mathbf{r}' = \mathbf{r}^C$ belonging to C,

$$\mathbf{R} = \mathbf{r} - \mathbf{r}', \quad R = |\mathbf{r} - \mathbf{r}'|, \tag{3.47}$$

and Ω is the solid angle subtended at $\mathbf{r}$ by the surface S resting on the line C,

$$\Omega = \int_S dS \frac{\mathbf{n}^S\cdot\mathbf{R}}{R^3}. \tag{3.48}$$

(here $\mathbf{r}' = \mathbf{r}^S$ belongs S). By definition this solid angle depends only on the configuration of the line C bounding the surface S being insensitive to a shape of S which shows again that a dislocation is a line defect. On the other hand, $\Omega(\mathbf{r})$ is a discontinuous function across the S surface changing by 4π on linking the curve C. Accordingly solution (3.46) automatically satisfies eq. (3.1),

$$\oint_\gamma du_i = -\frac{\mathbf{b}}{4\pi}\oint_\gamma d\Omega = \mathbf{b}. \tag{3.49}$$

An extension of Burgers' solution (3.46) for elastic media of an arbitrary anisotropy has been done by Indenbom & Orlov [149] (see also [150]). In contrast to isotropic expression (3.46), its absolutely nontrivial extension [149] has proved to be practically unknown to scientific community, at least until the publication [150]. We shall not reproduce it here, but instead directly present the result of its further generalization valid for a considered in this chapter 4D dislocation in an arbitrary piezoelectric medium. The derivation is very similar to that used in [149, 150]. Thus, the extension of (3.46) in our case reads

$$\mathbf{U}(\mathbf{r}) = -\frac{\mathbf{b}^{(4)}\Omega}{4\pi} + \frac{1}{8\pi^2}\int_C dl \frac{1}{R}\int_0^{2\pi} d\varphi (n_\varphi n_\varphi)^{-1}(\mathbf{n}_\varphi, \xi^C \times \mathbf{n}_\varphi)\mathbf{b}^{(4)}, \quad (3.50)$$

where the 4×4 matrix $(\mathbf{n}_\varphi, \xi^C \times \mathbf{n}_\varphi)$ of the type $(\mathbf{c}, \mathbf{d}) \equiv (cd)$ is defined by a convolution similar to (3.33)

$$(\mathbf{c}, \mathbf{d})_{KM} \equiv (cd)_{KM} = c_i C_{iKMj} d_j. \quad (3.51)$$

As already was stated, the solid angle Ω (3.48) is a discontinuous function having a jump 4π at crossing the surface S. Accordingly, the total distortion $U_{K,l}$ does have a structure similar to (3.2),

$$U_{K,l} = U_{lK} + U^0_{lK}, \quad (3.52)$$

where U_{lK} and U^0_{lK} are given by (3.15), (3.25). In spite of (3.52), the "elastic distortion" can be found by a differentiation of eqs. (3.50) or (3.44) if to suppose there an occurrence of the cut along the surface S,

$$\mathbf{r} - \mathbf{r}^S \neq 0, \quad (3.53)$$

providing a nonsingular behavior of the function $\mathbf{U}(\mathbf{r})$. Here we must recall that the surface S is arbitrary. So, if we are interested in some particular point $\mathbf{r}$, which happened to situate at S, it is always possible to deform S so that $\mathbf{r}$ would be beyond the cut. In evaluating the elastic distortion U_{lK} it is more convenient to return to initial eq. (3.44), which leads to

$$U_{lK}(\mathbf{r}) = C_{iMNp} b_N^{(4)} \int_S dS_p G_{KM,il}(\mathbf{r} - \mathbf{r}^S). \quad (3.54)$$

It is easily checked that the above surface integral is equivalent to the line integral along the dislocation (compare with the corresponding Mura representation [151] for a pure elasticity),

$$U_{lK}(\mathbf{r}) = -C_{iMNp} b_N^{(4)} \varepsilon_{jpl} \oint_C dl_j G_{KM,i}(\mathbf{r} - \mathbf{r}^C), \quad (3.55)$$

where

$$dl_j = \xi_j^C dl \quad (3.56)$$

and ε_{ijk} is the permutation tensor defined by

$$\begin{aligned} \varepsilon_{ijk} &= 1 \text{ if } ijk = 123, 231 \text{ or } 312; \\ &= -1 \text{ if } ijk = 132, 213 \text{ or } 321; \quad (3.57) \\ &= 0 \text{ if } ijk \text{ contain any repeated indices.} \end{aligned}$$

This tensor is connected with the Kronecker delta by the identity

$$\varepsilon_{ijk}\varepsilon_{imn} = \delta_{jm}\delta_{kn} - \delta_{jn}\delta_{km}. \tag{3.58}$$

The equivalence between (3.54) and (3.55) is proved using the Stokes theorem

$$\varepsilon_{ijk} \int_S dS_i A_{\dots,j} = \oint_C A_{\dots} dl_k \tag{3.59}$$

which by (3.58) is transformed into

$$\int_S dS_j A_{\dots,k} - \int_S dS_k A_{\dots,j} = \varepsilon_{ijk} \oint_C A_{\dots} dl_i. \tag{3.60}$$

Indeed, combining (3.54) with (3.60) and taking into account that at the surface S by eqs. (3.28), (3.53)

$$C_{iMNp} G_{KM,ip} = -\delta_{NK}\delta(\mathbf{r} - \mathbf{r}^S) = 0, \tag{3.61}$$

one immediately obtains eq. (3.55).

By a similar reasoning it can be found also the alternative representation to (3.55) extending the corresponding result of Indenbom & Orlov [47]

$$U_{lK}(\mathbf{r}) = C_{iMNp} b_N^{(4)} \varepsilon_{jpr} \oint_C dl_j (x_r - x_r^C) G_{KM,is}(\mathbf{r} - \mathbf{r}^C). \tag{3.62}$$

Starting from the expression of this type Indenbom and Orlov [47] have derived their famous formula expressing the elastic distortion of an arbitrary dislocation loop in terms of elastic fields of straight dislocations going from the observation point $\mathbf{r}$ to the points $\mathbf{r}^C$ at the dislocation line C. Brown [152] has independently found an analogous formula for the particular case of a plane dislocation loop. We shall again omit derivations, which are rather cumbersome, and represent only the general result [47] extended to the case of an arbitrary piezoelectric medium,

$$U_{lK}(\mathbf{r}) = -\frac{1}{2} \oint_C dl \xi_p^C \xi_q^C \frac{\partial^2}{\partial x_p \partial x_q} U_{lK}(\xi^C, \mathbf{r} - \mathbf{r}^C). \tag{3.63}$$

Here $U_{lK}(\xi^C, \mathbf{r}-\mathbf{r}^C)$ is the electro-elastic "distortion" field created at the point ξ^C by the straight 4D dislocation having the same Burgers vector $\mathbf{b}^{(4)}$ as the loop C and directed along $\mathbf{r} - \mathbf{r}^C$. Multiplying both sides of (3.63) by C_{iJKl}, eq. (3.20), we arrive at the same relation between "stresses" Σ_{iJ} (3.16) for the dislocation loop and for the appropriate set of straight dislocations. Thus, eq. (3.63) directly shows how 3D fields of an arbitrary dislocation can be constructed from much more simple 2D fields of straight dislocations. This provides a useful algorithm for a computing of complex dislocation fields. Willis [153] and Steeds & Willis [154] have already obtained important results concerning a realization of mentioned program for a pure elasticity.

(c) Energy of a 4D dislocation and its interaction with external fields

Consider an unbounded piezoelectric medium excited by the 4D dislocation loop (Fig. 3.1) and by some other sources. It will be convenient for further calculations to introduce the imaginary cut along the arbitrary surface S resting on the dislocation line C. As we know, such a cut excluding the surface S from the body does not change any physical results, however it provides a single-valued behavior of the "displacement" 4-vector $\mathbf{U} = (\mathbf{u}, \varphi)^t$ and excludes the "plastic distortion" $\mathbf{U}^0 = (\boldsymbol{\beta}^0, \mathbf{E}^0)^t$, eqs. (3.25), (3.15), which vanishes everywhere beyond the surface S. This allows putting in all equations

$$U_{iK} = U_{K,i}\ , \quad \mathbf{E}' = \mathbf{E}. \tag{3.64}$$

As was shown by Eshelby [155], the elastic energy of the body deformed under isothermal conditions is thermodynamically described by the Gibbs free energy potential

$$F = U - TS. \tag{3.65}$$

For a piezoelectric medium we should take for the same aims the Gibbs electric enthalpy (see (1.13), (1.15))

$$H^G(\boldsymbol{\beta}, \mathbf{E}) = F - \mathbf{D} \cdot \mathbf{E}. \tag{3.66}$$

As a ready result we can use the isothermal variant of eq. (1.20) taken at $\theta = 0$ and counted from the ground state,

$$H^G(\boldsymbol{\beta}, \mathbf{E}) = \frac{1}{2} c_{ijkl}\beta_{ij}\beta_{kl} - e_{kij}E_k\beta_{ij} - \frac{1}{2}\varepsilon_{kl}E_kE_l. \tag{3.67}$$

With constitutive equations (3.18), (3.19), definitions (3.12), $(3.15)_1$, (3.16) and relations (3.64) we obtain

$$H^G = \frac{1}{2}(\sigma_{ij}u_{i,j} + D_k\varphi_{,k}) = \frac{1}{2}\Sigma_{iJ}U_{J,i}. \tag{3.68}$$

So, the internal energy of a piezoelectric body, deformed under the isothermal conditions, is

$$W = \frac{1}{2}\int_V dV\, \Sigma_{iJ}U_{J,i}. \tag{3.69}$$

Here V is supposed to be the whole space, except the space in the cut.

Let us divide now the electro-elastic fields $\mathbf{U}(\mathbf{r})$ and $\Sigma(\mathbf{r})$, each into two parts related to the considered dislocation and to the other sources of any origin, except of body forces and body charges. We shall mark the first by the superscript d, and the second - by the superscript o :

$$\mathbf{U} = \mathbf{U}^d + \mathbf{U}^o, \tag{3.70}$$

$$\boldsymbol{\Sigma} = \boldsymbol{\Sigma}^d + \boldsymbol{\Sigma}^o. \tag{3.71}$$

Substituting eqs. (3.70) and (3.71) into (3.69) one obtains

$$W = \frac{1}{2}\int_V dV\,\Sigma^d_{iJ}U^d_{J,i} + \frac{1}{2}\int_V dV\,\Sigma^o_{iJ}U^o_{J,i} + \frac{1}{2}\int_V dV(\Sigma^d_{iJ}U^o_{J,i} + \Sigma^o_{iJ}U^d_{J,i}) \quad (3.72)$$

Here the first term describes the self-energy of the 4D dislocation

$$W^{self} = \frac{1}{2}\int_V dV\,\Sigma^d_{iJ}U^d_{J,i}. \quad (3.73)$$

The second term is the analogous self-energy of the other sources, and the third term - the interaction energy between the dislocation and the external field. In view of (3.17'), (3.20) and $(3.64)_1$, the latter interaction energy reads

$$W^{int} = \int_V dV\,\Sigma^o_{iJ}U^d_{J,i}. \quad (3.74)$$

The integral (3.73) is divergent close to the dislocation line, therefore for a calculation of a dislocation self-energy it is customary to introduce the cut-off corresponding to the radius of the dislocation core of order of the Burgers vector **b**. In [46, 150, 156] one can find very rigorous consideration of this problem for a pure elastic anisotropic medium. In contrast, the interaction energy (3.74) is a well-defined quantity and can be further transformed. Taking into account that in absence of body forces and body charges the divergence of $\mathbf{\Sigma}^0$ vanishes, $\Sigma^o_{iJ,i} = 0$, eq. (3.22), the integrand in (3.74) can be reduced to

$$\Sigma^o_{iJ}U^d_{J,i} = (\Sigma^o_{iJ}U^d_J)_{,i} - \Sigma^o_{iJ,i}U^d_J = (\Sigma^o_{iJ}U^d_J)_{,i}. \quad (3.75)$$

Hence, by Gauss' theorem, integral (3.74) is transformed to a surface integral over the both sides of the cut, S^+ and S^-, the first with the outward normal $-\mathbf{n}^s$, and the second with the outward normal $\mathbf{n}^S$

$$W^{int} = \int_{S^+ + S^-} dS_i\,\Sigma^o_{iJ}U^d_J = b^{(4)}_J \int_{S=S^-} dS_i\,\Sigma^o_{iJ}. \quad (3.76)$$

In (3.76) we have taken into account that by the third definition of a dislocation, eq. (3.4),

$$\mathbf{U}(\mathbf{r}^{S^-}) - \mathbf{U}(\mathbf{r}^{S^+}) = \mathbf{b}^{(4)}. \quad (3.77)$$

With eq. (3.76) we are prepared to extend the familiar Peach-Koehler formula [157] determining the force on the dislocation from the external fields. Of course, a force in the given case should be treated in a thermodynamic sense. It has nothing to do with any interatomic forces; rather it is similar to a force on a line of charge from an electric field or on a conductor with an electric current from a magnetic field. In our case the distributed force **F** per unit length of the dislocation from the other sources of electro-elastic fields is determined by the variation of the interaction energy δW^{int} due to a virtual displacement δx_p in the configuration of the dislocation line C. By definition [46],

$$\delta W^{int} = -\oint_C ds F_p \delta x_p. \tag{3.78}$$

On the other hand, the same variation of (3.76) gives

$$\delta W^{int} = \varepsilon_{ipq} b_J^{(4)} \oint_C ds \xi_q^C \delta x_p \Sigma_{iJ} \tag{3.79}$$

where for brevity we have omitted the superscript o at σ_{iJ}. Combining the last two equations with recognition that the variation δx_p is arbitrary, one finds

$$F_p = \varepsilon_{ipq} \Sigma_{iJ} b_J^{(4)} \xi_q^C, \tag{3.80}$$

which differs from the initial Peach-Koehler formula by a natural replacement

$$\sigma_{ij} b_j \rightarrow \Sigma_{iJ} b_J^{(4)} = \sigma_{ij} b_j + D_i \Delta\varphi, \tag{3.81}$$

where σ_{ij} and D_i are external fields of stress and electric displacement. We can see that due to electro-elastic coupling a nonvanishing force on a dislocation may arise due to external electric field $\mathbf{E}$ even in externally nondeformed piezoelectric media ($\beta^o = 0$) and even when we deal with an ordinary 3D dislocation ($\Delta\varphi = 0$). Indeed, in the latter case by eqs. (3.13), (3.16) and (3.18)

$$\Sigma_{iJ} b_J^{(4)} = \sigma_{ij} b_j = -e_{kij} E_k b_j, \tag{3.82}$$

and

$$F_p = -\varepsilon_{ipq} \xi_q^C E_k e_{kij} b_j. \tag{3.83}$$

As is well known, dislocations in crystals can move conservatively only in so-called slip planes, containing their Burgers vectors. If $\mathbf{n}$ is the normal to such a plane, then $\mathbf{n} \times \xi^C$ is the direction of a possible glide of dislocation. So, as a rule, the most physical interest is concerned with the projection of $\mathbf{F}$ (3.80) on $\mathbf{n} \times \xi^C$,

$$F = \mathbf{F} \cdot (\mathbf{n} \times \xi^C) = \varepsilon_{ipq} \varepsilon_{prs} \Sigma_{iJ} b_J^{(4)} \xi_q^C \xi_s^C n_r = n_i \Sigma_{iJ} b_J^{(4)}. \tag{3.84}$$

The scalar quantity F (3.84) is an extended resolved force. It is customary also to introduce the so-called resolved shear stress $\tau = F/b$.

When the external force on a dislocation with the Burgers vector $\mathbf{b}$ is provided by the electric field $\mathbf{E}$, eq. (3.84) with (3.82) is simplified to

$$F_{el} = -E_k e_{kij} n_i b_j. \tag{3.84a}$$

3.3 Electro-elastic fields of straight 4D dislocations

In this subsection we shall consider the very important particular case of straight dislocations in infinite piezoelectric bodies. Their fields are themselves of essential physical interest. However, as we have seen, the basic characteristics of fields excited by curvilinear dislocations of arbitrary shape can be expressed in terms of corresponding fields of straight dislocations with the same Burgers vector. This remarkably increases a significance of the forthcoming material. In fact, there is one more reason for paying so much attention in this chapter to straight-line sources. It is concerned with the role of this subsection (especially its (c) part) as a theoretical introduction for chapters 7-9 of the next part II, which will be devoted to an analysis of 2D fields excited by line sources in bounded piezoelectric media.

(a) The integral representations

Consider an infinite piezoelectric medium with a straight 4D dislocation having the direction $\xi^C = \mathbf{t}$(z-axis), the Burgers vector $\mathbf{b}^{(4)}$, the cut surface S(xz-plane) with the normal $\mathbf{n}$ (y-axis) and the in-plane normal $\mathbf{m}$ to the dislocation line (x-axis), see Fig. 3.3. In this case the contour C of the integration in eq. (3.55) is reduced to the $z-$axis ($-\infty < z < \infty$) with $\mathbf{r}^C = z\mathbf{t}, dl_j = t_j dz$, so that one obtains

$$U_{lK}(\mathbf{r}) = -C_{qMNp} b_N^{(4)} \varepsilon_{jpl} t_j \int\limits_{-\infty}^{\infty} dz G_{KM,q}(\mathbf{r} - z\mathbf{t}) \tag{3.85}$$

Substituting here (3.29) and (3.34) we have

$$U_{lK}(\mathbf{r}) =$$

$$= -iC_{qMNp} b_N^{(4)} \varepsilon_{jpl} t_j \frac{1}{(2\pi)^3} \int d^3\mathbf{k} e^{i\mathbf{k}\cdot\mathbf{r}} \frac{k_q}{k^2} [(\kappa\kappa)^{-1}]_{KM} \int\limits_{-\infty}^{\infty} dz \exp(-ik_z z). \tag{3.86}$$

By definition, the last integral in (3.86) is $2\pi\delta(k_z)$, therefore the Fourier integral is reduced to a double integral in the plane $\mathbf{m}$, $\mathbf{n}$. We shall evaluate this integral in the polar coordinate system (k_ρ, φ) counting off the angle φ from the vector $\mathbf{n}$ (Fig. 3.3). In this system $\boldsymbol{\kappa} \equiv \mathbf{k}/k = \mathbf{n}(\varphi) \equiv \mathbf{n}_\varphi$ (Fig. 3.3), $\mathbf{k}\cdot\mathbf{r} = k_\rho \boldsymbol{\kappa}\cdot\boldsymbol{\rho}$, $\boldsymbol{\rho} = \mathbf{r} - z\mathbf{t}$.
Taking into account that [150]

$$\int\limits_0^{\infty} dk_\rho \exp(ik_\rho \mathbf{n}_\varphi \cdot \boldsymbol{\rho}) = i/\mathbf{n}_\varphi \cdot \boldsymbol{\rho}, \tag{3.87}$$

one can easily derive

$$U_{lK}(\boldsymbol{\rho}) = \frac{C_{qMNp} b_N^{(4)} \varepsilon_{jpl} t_j}{(2\pi)^2} \int\limits_0^{2\pi} d\varphi \frac{n_q(\varphi)[(n_\varphi n_\varphi)^{-1}]_{KM}}{\mathbf{n}_\varphi \cdot \boldsymbol{\rho}}, \tag{3.88}$$

where the integral is understood in a principal value sense. Eq. (3.88) represents the extension of the Barnett-Swanger [158] formula.

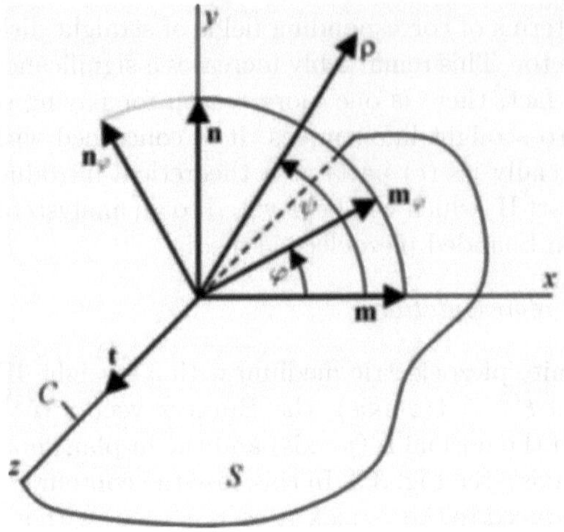

Fig. 3.3. Straight dislocation C in the coordinate system x, y, z and some vector geometry.

Let us slightly rearrange the latter result. Noting that by definition

$$\varepsilon_{jpl}t_j = m_p(\varphi))n_l(\varphi) - m_l(\varphi)n_p(\varphi), \tag{3.89}$$

and making use of the identity

$$[(n_\varphi n_\varphi)^{-1}]_{KM}(n_\varphi n\varphi)_{MN} = \delta_{KN} \ , \tag{3.90}$$

one arrives at the other form of (3.88)

$$U_{lK}(\boldsymbol{\rho}) = \frac{b_N^{(4)}}{(2\pi)^2}\int_0^{2\pi} d\varphi \frac{\{n_q(\varphi)C_{qMNp}[(n_\varphi n_\varphi)^{-1}]_{KM}n_l(\varphi) - \delta_{NK}\delta_{lp}\}\, m_p(\varphi)}{\mathbf{n}_\varphi \cdot \boldsymbol{\rho}} \tag{3.91}$$

With (3.91) the corresponding generalized stress given by eq. (3.20) reads

$$\Sigma_{iJ}(\boldsymbol{\rho}) = \frac{b_N^{(4)}}{(2\pi)^2}\int_0^{2\pi} d\varphi \frac{C^*_{iJNp}(\mathbf{n}_\varphi)m_p(\varphi)}{\mathbf{n}_\varphi \cdot \boldsymbol{\rho}}, \tag{3.92}$$

where we have introduced the plane tensor of our extended "elasticity" (3.20)

$$C^*_{iJNp}(\mathbf{n}) = C_{iJKl}n_l[(nn)^{-1}]_{KM}n_qC_{qMNp} - C_{iJNp} \ . \tag{3.93}$$

Eqs. (3.91), (3.92) extend the analogous results for pure elasticity obtained in [159].

In the ordinary elasticity the tensor analogous to (3.93) has quite a series of useful properties [139, 150]. Some of them hold also for piezoelectrics. For instance, the projection of $C^*_{iJNp}(\mathbf{n})$ on $\mathbf{n}$ from both sides identically vanishes,

$$n_i C^*_{iJNp}(\mathbf{n}) = 0, \quad C^*_{iJNp}(\mathbf{n}) n_p = 0. \tag{3.94}$$

This tensor plays its main role in the theory of internal elastic fields excited by 1D distributions of sources described by a plastic distortion tensor dependent on only one coordinate, $\beta^0 = \beta^0(\mathbf{r} \cdot \mathbf{n})$. In the latter case the corresponding stress field is given by [139]

$$\sigma_{ij}(\mathbf{r} \cdot \mathbf{n}) = C^*_{ijnp}(\mathbf{n}) \beta^0_{pn}(\mathbf{r} \cdot \mathbf{n}). \tag{3.95}$$

Let us prove that the similar solution is valid also for a considered generalized elasticity of piezoelectrics. Consider an infinite piezoelectric medium with a given 1D distribution of distortions $\mathbf{U}^0(\mathbf{r} \cdot \mathbf{n})$. In terms of eq. (3.27) this is equivalent to the body forces

$$F_M(\mathbf{r} \cdot \mathbf{n}) = -C_{qMNp} U^0_{pN,q}(\mathbf{r} \cdot \mathbf{n}) = -n_q C_{qMNp} \frac{\partial}{\partial(\mathbf{r} \cdot \mathbf{n})} U^0_{pN}(\mathbf{r} \cdot \mathbf{n}). \tag{3.96}$$

The generalized stress distribution determined by eq. (3.20) in the given case is

$$\Sigma_{iJ}(\mathbf{r} \cdot \mathbf{n}) = C_{iJKl}[U_{K,l}(\mathbf{r} \cdot \mathbf{n}) - U^0_{lK}(\mathbf{r} \cdot \mathbf{n})]. \tag{3.97}$$

For the considered 1D field $U_K(\mathbf{r} \cdot \mathbf{n})$ the Green function is also one-dimensional, $G_{KM}(\mathbf{r} \cdot \mathbf{n})$, so that we can look for $U_{K,l}(\mathbf{r} \cdot \mathbf{n})$ in the form

$$U_{K,l}(\mathbf{r} \cdot \mathbf{n}) = \int_{-\infty}^{\infty} ds G_{KM,l}(\mathbf{r} \cdot \mathbf{n} - s) F_M(s). \tag{3.98}$$

Let us express the function $G_{KM,l}(\mathbf{r} \cdot \mathbf{n} - s)$ in (3.98) as a Fourier integral

$$G_{KM,l}(\mathbf{r} \cdot \mathbf{n} - s) = \frac{i n_l}{2\pi} \int_{-\infty}^{\infty} dk k G_{KM}(k) e^{ik(\mathbf{r} \cdot \mathbf{n} - s)}, \tag{3.99}$$

where the Fourier image of $G_{KM}(\mathbf{r} \cdot \mathbf{n})$ is found similarly to (3.34),

$$G_{KM}(k) = \frac{1}{k^2} [(nn)^{-1}]_{KM}. \tag{3.100}$$

Substituting (3.99), (3.100) into (3.98) and noting that

$$\int_{-\infty}^{\infty} ds F_M(s) e^{-iks} = F_M(k) = -i n_q C_{qMNp} U^0_{pN}(k) k, \tag{3.101}$$

one easily finds

$$U_{K,l}(\mathbf{r} \cdot \mathbf{n}) = n_l [(nn)^{-1}]_{KM} n_q C_{qMNp} U^0_{pN}(\mathbf{r} \cdot \mathbf{n}). \tag{3.102}$$

Combining eqs. (3.102) and (3.97) we finally obtain

$$\Sigma_{iJ}(\mathbf{r}\cdot\mathbf{n}) = C^{*}_{iJNp}(\mathbf{n})U^{0}_{pN}(\mathbf{r}\cdot\mathbf{n}), \tag{3.103}$$

where the tensor $C^{*}_{iJNp}(\mathbf{n})$ is defined by (3.93).

With (3.102) and (3.103) one could derive eqs. (3.91) and (3.92) much more easily basing on the Radon transformations

$$\Sigma_{iJ}(\boldsymbol{\rho}) = \int_{0}^{2\pi} d\varphi\Sigma_{iJ}(\boldsymbol{\rho}\cdot\mathbf{n}), \qquad U^{0}_{pN}(\boldsymbol{\rho}) = \int_{0}^{2\pi} d\varphi U^{0}_{pN}(\boldsymbol{\rho}\cdot\mathbf{n}), \tag{3.104}$$

having in mind that $\mathbf{r}\cdot\mathbf{n} = p\cdot\mathbf{n}$,

$$U^{0}_{pN}(\boldsymbol{\rho}) = -b^{(4)}_{N} n_p H(\boldsymbol{\rho}\cdot\mathbf{m})\delta(\boldsymbol{\rho}\cdot\mathbf{n}) = \frac{b^{(4)}_{N} n_p}{(2\pi)^2}\int_{0}^{2\pi}\frac{d\varphi}{(\boldsymbol{\rho}\cdot\mathbf{n}_\varphi)(\mathbf{n}\cdot\mathbf{m}_\varphi)}, \tag{3.105}$$

and making use of the identities $n_p = n_p(\varphi)(\mathbf{n}\cdot\mathbf{n}_\varphi) + m_p(\varphi)(\mathbf{n}\cdot\mathbf{m}_\varphi)$ and $(3.94)_2$.

(b) The matrix representations

The found expressions for dislocation fields, eqs. (3.88), (3.91), (3.92), contain in the denominators of the integrands the scalar product $\boldsymbol{\rho}\cdot\mathbf{n}_\varphi$, which provides a singularity at integration. Therefore, as was already noted, all above integrals are understood in principal value sense. This is not very convenient for numerical calculations and can be eliminated [160, 161]. Up to now our coordinate system was independent of the observation point $\boldsymbol{\rho}$. Let us abandon this invariance and choose the orientation of the vector $\mathbf{m}$ in Fig. 3.3 not in the S plane but along the vector $\boldsymbol{\rho}$, so that

$$\boldsymbol{\rho} = \rho\mathbf{m}, \quad \boldsymbol{\rho}\cdot\mathbf{n}_\varphi = \rho\mathbf{m}\cdot\mathbf{n}_\varphi = -\sin\varphi. \tag{3.106}$$

Substituting into (3.88) together with (3.106) the relations

$$n_q(\varphi) = -m_q\sin\varphi + n_q\cos\varphi, \tag{3.107}$$

$$\varepsilon_{jpl}t_j = m_p n_l - m_l n_p, \tag{3.108}$$

we have

$$\begin{aligned} U_{lK}(\boldsymbol{\rho}) = \frac{b^{(4)}_{N}}{(2\pi)^2}\Bigg\{&[n_l(mm)_{NM} - m_l(nm)_{NM}]\int_{0}^{2\pi} d\varphi(n_\varphi n_\varphi)^{-1}_{MK} - \\ &-[n_l(mn)_{NM} - m_l(nn)_{NM}]\int_{0}^{2\pi} d\varphi\frac{\cos\varphi}{\sin\varphi}(n_\varphi n_\varphi)^{-1}_{MK}\Bigg\}. \end{aligned} \tag{3.109}$$

Now let us derive some useful identity starting from the trivial

$$(nn_\varphi) \equiv (nn_\varphi). \tag{3.110}$$

Replace in the left-hand side of (3.110) n_φ by $-m\sin\varphi + n\cos\varphi$ and in the right-hand side n by $n_\varphi\cos\varphi + m_\varphi\sin\varphi$. After simple rearrangements the identity becomes

$$(nn)\frac{\cos\varphi}{\sin\varphi} \equiv (nm) + (m_\varphi n_\varphi) + (n_\varphi n_\varphi)\frac{\cos\varphi}{\sin\varphi}. \tag{3.111}$$

Multiplying both sides of (3.111) by $(nn)^{-1}$ from the left and by $(n_\varphi n_\varphi)^{-1}$ from the right, one finally obtains

$$(n_\varphi n_\varphi)^{-1}\frac{\cos\varphi}{\sin\varphi} \equiv$$
$$\equiv (nn)^{-1}\frac{\cos\varphi}{\sin\varphi} + (nn)^{-1}(nm)(n_\varphi n_\varphi)^{-1} + (nn)^{-1}(m_\varphi n_\varphi)(n_\varphi n_\varphi)^{-1}. \tag{3.112}$$

Noting that the principal value integration of the first term vanishes,

$$\int_0^{2\pi} d\varphi \frac{\cos\varphi}{\sin\varphi} = 0, \tag{3.113}$$

we see that the last integral in (3.109) is not singular anymore. Let us introduce two 4×4 matrices

$$\mathbf{Q} = -\frac{1}{2\pi}\int_0^{2\pi} d\varphi (n_\varphi n_\varphi)^{-1}, \quad \mathbf{S} = -\frac{1}{2\pi}\int_0^{2\pi} d\varphi (n_\varphi n_\varphi)^{-1}(n_\varphi m_\varphi). \tag{3.114}$$

In these terms eq. (3.109) takes the form.

$$\begin{aligned} U_{lK} = \frac{b_N^{(4)}}{2\pi\rho}\{&-m_l S_{NK} + n_l\{(mn)(nn)^{-1}\mathbf{S}+ \\ &+ [(mn)(nn)^{-1}(nm) - (mm)]\mathbf{Q}\}_{NK}\}. \end{aligned} \tag{3.115}$$

This representation looks rather bulky compared with eq. (3.88), however it has an important advantage being much more suitable for computing.

Let us calculate now self-energy (3.73) of a straight dislocation per its unit length. It is clear that the bulk integral in (3.73) considered per unit length is reduced to the integration over the plane **m**, **n**. It is convenient to evaluate the integral in the polar coordinates $\{\rho, \psi\}$ (Fig. 3.3)

$$W^{self} = \frac{1}{2}\int \rho d\rho \int d\psi \Sigma_{iJ}(\boldsymbol{\rho}) U_{iJ}(\boldsymbol{\rho}) \tag{3.116}$$

where for brevity we omitted the superscripts d and replaced $U_{J,i}$ by U_{iJ} having in mind that we do not introduce any cut along the S surface. In this case one can substitute into (3.116) $U_{iJ} = U_{J,i} - U^0_{iJ}$ and make use of identical vanishing of the first integral due to the Colonnetti theorem [139]. Taking into account the logarithmic divergence in (3.116) both at $\rho \to 0$ and at $\rho \to \infty$, the integration will be accomplished in the ring between $\rho = r_0$ and $\rho = R$, where r_0 is of order of the lattice parameter and R of order of the crystal size:

$$W^{self} = -\frac{1}{2}\int_{r_0}^{R} \rho d\rho \int_0^{2\pi} d\psi \Sigma_{iJ}(\rho) U^0_{iJ}(\rho). \tag{3.117}$$

Combining (3.117) with (3.92) and $(3.105)_1$ one obtains

$$W^{self} = \frac{b^{(4)}_J b^{(4)}_N}{8\pi^2} \ln\left(\frac{R}{r_0}\right) \int_0^{2\pi} d\varphi m_i(\varphi) C^*_{iJNp}(\mathbf{n}_\varphi) m_p(\varphi) I, \tag{3.118}$$

where I is the integral

$$I = \sin\varphi \int_0^{2\pi} d\psi \frac{\delta(\sin\psi) H(\cos\psi)}{\sin(\psi - \varphi)} = -1. \tag{3.119}$$

Summerizing, the self-energy per unit length of a straight dislocation is given by

$$W^{self} = E_0 \ln\left(\frac{R}{r_0}\right) \tag{3.120}$$

with the prelogarithmic energy factor E_0 defined by the convolution

$$E_0 = \frac{1}{4\pi} b^{(4)}_I B_{IJ} b^{(4)}_J \tag{3.121}$$

where the 4×4 matrix $\mathbf{B}$ is introduced according to

$$B_{JN} = -\frac{1}{2\pi}\int_0^{2\pi} d\varphi m_i(\varphi) C^*_{iJNp}(\mathbf{n}_\varphi) m_p(\varphi). \tag{3.122}$$

In view of (3.94) the latter matrix can be presented in an invariant form similar to (3.114)

$$\mathbf{B} = -\frac{1}{2\pi}\int_0^{2\pi} d\varphi [(m_\varphi n_\varphi)(n_\varphi n_\varphi)^{-1}(n_\varphi m_\varphi) - (m_\varphi m_\varphi)]. \tag{3.123}$$

Thus, as we have seen, all basic characteristics of electro-elastic fields of a straight dislocation can be expressed in terms of three matrices $\mathbf{Q}$, $\mathbf{S}$ and $\mathbf{B}$, eqs. (3.114), (3.123), which depend only on an orientation of the dislocation line in the given piezoelectric medium. In ordinary theory of elasticity the applied method is known as the matrix (or integral) formalism [46, 150]. Eqs. (3.114), (3.115) and (3.120)-(3.123) extend the most important results of this theory for the case of piezoelectric media.

(c) The Stroh-like algebraic representations

In this subsection we shall discuss an alternative approach to a description of 2D electro-elastic fields excited by straight-line sources. This approach is based on the Stroh-like formalism [41, 42] extended for piezoelectrics by Jergan [60] and Barnett & Lothe [61]. It reduces the problem to a purely algebraic procedure of finding eigenvalues and eigenvectors of some 8×8 matrix. As will be shown, it allows considering in a unified manner the 8D general line source, which consists of the above 4D straight dislocation, the line of body forces

$$\mathbf{f}(\boldsymbol{\rho}) = \mathbf{f}\delta(\boldsymbol{\rho}\cdot\mathbf{n})\delta(\boldsymbol{\rho}\cdot\mathbf{m}), \tag{3.124}$$

and the line of body charges

$$q(\boldsymbol{\rho}) = q\delta(\boldsymbol{\rho}\cdot\mathbf{n})\delta(\boldsymbol{\rho}\cdot\mathbf{m}), \tag{3.125}$$

all combined in the same straight line. Thus, equilibrium equation (3.22) is transformed into

$$\Sigma_{iJ,i} = -f_J^{(4)}\delta(\boldsymbol{\rho}\cdot\mathbf{n})\delta(\boldsymbol{\rho}\cdot\mathbf{m}), \tag{3.126}$$

where $\mathbf{f}^{(4)}$ is the generalized 4D force

$$\mathbf{f}^{(4)} = (\mathbf{f}, -q)^t. \tag{3.127}$$

Let us introduce the "stress"

$$\boldsymbol{\Sigma}^0(\boldsymbol{\rho}) = \mathbf{m}\otimes\mathbf{f}^{(4)}\delta(\boldsymbol{\rho}\cdot\mathbf{n})H(\boldsymbol{\rho}\cdot\mathbf{m}). \tag{3.128}$$

It is easily checked that in these terms eq. (3.126) is reduced to (see Fig. 3.3)

$$(\boldsymbol{\Sigma}+\boldsymbol{\Sigma}^0)_{xJ,x} + (\boldsymbol{\Sigma}+\boldsymbol{\Sigma}^0)_{yJ,y} = 0. \tag{3.129}$$

Eq. (3.129) allows us to introduce a stress function 4-vector $\mathbf{V}(x,y)$,

$$\mathbf{V}(x,y) = \begin{pmatrix} \mathbf{J}(x,y) \\ D(x,y) \end{pmatrix} \tag{3.130}$$

so that

$$(\boldsymbol{\Sigma}+\boldsymbol{\Sigma}^0)_{xJ} = V_{J,y}, \qquad (\boldsymbol{\Sigma}+\boldsymbol{\Sigma}^0)_{yJ} = -V_{J,x}. \tag{3.131}$$

With this substitution, eq. (3.129) is identically satisfied and constitutive equation (3.20) with (3.23) is transformed to the system

$$-V_{J,x} = (nm)_{JK}U_{K,x} + (nn)_{JK}(U_{K,y} - U^0_{yK}), \tag{3.132}$$

$$V_{J,y} = (mm)_{JK}U_{K,x} + (mn)_{JK}(U_{K,y} - U^0_{yK}) + \Sigma^0_{xJ}, \tag{3.133}$$

where by eqs. $(3.105)_1$ and (3.128)

$$U^0_{yK} = -b^{(4)}_K \mathrm{H}(x)\delta(y), \quad \Sigma^0_{xJ} = f^{(4)}_J \mathrm{H}(x)\delta(y). \tag{3.134}$$

After some rearrangements in eqs. (3.132) and (3.133), this system can be transformed to the single eight-dimensional equation:

$$(\mathbf{I}\partial_y - \mathbf{N}\partial_x)\boldsymbol{\eta}(x,y) = -\mathbf{g}\mathrm{H}(x)\delta(y), \tag{3.135}$$

where $\boldsymbol{\eta}(x,y)$ is the unknown eight-component vector function

$$\boldsymbol{\eta}(x,y) = [\mathbf{U}(x,y), \mathbf{V}(x,y)]^t, \tag{3.136}$$

$\mathbf{g}$ is the strength of the general line source,

$$\mathbf{g} = [\mathbf{b}, \Delta\varphi, -\mathbf{f}, q]^t, \tag{3.137}$$

I is the unit matrix and $\mathbf{N}$ is the 8×8 constant matrix:

$$\mathbf{N} = -\begin{pmatrix} (nn)^{-1}(nm) & (nn)^{-1} \\ (mn)(nn)^{-1}(nm) - (mm) & (mn)(nn)^{-1} \end{pmatrix}, \tag{3.138}$$

depending only on the orientation of the frame $\mathbf{m}$, $\mathbf{n}$ and on the material constants C_{iJKl} (see eq. (3.51)). The matrix $\mathbf{N}$ is an extension [60, 61] of the 6×6 Stroh matrix initially introduced [41, 42] for purely elastic media.

Note that eq. (3.135) is equivalent to a system of eight differential equations of the first order, in contrast to the standard system of four differential equations of the second order given by eq. (3.24). As a result, the right-hand side of (3.135) contains the Heaviside function $\mathrm{H}(x)$, instead of the delta-function $\delta(x)$ in (3.126). The next natural step in our analysis is a Fourier transformation of the unknown function $\boldsymbol{\eta}(x,y)$ and the Heaviside function in the right-hand side of eq. (3.135) with respect to x :

$$\begin{aligned} \boldsymbol{\eta}(x,y) &= \int_{-\infty}^{\infty} dk \exp(ikx)\boldsymbol{\eta}(k,y), \\ \boldsymbol{\eta}(k,y) &= \frac{1}{2\pi}\int_{-\infty}^{\infty} dx \exp(-ikx)\boldsymbol{\eta}(x,y), \end{aligned} \tag{3.139}$$

$$\mathrm{H}(x) = \frac{1}{2\pi i}\int_{-\infty}^{\infty} \frac{dk}{k}\exp(ikx). \tag{3.140}$$

In the Fourier representation of $H(x)$, as usual, for the sake of convergence in (3.140), k was given an infinitesimally small negative imaginary part

$$k = \mathrm{Re}k - i\varepsilon, \qquad \varepsilon \to +0. \tag{3.141}$$

At $x > 0$ the integration in (3.140) can be reduced to the integration over the closed path C_R shown in Fig. 3.4, a. Calculating the residue at the only pole

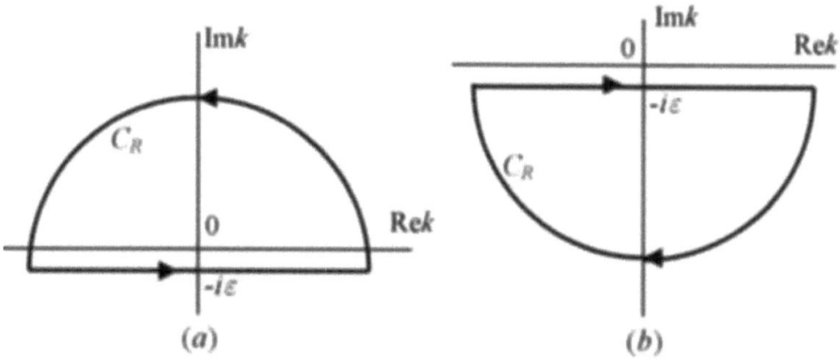

Fig. 3.4. Straight dislocation C in the coordinate system x, y, z and some vector geometry.

$k = 0$ one easily obtains $H(x > 0) = 1$. Similarly, at $x < 0$ the integration path in (3.140) can be deformed into the contour shown in Fig. 3.4, b with no poles inside which leads to the result $H(x < 0) = 0$. Naturally, (3.141) will be implied in all further equations. After the Fourier transformation eq. (3.135) takes the form

$$(\mathbf{I}\partial_y - ik\mathbf{N})\boldsymbol{\eta}(k, y) = -\mathbf{g}(2\pi ik)^{-1}\delta(y). \tag{3.142}$$

We stress that the matrix $\mathbf{N}$ in (3.142) is independent of k (see eq. 3.128)). It is convenient at this stage to introduce eigenvectors $\boldsymbol{\xi}_\alpha$ and eigenvalues p_α of the matrix $\mathbf{N}$:

$$\mathbf{N}\boldsymbol{\xi}_\alpha = p_\alpha \boldsymbol{\xi}_\alpha. \tag{3.143}$$

According to [61, 63, 64], they form four pairs of complex conjugates. With an appropriate choice of numbering,

$$\boldsymbol{\xi}_{\alpha+4} = \boldsymbol{\xi}^*_\alpha, \qquad p_{\alpha+4} = p^*_\alpha, \qquad \alpha = 1, ..., 4. \tag{3.144}$$

$$\text{Im}\, p_\alpha > 0 \quad \text{for} \quad \alpha = 1, ..., 4; \qquad \text{Im}\, p_\alpha < 0 \quad \text{for} \quad \alpha = 5, ..., 8. \tag{3.145}$$

The Stroh eigenvectors $\boldsymbol{\xi}_\alpha$ are orthogonal and complete [61, 63, 64]:

$$\boldsymbol{\xi}_\alpha \cdot \mathbf{T}\boldsymbol{\xi}_\beta = \delta_{\alpha\beta}, \tag{3.146}$$

$$\sum_{\alpha=1}^{8} \boldsymbol{\xi}_\alpha \otimes \mathbf{T}\boldsymbol{\xi}_\alpha = \mathbf{I}, \tag{3.147}$$

where the 8×8 matrix $\mathbf{T}$ has the following block form:

$$\mathbf{T} = \begin{pmatrix} \mathbf{0} & \mathbf{I} \\ \mathbf{I} & \mathbf{0} \end{pmatrix}. \tag{3.148}$$

Now let us return to eq. (3.142). We shall look for a solution of this equation in a customary form of the sum of a general solution obeying the corresponding homogeneous equation,

$$(\mathbf{I}\partial_y - ik\mathbf{N})\boldsymbol{\eta}(k, y) = \mathbf{0}, \qquad (3.149)$$

and a partial solution satisfying the initial eq. (3.142). The formal solution of homogeneous eq. (3.149) is written as

$$\boldsymbol{\eta}(k, y) = \exp[ik\mathbf{N}y]\mathbf{X}, \qquad (3.150)$$

where $\mathbf{X}$ is an arbitrary eight-component vector independent of y (but not of k). Calculating the scalar product of this vector with eq. (3.147) from the right and substituting the result into (3.150), one obtains the general solution of eq. (3.149)

$$\boldsymbol{\eta}_{gen}(k, y) = \sum_{\alpha=1}^{8} (\mathbf{X} \cdot \mathbf{T}\boldsymbol{\xi}_\alpha) \exp[ik\mathbf{N}y]\boldsymbol{\xi}_\alpha = \sum_{\alpha=1}^{8} c_\alpha \boldsymbol{\xi}_\alpha \exp[ikp_\alpha y] \qquad (3.151)$$

with a set of unknown coefficients c_α, $\alpha = 1, ..., 8$, replacing the unknown 8-vector $\mathbf{X}$. Adding to (3.151) the partial solution of eq. (3.142)

$$\boldsymbol{\eta}_{part}(k, y) = -[2\pi ik]^{-1} H(y) \sum_{\alpha=1}^{8} \boldsymbol{\xi}_\alpha (\boldsymbol{\xi}_\alpha \cdot \mathbf{Tg}) \exp(ikp_\alpha y) \qquad (3.152)$$

and transforming the result to x-space, by eq. (3.139) with (3.141), we have

$$\boldsymbol{\eta}(x, y) = \frac{1}{2\pi i} \sum_{\alpha=1}^{8} \int_{-\infty}^{\infty} \frac{dk}{k} \boldsymbol{\xi}_\alpha \exp\{ik(x + p_\alpha y)\}[b_\alpha(k) - (\boldsymbol{\xi}_\alpha \cdot \mathbf{Tg})\mathrm{H}(y)]. \qquad (3.153)$$

Here the 8 unknown coefficients b_α have appeared from the substitution to (3.151)

$$c_\alpha = b_\alpha [2\pi ik]^{-1} \exp(ikp_\alpha y). \qquad (3.154)$$

Let us find these coefficients from the requirement of a convergence of the integral in (3.153). The crucial role belongs here to a sign of the product $\mathrm{Im}(p_\alpha)ky$. Bearing in mind (3.145), we can conclude that the function $b_\alpha(k)$ must be constructed in such a way, in order to exclude from the integration the terms relating to $ky < 0$ for $\alpha = 1, ..., 4$, and to $ky > 0$ for $\alpha = 5, ..., 8$. One can easily check that the only variant satisfying the above criterion is provided by the choice

$$b_\alpha(k) = (\boldsymbol{\xi}_\alpha \cdot \mathbf{Tg}) H(-k\mathrm{Im}p_\alpha). \qquad (3.155)$$

Substitution of (3.155) into (3.153) leads to

$$\boldsymbol{\eta}(x, y) = -\frac{\mathrm{sgn}(y)}{2\pi i} \sum_{\alpha=1}^{8} (\boldsymbol{\xi}_\alpha \otimes \mathbf{T}\boldsymbol{\xi}_\alpha)\mathbf{g} \int_{-\infty}^{\infty} \frac{dk}{k} e^{ik(x+p_\alpha y)} \mathrm{H}(ky\mathrm{Im}p_\alpha), \qquad (3.156)$$

where we have made use of the identity

$$\mathrm{H}(a) - \mathrm{H}(-b) = \ \mathrm{sgn}(a)\mathrm{H}(ab) = \mathrm{sgn}(b)\mathrm{H}(ab). \tag{3.157}$$

Now one can see that the integrand in (3.156) is exponentially small at $k \to \pm\infty$ for all numbers α and independently of a sign of y. However, there is some problem with the integration along the path (3.141) close to the pole $k = 0$. Fortunately, as shows a rigorous analysis, this pole contributes to $\boldsymbol{\eta}(x,y)$ (3.156) just a large constant term $\propto \ln\varepsilon$. Such contribution is unimportant, because, as we know, all components of the function $\boldsymbol{\eta}(x,y) = [\mathbf{U}(x,y), \mathbf{V}(x,y)]^t$ are defined to the accuracy of an arbitrary constant. With this remark, after the integration in (3.156) we obtain

$$\boldsymbol{\eta}(x,y) = \frac{1}{2\pi i}\sum_{\alpha=1}^{8} \mathrm{sgn}(\mathrm{Im}p_\alpha)[\boldsymbol{\xi}_\alpha \otimes \mathbf{T}\boldsymbol{\xi}_\alpha]\mathbf{g}\ln(x + p_\alpha y), \tag{3.158}$$

which is very similar to the classic result of the theory of elasticity [41, 43, 150].

It is convenient along with decomposition (3.136) to do the same with the eigenvectors $\boldsymbol{\xi}_\alpha$,

$$\boldsymbol{\xi}_\alpha = \begin{pmatrix} \mathbf{U}_\alpha \\ \mathbf{V}_\alpha \end{pmatrix}. \tag{3.159}$$

In these terms the "displacement" 4-vector $\mathbf{U}(\boldsymbol{\rho})$ of our generalized dislocation is

$$\mathbf{U}(\boldsymbol{\rho}) = \frac{1}{2\pi i}\sum_{\alpha=1}^{8} \mathrm{sgn}(\mathrm{Im}p_\alpha)\mathbf{U}_\alpha(\mathbf{V}_\alpha \cdot \mathbf{b}^{(4)})\ln(\mathbf{m}\cdot\boldsymbol{\rho} + p_\alpha\mathbf{n}\cdot\boldsymbol{\rho}), \tag{3.160}$$

where we have returned to an invariant presentation. In order to eliminate the multivaluedness of the logarithm let us again introduce the cut along the surface S in Fig. 3.3. In this case the total and "elastic" distortions coincide, so that

$$U_{K,l}(\boldsymbol{\rho}) = U_{lK}(\boldsymbol{\rho}) = \frac{1}{2\pi i}\sum_{\alpha=1}^{8} \mathrm{sgn}(\mathrm{Im}\ p_\alpha)U_{K\alpha}\mathbf{V}_\alpha \cdot \mathbf{b}^{(4)}\frac{m_l + p_\alpha n_l}{\mathbf{m}\cdot\boldsymbol{\rho} + p_\alpha\mathbf{n}\cdot\boldsymbol{\rho}}. \tag{3.161}$$

The corresponding "stress" field is

$$\Sigma_{iJ}(\boldsymbol{\rho}) = \frac{1}{2\pi i}\sum_{\alpha=1}^{8} \mathrm{sgn}(\mathrm{Im}p_\alpha)U_{K\alpha}\mathbf{V}_\alpha \cdot \mathbf{b}^{(4)}\frac{C_{iJKl}(m_l + p_\alpha n_l)}{\mathbf{m}\cdot\boldsymbol{\rho} + p_\alpha\mathbf{n}\cdot\boldsymbol{\rho}}. \tag{3.162}$$

Identical equivalence between eqs. (3.161), (3.162) and (3.91), (3.92), respectively, can be proved quite similarly as it was done in [150] for ordinary elasticity.

In the Part II we shall return to a theory of 2D coupled fields excited by straight-line sources in piezoelectric structures with surfaces and interfaces. Our theoretical approach will be based on the same algebraic Stroh-like method, which was presented in this subsection.

4

Thermo-Electro-Elastic Fields Accompanying Plastic Deformation

In the previous chapter we have developed a static theory of generalized dislocations in piezoelectric media of unrestricted anisotropy. Now we are going to consider coupled fields excited by an arbitrary dynamic dislocation ensemble determined by the corresponding tensor of plastic distortion $\boldsymbol{\beta}^0(\mathbf{r}), \mathbf{t})$, which will be considered as a known source of thermo-electro-elastic fields.

4.1 Thermodynamics of plastic deformation in thermoelastic piezoelectrics

(a) Some introductory remarks

As was shown in chapter 1, in piezoelectric media electroelastic fields are generally coupled with a thermal field due to thermoelasticity and thermal conductivity. This coupling is not essential for static problems and homogeneous thermal fields but in dynamics the situation becomes completely different. It is very well known that plastic deformation can heat bodies. This is just a reflection of the fact that a moving dislocation can effectively excite thermal fluxes. On the other hand, a generation of heat by dislocations is usually accompanied by reversible thermoelastic processes. For instance, in our paper [73] such reversible processes at plastic deformation have been generally considered for purely thermoelastic media without piezoeffect. In this chapter we shall extend the consideration [73] for a more complex medium with thermo-electro-elastic couplings.

In earlier papers [71,72], it has been shown that in piezoelectrics without thermoelasticity the mechanical and electrical fields produced by a distribution of moving dislcocations can be expressed entirely in terms of the dislocation-density and dislocation-flux tensors. On the other hand, the electro-mechanical fields produced by moving dislocations in a magneto-elastic solid under a large magnetic field depend explicitly on the tensor of plastic distortion [162]. In [73] these results were interpreted as a consequence of the fact that in the former case the piezoelectric effect is related to the elastic distortion, while the Lorentz force is incorporated with the total distortion. According to [73], this situation is similar to the case of the thermo-mechanical fields produced by moving dislocations. Since the irreversible generation of heat is caused by the plastic work done by the stress on the materials, the fields depend explicitly on the tensor of plastic distortion. On the contrary, the reversible generation of field is incorporated with the elastic distortion, and therefore the reversible thermo-mechanical fields can be expressed entirely in terms of the dislocation-density and flux tensors. Below we shall not return to this aspect of the problem.

The investigation of thermoelastic fields accompanying a plastic deformation of purely elastic isotropic bodies was also accomplished by Emaletdinov [163]. However in his analysis there was missed the important term in the basic equation of motion, namely the term linear in the gradient of plastic distortions (see the last term in the first equation of the system (4.1)). Therefore the results of [163] have the meaning of some preliminary qualitative study.

(b) Mathematical formulation of the problem and its solution

We shall start from the equations of motion (1.40)-(1.42) (with $\mathbf{f} = q = W = 0$) and the constitutive equations (1.21)-(1.23). Combining these equations one obtains the system

$$\begin{aligned} c_{ipqj}u_{j,qp} + e_{pqi}\varphi_{,qp} - \gamma_{ip}\theta_{,p} &= \rho\ddot{u}_i + c_{ipqj}\beta^0_{qj,p}, \\ e_{pqj}u_{j,qp} - \varepsilon_{pq}\varphi_{,qp} + g_p\theta_{,p} &= e_{pqj}\beta^0_{qj,p}, \\ \gamma_{pj}\dot{u}_{j,p} - g_p\dot{\varphi}_{,p} + \frac{c_v}{\bar{T}}\dot{\theta} &= \frac{\kappa_{pq}}{\bar{T}}\theta_{,qp} + \gamma_{pj}\dot{\beta}^0_{qj}. \end{aligned} \tag{4.1}$$

It is convenient to look for a solution of this system in the form of the Fourier expansion

$$\begin{pmatrix} \mathbf{u}(\mathbf{r},t) \\ \varphi(\mathbf{r},t) \\ \theta(\mathbf{r},t) \\ \boldsymbol{\beta}^0(\mathbf{r},t) \end{pmatrix} = \frac{1}{(2\pi)^4}\int d\mathbf{k}d\omega \begin{pmatrix} \bar{\mathbf{u}}(\mathbf{k},\omega) \\ \bar{\varphi}(\mathbf{k},\omega) \\ \bar{\theta}(\mathbf{k},\omega) \\ \bar{\boldsymbol{\beta}}^0(\mathbf{k},\omega) \end{pmatrix} \exp\left[i(\mathbf{k}\cdot\mathbf{r} - \omega t)\right]. \tag{4.2}$$

After the substitution of eq. (4.2) into initial system (4.1) the latter takes the form

$$\begin{aligned} [\rho\omega^2\delta_{ij} - k_p c_{pijq}k_q]\bar{u}_j - k_p e_{piq}k_q\bar{\varphi} - ik_p\gamma_{pi}\bar{\theta} &= ik_p c_{pijq}\bar{\beta}^0_{qj}, \\ -k_p e_{pjq}k_q\bar{u}_j + k_p\varepsilon_{pq}k_q\bar{\varphi} + ik_p g_p\bar{\theta} &= ik_p e_{pjq}\bar{\beta}^0_{qj}, \\ ik_p\gamma_{pj}\bar{u}_j - ik_p g_p\bar{\varphi} + \frac{1}{\bar{T}}\left(c_v + \frac{k_p\kappa_{pq}k_q}{-i\omega}\right)\bar{\theta} &= \gamma_{qj}\bar{\beta}^0_{qj}. \end{aligned} \tag{4.3}$$

This is a linear algebraic system of equations with respect to three unknown Fourier amplitudes $\bar{\mathbf{u}}(\mathbf{k},\omega)$, $\bar{\varphi}(\mathbf{k},\omega)$ and $\bar{\theta}(\mathbf{k},\omega)$. Taking into account that $\bar{\mathbf{u}}$ is a 3-component vector, we have got five equations with five unknown values.

For further conveniences in solving the problem and for the sake of more compact final results let us reformulate the system (4.3) into a matrix form. In analogy with our 4-dimensional approach to the problem in Ch. 3, now it is convenient to develop an appropriate 5-dimensional formalism. Introduce the unknown 5-vector function as an extension of (3.14):

$$\bar{U}_J(\mathbf{k},\omega) = \begin{cases} \bar{u}_j(\mathbf{k},\omega), & J = j = 1,2,3, \\ \bar{\varphi}(\mathbf{k},\omega), & J = 4, \\ \bar{\theta}(\mathbf{k},\omega), & J = 5, \end{cases} \tag{4.4}$$

and the two tensors:

$$\Gamma_{Iqj}(\mathbf{k}) = c \begin{cases} ik_p c_{pijq}, & I = i = 1,2,3, \\ ik_p e_{pjq}, & I = 4, \\ -\gamma_{qj}, & I = 5, \end{cases} \tag{4.5}$$

and

$$C_{IJ}(\mathbf{k},\omega) = \begin{cases} \rho\omega^2\delta_{ij} - k_p c_{pijq} k_q, & I = i = 1,2,3,\ J = j = 1,2,3, \\ -k_p e_{piq} k_q, & I = i = 1,2,3, \quad J = 4, \\ -ik_p\gamma_{pi}, & I = i = 1,2,3, \quad J = 5, \\ k_p\varepsilon_{pq}k_q, & I = 4, \quad J = 4, \\ ik_p g_p, & I = 4, \quad J = 5, \\ -\dfrac{1}{\bar{T}}\left(c_v - \dfrac{k_p\kappa_{pq}k_q}{i\omega}\right), & I = 5, \quad J = 5. \end{cases} \tag{4.6}$$

The latter 5×5 tensor is symmetric by definition, $C_{IJ} = C_{JI}$, therefore there is no need to define its other components. The more explicit alternative form of this tensor is given by the matrix

$$\mathbf{C}(\mathbf{k},\omega) = \begin{pmatrix} \rho\omega^2\delta_{ij} - k_p c_{pijq} k_q & -k_p e_{piq} k_q & -ik_p\gamma_{pi} \\ -k_p e_{pjq} k_q & k_p\varepsilon_{pq}k_q & ik_p g_p \\ -ik_p\gamma_{pj} & ik_p g_p & \dfrac{1}{\bar{T}}\left(-c_v + \dfrac{k_p\kappa_{pq}k_q}{i\omega}\right) \end{pmatrix}, \tag{4.7}$$

where $i,j = 1,2,3$.

In terms of the introduced five-dimensional objects the system (4.3) acquires a very compact form:

$$C_{IJ}\bar{U}_J = \Gamma_{Ilq}\bar{\beta}^0_{ql}, \quad I,J = 1,\ldots,5. \tag{4.8}$$

The solution of eq. (4.8) is

$$\bar{U}_J(\mathbf{k},\omega) = \frac{\tilde{C}_{JI}(\mathbf{k},\omega)\Gamma_{Ilq}(\mathbf{k})\bar{\beta}^0_{ql}(\mathbf{k},\omega)}{\det|\mathbf{C}(\mathbf{k},\omega)|}, \tag{4.9}$$

where $\tilde{C}_{JI}$ is a IJ-cofactor of the matrix $\mathbf{C}$. Taking into account the definition (4.4) and the relations (4.2) it is easy to find the final expressions for the coupled fields of displacements $u_i(\mathbf{r},t)$, electric potential $\varphi(\mathbf{r},t)$ and temperature $\theta(\mathbf{r},t)$:

$$\begin{aligned} u_j(\mathbf{r},t) &= \frac{1}{(2\pi)^4}\int d\mathbf{k}d\omega \frac{\tilde{C}_{jI}(\mathbf{k},\omega)\Gamma_{Ilq}(\mathbf{k})\bar{\beta}^0_{ql}(\mathbf{k},\omega)}{\det|\mathbf{C}(\mathbf{k},\omega)|} e^{i(\mathbf{k}\cdot\mathbf{r}-\omega t)}, \\ \varphi(\mathbf{r},t) &= \frac{1}{(2\pi)^4}\int d\mathbf{k}d\omega \frac{\tilde{C}_{4I}(\mathbf{k},\omega)\Gamma_{Ilq}(\mathbf{k})\bar{\beta}^0_{ql}(\mathbf{k},\omega)}{\det|\mathbf{C}(\mathbf{k},\omega)|} e^{i(\mathbf{k}\cdot\mathbf{r}-\omega t)}, \\ \theta(\mathbf{r},t) &= \frac{1}{(2\pi)^4}\int d\mathbf{k}d\omega \frac{\tilde{C}_{5I}(\mathbf{k},\omega)\Gamma_{Ilq}(\mathbf{k})\bar{\beta}^0_{ql}(\mathbf{k},\omega)}{\det|\mathbf{C}(\mathbf{k},\omega)|} e^{i(\mathbf{k}\cdot\mathbf{r}-\omega t)}. \end{aligned} \tag{4.10}$$

These coupled fields created by arbitrary dislocation motions in a thermoelastic piezoelectric medium extend our results [73] obtained for thermoelastic bodies without electromechanical coupling. The corresponding expressions for the displacement and temperature distributions, $u_j(\mathbf{r},t)$, and $\theta_j(\mathbf{r},t)$ in a particular case considered in [73] can be easily obtained from eqs. $(4.10)_1$ and $(4.10)_3$ after replacements there the tensors $\Gamma(\mathbf{r},t)$, and $\mathbf{C}(\mathbf{r},t)$ by their four-dimensional counterparts

$$\Gamma^{th}_{Iqj}(\mathbf{k}) = \begin{cases} ik_p c_{pijq}, & I = i = 1,2,3, \\ -\gamma_{qj}, & I = 4, \end{cases} \tag{4.11}$$

and

$$\mathbf{C}^{th}(\mathbf{k},\omega) = \begin{pmatrix} \rho\omega^2\delta_{ij} - k_p c_{pijq}k_q & -ik_p\gamma_{pi} \\ -ik_p\gamma_{pj} & \frac{1}{\bar{T}}\left(-c_v + \frac{k_p\kappa_{pq}k_q}{i\omega}\right) \end{pmatrix}. \tag{4.12}$$

Of course, one should additionally replace in $(4.10)_3$ $\tilde{C}_{5I}$ by $\tilde{C}^{th}_{4I}$, where now $I = 1,\ldots,4$.

4.2 Coupled fields produced by a moving straight dislocation

(a) Uniformly moving straight dislocation, general case

Let us consider the above solutions (4.10) for a particular case of a steady state motion of an infinite straight dislocation. It is convenient to use the same coordinate system as in Fig. 3.3: dislocation is parallel to the z-axis and moves in the xz slip plane with the velocity v along the x-axis (see Fig. 4.1). Accordingly by eq. (3.105) the plastic distortion $\beta^0_{ql}(\mathbf{r}, t)$ is equal

$$\beta^0_{ql}(\mathbf{r}, t) = -b_l n_q \delta(y) \mathrm{H}(x - vt), \tag{4.13}$$

and its Fourier transform $\bar{\beta}^0_{ql}(\mathbf{k}, \omega)$ is given by

$$\bar{\beta}^0_{ql}(\mathbf{k}, \omega) = \frac{(2\pi)^2 i b_l n_q}{k_x} \delta(k_z) \delta(\omega - k_x v). \tag{4.14}$$

Substituting (4.14) into (4.10) we obtain in a 5D formulation:

$$U_J(x - vt, y) =$$

$$= \frac{ib_p}{(2\pi)^2} \int\limits_{-\infty}^{\infty} \frac{dk_1}{k_1} e^{ik_1(x-vt)} \int\limits_{-\infty}^{\infty} dk_2 e^{ik_2 y} \frac{\tilde{C}_{JI}(\mathbf{k}, k_1 v) \Gamma_{Ip2}(\mathbf{k})}{\det |\mathbf{C}(\mathbf{k}, k_1 v)|} \Bigg|_{k_3=0}, \tag{4.15}$$

where for the sake of convergence the integration with respect to k_1 is supposed to be accomplished along the contour (3.141), (Fig. 3.4)).

We stress that the found solution does not represent a steady-state field, in spite of the form of the first argument in (4.15). Indeed, even in the coordinate

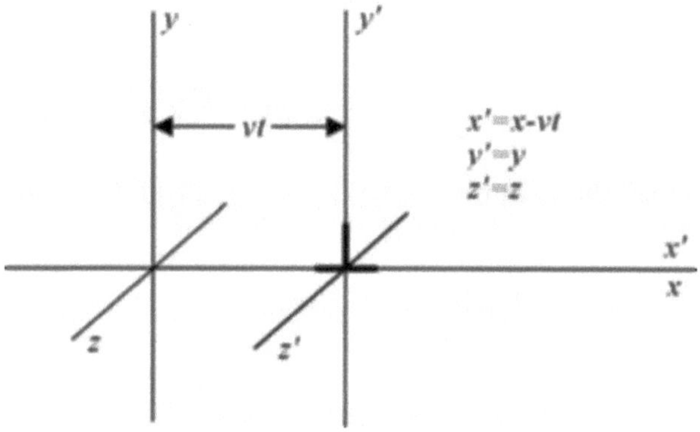

Fig. 4.1. Moving straight dislocation directed along z'-axis and the coordinate systems $\{x, y, z\}$ and $\{x', y', z'\}$

system $x' = x - vt$, $y' = y$ moving together with the dislocation, the corresponding 5D coupled field $U_J(x', y')$ retains the dependence on the velocity v and does not coincide with the static field. Let us consider the found fields at the formal limit $v \to 0$. In this limiting case the matrix element C_{55} in eq. (4.7) taken at $\omega = k_1 v$ becomes much larger than the other elements. Accordingly, the determinant $\det|\mathbf{C}|$ tends to the value $C_{55} \det|C^{piezo}|$, where $\mathbf{C}^{piezo}$ is the 4×4 sub-matrix obtained from (4.7) by eliminating there 5th line and 5th column. Quite similarly the cofactors $\tilde{C}_{JI}$ in (4.15) with $I, J \neq 5$ in the same limit reduce to $C_{55}\tilde{C}_{JI}^{piezo}$. As a result, the singular factor C_{55} disappears and the corresponding electro-elastic field of static dislocation is described by the same eq. (4.15) where one should replace $\tilde{C}_{JI}/\det|C|$ by $\tilde{C}_{JI}^{piezo}/\det|\mathbf{C}^{piezo}|$ and put everywhere $v = 0$. Of course, this integral representation can be reduced to known forms derived in Chapter 3.

The fifth line in eq. (4.15) describing a thermal field contains in the integrand the nonsingular cofactors $\tilde{C}_{JI}^{piezo}$, which remain finite at $v \to 0$. Therefore in this case the large factor C_{55} in the denominator does not disappear and the integrand together with the thermal field θ must vanish at $v \to 0$. So the static limit of the fields (4.15) gives quite reasonable results. On the other hand, it appears to be of interest to look at the problem in the opposite limiting case of large enough speeds v of a dislocation when one can expect less trivial coupling effects, in particular, non-vanishing thermal fields excited by a moving dislocation. Below we shall study such temperature fields in a particular case of a thermo-elastic medium without piezoelectric coupling (see [73]).

b) Reversible thermal field around moving dislocation

In accordance with our comments in the end of Sec. 4.1, the temperature field produced by a dislocation motion in a thermo-elastic medium is described by the integral

$$\theta(x', y') =$$

$$= \frac{ib_q}{(2\pi)^2} \int_{-\infty}^{\infty} \frac{dk_1}{k_1} e^{ik_1 x'} \int_{-\infty}^{\infty} dk_2 e^{ik_2 y'} \left. \frac{\tilde{C}_{4I}^{th}(\mathbf{k}, k_1 v) \Gamma_{Iq2}^{th}(\mathbf{k})}{\det|\mathbf{C}^{th}(\mathbf{k}, k_1 v)|} \right|_{k_3=0} \tag{4.16}$$

Bearing in mind the definitions (4.11) and (4.12) of Γth and $\mathbf{C}^{th}$ one can easily transform the integrand of (4.16) to the form

$$\left. \frac{\tilde{C}_{4I}^{th}(\mathbf{k}, k_1 v) \Gamma_{Iq2}^{th}(\mathbf{k})}{\det|\mathbf{C}^{th}(\mathbf{k}, k_1 v)|} \right|_{k_3=0} =$$

$$= \left. \frac{-ik_p c_{piq2} \tilde{C}_{4i}^{th} + \gamma_{q2} \tilde{C}_{44}^{th}}{ik_p \gamma_{pi} \tilde{C}_{4i}^{th} + \frac{1}{\bar{T}} \left[c_v - \frac{k_i \kappa_{ij} k_j}{ik_1 v} \right] \tilde{C}_{44}^{th}} \right|_{k_3=0} . \tag{4.17}$$

Let us introduce the real 4×4 matrix

$$\mathbf{A} = -\begin{pmatrix} B_{ij} & \bar{k}_p\gamma_{pi} \\ \bar{k}_p\gamma_{pj} & -c_v/\bar{T} \end{pmatrix}, \tag{4.18}$$

where $i,j = 1,2,3,\ \bar{k}_p = k_p/k_1,\ p = 1,2$, and $\mathbf{B} = \{B_{ij}\}$ is a 3×3 submatrix with the components

$$B_{ij} = \bar{k}_p c_{pijq} \bar{k}_q - \rho v^2 \delta_{ij}. \tag{4.19}$$

It is easily checked that in the new notation

$$i\tilde{C}^{th}_{4i} = \tilde{A}_{4i} k_1^5, \qquad \tilde{C}^{th}_{44} = -\tilde{A}_{44} k_1^6. \tag{4.20}$$

Accordingly, the integrand (4.17) is transformed to

$$\left.\frac{\tilde{C}^{th}_{4I}(\mathbf{k}, k_1 v)\Gamma^{th}_{Iq2}(\mathbf{k})}{\det|\mathbf{C}^{th}(\mathbf{k}, k_1 v)|}\right|_{k_3=0} = v\frac{H_q}{Av + ik_1 KB/\bar{T}}, \tag{4.21}$$

where

$$K = \bar{k}_p \kappa_{pq} \bar{k}_q, \qquad B = \tilde{A}_{44} = \det \mathbf{B}, \tag{4.22}$$

whereas the scalar $A = \det \mathbf{A}$ and the vector H_q are expressed as linear combinations of the determinant B and its cofactors $\tilde{B}_{ij}$:

$$A = \frac{c_v}{\bar{T}} B - \bar{k}_p \gamma_{pi} \tilde{B}_{ij} \gamma_{jq} \bar{k}_q, \quad H_q = \gamma_{2q} B + \bar{k}_l \gamma_{li} \tilde{B}_{ij} c_{jp2q} \bar{k}_p. \tag{4.23}$$

In the right-hand side of eq. (4.21) all the functions A, B, H_q and K depend only on $\bar{k}_p = k_p/k_1,\ p = 1,2$, i.e. on k_2/k_1. Therefore it is convenient to change in the double integral (4.16) the variables from (k_1, k_2) to (k_1, s) with $s \equiv k_2/k_1$. After some manipulations one obtains

$$\theta(x', y') = \bar{T}\frac{vb_q}{2\pi^2} \int\limits_{-\infty}^{\infty} ds \frac{H_q(s)}{K(s)B(s)} \exp\left[-c(x' + sy')\right] \mathrm{Ei}\left[c(x' + sy')\right], \tag{4.24}$$

where

$$c = \frac{v\bar{T}A(s)}{K(s)B(s)} \tag{4.25}$$

and $\mathrm{Ei}(x)$ is the exponential integral function defined as

$$\mathrm{Ei}(x) = -\int\limits_{-x}^{\infty} \frac{dt}{t} \exp(-t). \tag{4.26}$$

Of course, at $x > 0$ the integral (4.26) is understood in a principle value sense.

Let us consider the temperature field (4.24) for the particular case of a cubic crystal, when the material tensors have a very simple form:

$$\gamma_{ij} = \gamma\delta_{ij}, \quad \kappa_{ij} = \kappa\delta_{ij} \quad \{c_{\alpha\beta}\} = \begin{pmatrix} c_{11} & c_{12} & c_{12} & 0 & 0 & 0 \\ c_{12} & c_{11} & c_{12} & 0 & 0 & 0 \\ c_{12} & c_{12} & c_{11} & 0 & 0 & 0 \\ 0 & 0 & 0 & c_{44} & 0 & 0 \\ 0 & 0 & 0 & 0 & c_{44} & 0 \\ 0 & 0 & 0 & 0 & 0 & c_{44} \end{pmatrix}. \tag{4.27}$$

Accordingly,

$$K = \kappa(s^2 + 1),$$

$$\mathbf{B} = \begin{pmatrix} c_{11} + c_{44}\left(s^2 - \frac{v^2}{c_t^2}\right) & (c_{44} + c_{12})s & 0 \\ (c_{44} + c_{12})s & c_{11}s^2 + c_{44}\left(1 - \frac{v^2}{c_t^2}\right) & 0 \\ 0 & 0 & c_{44}\left(s^2 + 1 - \frac{v^2}{c_t^2}\right) \end{pmatrix} \tag{4.28}$$

$$\begin{aligned} B = B_{33}\tilde{B}_{33} = B_{33}\Bigg\{\left[c_{11} + c_{44}\left(s^2 - \frac{v^2}{c_t^2}\right)\right] \times \\ \times \left[c_{11}s^2 + c_{44}\left(1 - \frac{v^2}{c_t^2}\right)\right] - (c_{44} + c_{12})^2 s^2\Bigg\}, \end{aligned} \tag{4.29}$$

$$A = B_{33}\tilde{A}_{33} = B_{33}\left\{\frac{C_v}{\bar{T}}\tilde{B}_{33} - \gamma^2 c_{44}\left[(s^2 - 1)^2 - \frac{v^2}{c_t^2}(s^2 + 1)\right]\right\}, \tag{4.30}$$

$$H_1 = B_{33}H_1' = B_{33}\gamma c_{44}s\left[(c_{11} - c_{12})(s^2 + 1) - 2c_{44}\frac{v^2}{c_t^2}\right], \quad H_3 = 0, \tag{4.31}$$

where $c_t = (c_{44}/\rho)^{1/2}$ is the phase speed of the transverse sound wave.

Below we shall be interested by fast dislocation motions with velocities v of order of the sound speed c_t. Such motions could be possible only in a regime of conservative glide of a dislocation in a slip plane xz. This means that the Burgers vector b_q must belong to the same plane xz. Its screw component b_3 creates only shear strains, which cannot generate thermal fluxes. Indeed, this physical argument is in a complete accordance with the found vanishing component $H_3 = 0$ (4.31) (see also eq. (4.24)). Only the edge component

b_1 creates dilatational strains and nonvanishing temperature field θ (4.24). Let us analytically estimate this field basing on the found expressions for the functions $A(s), B(s), H_1(s)$ and $K(s)$, eqs. (4.28)-(4.31).

4.3 The temperature distribution around a moving edge dislocation

(a) Analytical study of thermal field close to the dislocation

To begin with, we note that the function $H_1(s)/K(s)B(s)$ in (4.24) is odd. With this observation one can immediately conclude that

in the slip plane xz the temperature θ must vanish:

$$\theta(x', 0) = 0. \tag{4.32}$$

Indeed, at $y' = 0$ the integrand in (4.24) becomes uneven, and its integration between symmetrical limits gives automatically (4.32). For the further analysis it is convenient to re-arrange eq. (4.24) to the form

$$\theta(x', y') = \bar{T}\frac{vb_1}{2\pi^2}\int_0^\infty ds \frac{H_1(s)}{K(s)B(s)} F(s|x', y'), \tag{4.33}$$

where

$$F(s|x', y') =$$
$$= \exp\left(-c(x' + sy')\right) \mathrm{Ei}\left(c(x' + sy')\right) - \exp\left(-c(x' - sy')\right) \mathrm{Ei}\left(c(x' - sy')\right). \tag{4.34}$$

Taking into account that identically

$$F(s|x', -y') = -F(s|x', y'), \tag{4.35}$$

we can state that generally

the temperature field $\theta(x', y')$ is anti-symmetric with respect to the slip plane xz:

$$\theta(x', -y') = -\theta(x', y'). \tag{4.36}$$

In fact, the property (4.32) is just a consequence of theorem (4.36).

Now let us analyze the temperature distribution (4.33) close to the dislocation. Introduce the polar coordinates $(r, \varphi) : x' = r\cos\varphi, y' = r\sin\varphi$ and consider the function $F(s|r, \varphi)$ for small enough r, so that the arguments in (4.34) would be small:

$$|cr(\cos\varphi \pm s\sin\varphi)| \ll 1. \tag{4.37}$$

Bearing in mind that the parameter c (4.25) is a function of s, let us look at this function with more attention:

$$c(s) = \frac{v\bar{T}A(s)}{K(s)B(s)} = \frac{vc_v}{K(s)}\left[1 - \frac{\bar{T}}{c_v}\bar{k}_p\gamma_{pi}B_{ij}^{-1}\gamma_{jq}\bar{k}_q\right] \approx \frac{c_0}{s^2+1}, \tag{4.38}$$

where

$$c_0 = \frac{vc_v}{\kappa}. \tag{4.39}$$

It is taken into account in (4.38) that the second term in the square brackets is of order of the thermoelastic coupling coefficient $\varepsilon_\theta = \gamma^2\bar{T}/c_v c_{44}$, which is supposed to be small (at $\bar{T} = 300K$ for copper $\varepsilon_\theta \approx 0,083$, for NaCl $\varepsilon_\theta \approx 1.33 \cdot 10^{-3}$). So, the condition (4.37) is fulfilled when $r \ll c_0^{-1}$ (at $v \sim 10^{-1}c_t$ for copper $c_0^{-1} \sim 10^3 b$, for NaCl $c_0^{-1} \sim 10b$).

In the limiting case (4.37) the both exponents in (4.34) may be replaced by the unities. In the same approximation the exponential integral function

$$\mathrm{Ei}(x) = C + \ln|x| + \sum_{k=1}^{\infty}\frac{x^k}{k \cdot k!} \tag{4.40}$$

can be represented by its two first terms, $\mathrm{Ei}(x) \approx \boldsymbol{C} + \ln|x|$, where $\boldsymbol{C}$ is the Euler number ($\boldsymbol{C} \approx 0,577\ldots$). In these terms function (4.34) is reduced to

$$F(s|r,\varphi) \approx \ln\left|\frac{1 + s\tan\varphi}{1 - s\tan\varphi}\right| \equiv F(s|\varphi). \tag{4.41}$$

In other words, we have come to the following very important conclusion:

close to the dislocation the reversible temperature field $\theta(r,\varphi)$ excited by its motion becomes independent of r being described by the universal function of azimuth $\theta(\varphi)$,

$$\theta(\varphi) = \bar{T}\frac{vb_1}{2\pi^2}\int_0^\infty ds\frac{H_1(s)}{K(s)B(s)}\ln\left|\frac{1 + s\tan\varphi}{1 - s\tan\varphi}\right|. \tag{4.42}$$

This function certainly retains properties (4.32), (4.36): $\theta(0) = 0$ and $\theta(-\varphi) = -\theta(\varphi)$. However, it acquires an additional symmetry,

$$\theta(\varphi) = -\theta(\pi - \varphi) \tag{4.43}$$

and an additional zero direction,

$$\theta(\pi/2) = 0. \tag{4.44}$$

Thus, one can state that:

$$\begin{gathered} \textit{the limiting temperature distribution } \theta(\varphi) \textit{ is anti}-\textit{symmetric} \\ \textit{with respect to a reflection both in the horizontal plane } x'z' \\ \textit{and in the vertical plane } y'z'. \end{gathered} \tag{4.45}$$

But then

this distribution must also possess a center of symmetry:

$$\theta(\varphi) = \theta(\pi + \varphi). \tag{4.46}$$

The latter statement is also directly follows from eq. (4.42).

The integral in (4.42) is well convergent. The logarithm function only improves the behavior of the integrand at small and large s :

$$\ln\left|\frac{1+s\tan\varphi}{1-s\tan\varphi}\right| \approx \begin{cases} 2s\tan\varphi, & s\tan\varphi \ll 1 \\ 2/s\tan\varphi, & s\tan\varphi \gg 1. \end{cases} \tag{4.47}$$

In particular, this function becomes essential for a convergence of the integral at the "relativistic" limit $v \to c_t$ when in the denominator we obtain $B(s) \propto s^2$. Thus, we can see that due to (4.47) even in this situation the integrand remains finite at $s \to 0$ and for $v \approx c_t$ there arise no "relativistic" singularities in the temperature field $\theta(\varphi, v)$.

Remarkable simplifications of the integrals (4.33) and (4.42) arise in the particular case of an isotropic medium, when the three elastic constants c_{11}, c_{44} and c_{12} present in eqs. (4.29)-(4.31) can be expressed in terms of only two elastic moduli, e.g. the Lame coefficients λ and μ,

$$c_{11} = \lambda + 2\mu, \qquad c_{44} = \mu, \qquad c_{12} = \lambda. \tag{4.48}$$

In these terms

$$\begin{aligned} \frac{H_1}{KB} &= \frac{2\gamma}{\kappa(\bar{\lambda}+2)} \frac{s}{(s^2+1)(s^2+1-\varepsilon_v)} = \\ &= \frac{2\gamma}{\kappa(\bar{\lambda}+2)} \frac{s}{(s^2+1)^2} \left[1 + \sum_{n=1}^{\infty} \left(\frac{\varepsilon_v}{s^2+1}\right)^n\right], \end{aligned} \tag{4.49}$$

where the two dimensionless parameters are introduced:

$$\bar{\lambda} = \lambda/\mu, \quad \varepsilon_v = \frac{v^2/c_t^2}{\bar{\lambda}+2}. \tag{4.50}$$

Supposing ε_v to be a small parameter, one obtains, instead of (4.42),

$$\theta(\varphi) \approx \bar{T} \frac{v b_1 \gamma}{\pi^2 \kappa(\bar{\lambda}+2)} \int_0^\infty ds \frac{s}{(s^2+1)^2} \ln\left|\frac{1+s\tan\varphi}{1-s\tan\varphi}\right|. \tag{4.51}$$

The convergent integral in (4.51) can be evaluated analytically. The corresponding indefinite integral is characterized by the primitive function

$$\Phi(s) = \frac{\cos^2\varphi}{2} \left\{ \frac{s^2\tan^2\varphi - 1}{s^2+1} \ln\left|\frac{1+s\tan\varphi}{1-s\tan\varphi}\right| + 2\tan\varphi \arctan s \right\} \tag{4.52}$$

which has no singularities, even at $s\tan\varphi \to \pm 1$. So the definite integral is equal

$$\Phi(\infty) - \Phi(0) = \frac{\pi}{4} \sin 2\varphi. \tag{4.53}$$

Accordingly, eq. (4.51) gives

$$\theta(\varphi) \approx \bar{T} \frac{v b_1 \gamma}{4\pi\kappa(\bar{\lambda}+2)} \sin 2\varphi. \tag{4.54}$$

It is convenient to present this relation in the form

$$\theta(\varphi) \approx g \frac{v}{c_t} \bar{T} \sin 2\varphi, \tag{4.55}$$

where the dimensionless factor g is given by

$$g = \frac{c_t b_1 \gamma}{4\pi\kappa(\bar{\lambda}+2)} \tag{4.56}$$

Fig. 4.2 shows a polar plot $\theta(\varphi)$ related to the found analytical temperature distribution (4.54), which provides a visible illustration of the established above general symmetry of properties of this coupled field close to the dislocation.

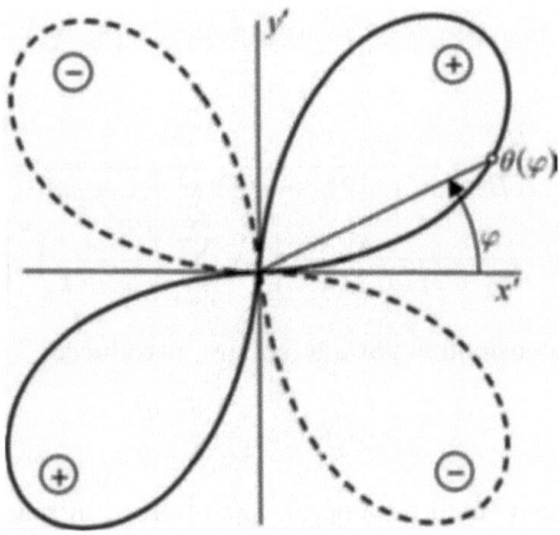

Fig. 4.2. Polar plot of the temperature distribution close to the moving edge dislocation; the solid lines indicate an increased temperature $\theta > 0$, the dashed lines correspond to a decreased temperature $\theta < 0$.

(b) Analytical and computer analysis of the thermal field $\theta(r, \varphi)$

Unfortunately, the found additional nice symmetry features of the limiting thermal pattern around the dislocation close to the latter ($rc_0 \ll 1$) do not retain for distances from the dislocation line not obeying the condition (4.37).

Of course, the general properties (4.32) and (4.36) remain valid at any distances r, however not the anti-symmetry with respect to a reflection in the plane $y'z'$ (4.45). Accordingly, we cannot expect an occurrence of the symmetry center (4.46) for a general pattern $\theta(r,\varphi)$ even in its isotropic simplified variant

$$\theta(r,\varphi) = \bar{T}\frac{vb_1\gamma}{\pi^2\kappa(\bar{\lambda}+2)}\int_0^\infty ds\frac{s}{(s^2+1)(s^2+1-\varepsilon_v)}F(s|r,\varphi). \tag{4.57}$$

Let us look at this temperature distribution basing on some additional approximate analytical analysis of eq.(4.57). As before, for the sake of brevity of final expressions, we shall suppose ε_v (4.50) to be a small parameter. It is easily checked that extending the study of the close field $rc_0 < 1$ by a transition from the used zero approximation of the function $\mathrm{Ei}(x)$ (4.40) to the linear approximation

$$\mathrm{Ei}\ (x) \approx \boldsymbol{C} + \ln|x| + x \tag{4.58}$$

leads to a much less symmetrical expression,

$$\theta(\varphi) \approx \bar{T}\frac{vb_1\gamma}{4\pi\kappa(\bar{\lambda}+2)}\left(\sin 2\varphi + \frac{1}{2}c_0 r\sin\varphi\right), \tag{4.59}$$

instead of (4.54). Thus, as we expected, the first line of zero field $y' = 0$ (4.32), described in the polar coordinates as $\sin\varphi = 0$, remains at its place (Fig. 4.3). However, the other line of vanishing θ, which in the zero approximation coincided with the axis $y'(\cos\varphi = 0)$, now becomes a curve

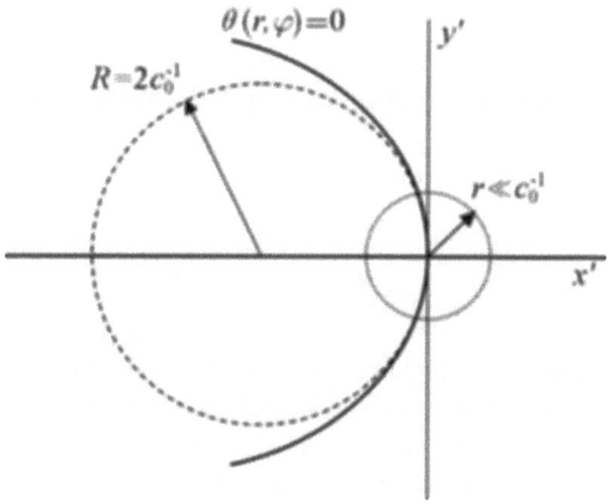

Fig. 4.3. Two lines of solutions of the equation $\theta(x',y') = 0$ (bold lines, one coinciding with x').

$$\cos\varphi = -\frac{1}{4}c_0 r. \tag{4.60}$$

One can readily verify that eq. (4.60) describes the circle

$$\left(x' + \frac{2}{c_0}\right)^2 + y'^2 = \left(\frac{2}{c_0}\right)^2 \tag{4.61}$$

with the radius $2/c_0$ and the center at the point $x' = -2/c_0$, $y' = 0$. Of course, only a small part of this circle close to the origin goes along the line $\theta(r,\varphi) = 0$ (Fig. 4.3). It is essential that deviation of this line from the ordinate axis y' occurs to the side opposite to the dislocation motion. For a given small r this deviation is increased with the growth of the speed v and with the decrease of the thermal conductivity, which provide an increase of c_0 (4.39) and a corresponding decrease of the line curvature at the origin. Therefore there is no doubt that such a deformation of the unperturbed symmetric pattern in Fig. 4.2 is directly related to a retarding thermal field from the moving dislocation. This field has been found numerically for the case of a copper in our paper [73]. Unfortunately, because of some misprints the iso-temperature lines $\theta(x', y') = \text{const}$ in the corresponding figure in [73] were supplied by wrong values of temperature $\theta \sim 10K$. Actually, in view of high thermal conductivity of copper the correct magnitudes for θ should be less by almost three orders of magnitude. On the other hand, in any dielectrics the considered effect must be much stronger than in metals. Say, in rather plastic dielectric crystals NaCl the thermal conductivity κ is smaller than in copper by the factor 346. Accordingly, the thermal amplitudes $\theta(x', y')$ in NaCl must be proportionally higher. One can compare the factors g (4.56) for these two materials

$$g_{Cu} \approx 5,53 \cdot 10^{-5}, \quad g_{NaCl} \approx 7,11 \cdot 10^{-2}. \tag{4.62}$$

Figure 4.4 represents some rearrangement of the above numerical results [73] organized so that the appropriate iso-temperature lines are universally valid for any crystal. It has proved to be possible due to a choice of dimensionless distances $\bar{c}_0 r$ and temperature θ/θ_0 with scales

$$\bar{c}_0 = \frac{c_t c_v}{\kappa}, \qquad \theta_0 \approx \frac{1}{8} g\bar{T} \tag{4.63}$$

(we remind that by (4.39) $c_0 = \bar{c}_0 v/c_t$). Bearing in mind the established anti-symmetry (4.36) of the thermal field with respect to its reflection in the plane $x'z'$, we present only the upper parts of thermal distributions at $y' > 0$. One can see, that very close to the dislocation the iso-lines have tendency to be anti-symmetric also with respect to the reflection in the plane $y'z'$, which is in accordance with the above analytical analysis (see theorem (4.45)). However in a more large scale of distances from the dislocation the temperature pattern loses this symmetry and manifests some lagging behind the dislocation, which, as was expected, is the more pronounced the faster is the dislocation motion (compare the lines of zero field $\theta = 0$ in Figs. 4.4

a, b, c and (4.3). On the other hand, one can see from Fig. 4.4c, where the condition $c_0 r \gg 1$ is practically fulfilled, that at these large distances the field $\theta(x', y')$ becomes inversely proportional to the distance (at least along y') being simultaneously less sensitive to the dislocation speed (compare Fig. 4.4b and c).

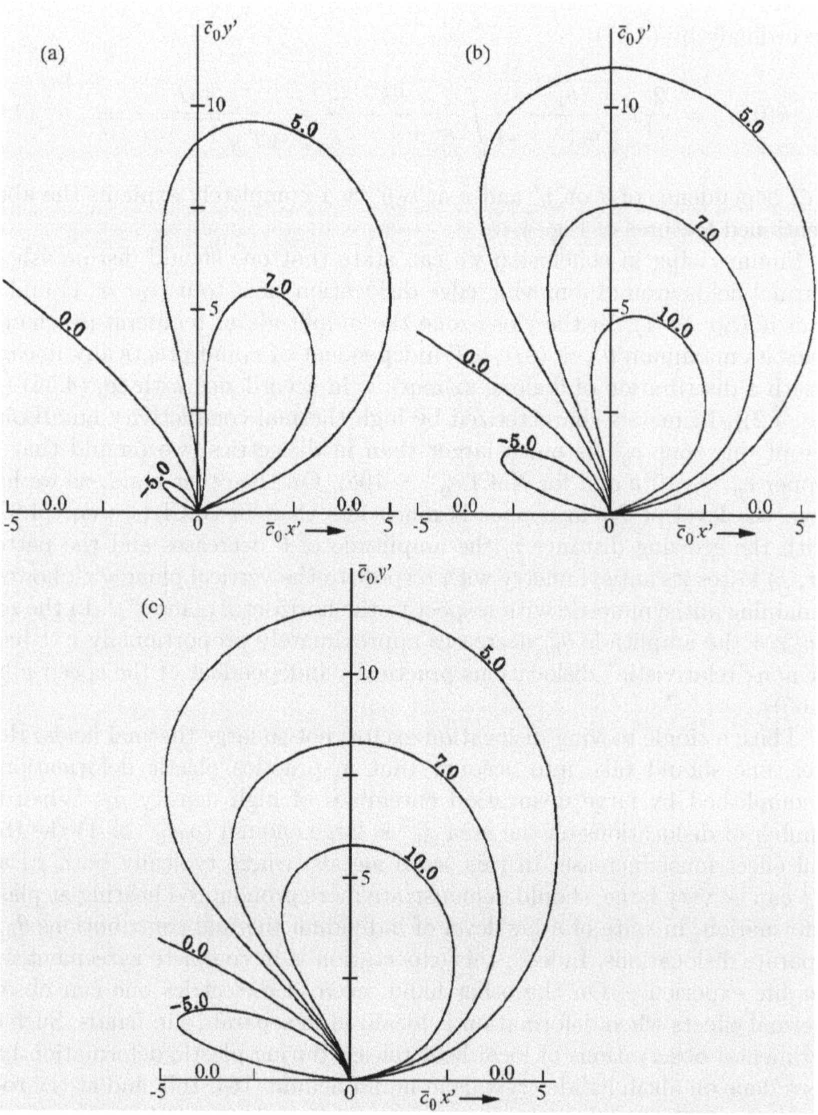

Fig. 4.4. The reversible thermal fields around a gliding edge dislocation; *(a)* for $v/c_t = 0,2$, *(b)* for $v/c_t = 0,4$, and *(c)* for $v/c_t = 0,8$. The contour values of temperature are given in the units θ_0 (4.63).

These observations can be easily understood analytically. Consider eq. (4.57) at $x' = 0$ for large distances $c_0 y' \gg 1$. In this case practically for all s contributed to the integral one can represent the function $F(s|r, \frac{\pi}{2}) = F(s|0, y')$ (4.34) in the asymptotical form

$$F(s|0, y') \approx \frac{2}{c_0 y'} \frac{s^2+1}{s}. \tag{4.64}$$

Accordingly by (4.57)

$$\theta(0, y') = \frac{2}{\pi^2} \bar{T} \frac{b_1 \gamma}{y' c_v (\bar{\lambda}+2)} \int_0^\infty \frac{ds}{s^2+1-\varepsilon_v} = \frac{4g\bar{T}}{\bar{c}_0 y' \sqrt{1-(v/c_t)^2}}. \tag{4.65}$$

This dependence of θ on y' and v at $c_0 y' \gg 1$ completely explains the above mentioned features of Fig. 4.4.

Summarizing in conclusion we can state that one should distinguish the thermal fields around a moving edge dislocation close to it ($c_0 r \ll 1$) and far from it ($c_0 r \gg 1$). In the close zone the amplitude of temperature changes takes its maximum $\theta_m = (v/c_t) g\bar{T}$ independent of r and practically linear in v with a distribution of θ along azimuth φ in accordance with eq. (4.55) (see Fig. 4.2)). In metals characterized by high thermal conductivity an absolute size of this zone c_0^{-1} is much larger than in dielectrics (we remind that for copper $c_0^{-1} \sim 10^3 b$ and for NaCl $c_0^{-1} \sim 10b$). On the other hand, as we have seen, the level of θ_m in copper is much less than in NaCl (see eq. (4.62)). With the growing distance r, the amplitude of θ decreases and the pattern $\theta(r, \varphi)$ losses its antisymmetry with respect to the vertical plane $y'z'$, however remaining antisymmetric with respect to the horizontal plane $x'z'$. In the zone $c_0 r \gg 1$ the amplitude θ_m decreases approximately proportionally r^{-1} being for non-"relativistic" dislocations practically independent of the speed v (see (4.65)).

Thus, a single moving dislocation excites not so large thermal fields. However, one should take into account that in practice plastic deformation is accomplished by large dislocation ensembles of high density ρ_d. When the number of dislocations on the area c_0^{-2} is large enough ($\rho_d c_0^{-2} \gg 1$) the thermal effect must increase. In this sense metals, where typically both ρ_d and c_0^{-1} can be very large, should demonstrate more pronounced heating at plastic deformation, in spite of a low level of individual thermal contributions θ_m of separate dislocations. Indeed, this expectation is in complete agreement with our life experience. On the other hand, even in dielectrics one can observe thermal effects when deformation is localized in separate slip bands. Such experimental observations of local heat release during plastic deformation have been done on alkali halide crystals in liquid helium [164, 165] and at the room temperature [166-168], with a registration of heat fluxes at the surface by helium boiling [165] or by means of a local change of color of a liquid crystal coating of the crystal face [166-168]. In particular, the experiments on crystals NaCl [166, 167] gave local temperature increase

$$\theta_{exp} \sim 0.1 - 0.5\text{K}. \qquad (4.66)$$

Let us compare these data with our estimate (4.65). Typically the density of dislocations in slip bands in such experiments can be roughly estimated as $\rho_d \sim 10^9 - 10^{10}\text{cm}^{-2}$. This allows us to put into (4.65)

$$y' \approx 1/\sqrt{\rho_d} \sim (1-3)10^{-5}\text{cm}.$$

The other necessary parameters for NaCl have been established before:

$$\bar{c}_0 \sim b^{-1} \sim 3 \cdot 10^7\text{cm}^{-1}, \quad g_{\text{NaCl}} \approx 7,11 \cdot 10^{-2}, \bar{T} \approx 300\text{K}.$$

And the resulting estimate is

$$\theta_{theor} \sim 0.1 - 0.3\text{K}, \qquad (4.67)$$

which provides really quite a good agreement with (4.66).

5

Some Extensions for Dislocations in Materials with Polarization Gradient

In the previous two chapters we dealt with dislocations and their coupled fields in ordinary classical media without Mindlin's effects. In this chapter we shall consider dislocations in a material with polarization gradient. Our purpose is to derive a new appearance of dislocation field including its electrical component. The material of this chapter is mainly based on the results obtained in the publications [112-114, 123].

5.1 Linear and surface definitions of a dislocation

(a) Basic equations and corresponding Green's functions

We shall start from the static variant of the Navier system of equations (2.113)- (2.115) with the dislocation strain (2.109), (3.5) as an only source of internal fields:

$$C_{ji}u_i + D_{ji}P_i = -f_j^{\mathrm{pl}}, \tag{5.1}$$

$$D_{ij}u_i + B_{ij}P_i + E_j = -E_j^{\mathrm{pl}}, \tag{5.2}$$

$$\varepsilon_0 E_{i,i} + P_{i,i} = 0. \tag{5.3}$$

Here the differential operators C_{ij}, D_{ij} and B_{ij} are defined by eqs. (2.110)-(2.112). The inhomogeneous terms in the right-hand side of (5.1) and (5.2) for the considered dislocation loop (Fig. 3.1), by eqs.(2.108), (2.109) and (3.5), are given by the expressions

$$F_j^{\mathrm{pl}}(\mathbf{r}) = \bar{\beta}_{ji}^{0(c)} \frac{\partial}{\partial x_i} \delta\left[(\mathbf{r} - \mathbf{r}^S) \cdot \mathbf{n}^S\right], \tag{5.4}$$

$$E_j^{\mathrm{pl}}(\mathbf{r}) = \bar{\beta}_{ji}^{0(d)} \frac{\partial}{\partial x_i} \delta\left[(\mathbf{r} - \mathbf{r}^S) \cdot \mathbf{n}^S\right], \tag{5.5}$$

where the notation is introduced:

$$\begin{aligned} \bar{\beta}_{ij}^{0(c)} &= c_{ijkl} b_k n_l = c_{12} \bar{\varepsilon}_{kk}^0 \delta_{ij} + 2 c_{44} \bar{\varepsilon}_{ij}^0, \\ \bar{\beta}_{ij}^{0(d)} &= d_{ijkl} b_k n_l = d_{12} \bar{\varepsilon}_{kk}^0 \delta_{ij} + 2 d_{44} \bar{\varepsilon}_{ij}^0, \end{aligned} \tag{5.6}$$

$$\bar{\varepsilon}_{ij}^0 = \frac{1}{2}\left(b_i n_j^S + b_j n_i^S\right) \tag{5.7}$$

Basing on the reciprocity theorem, Sec. 2.4(b), for infinite region and static isothermal variant (see also [112]) one can introduce the appropriate Green's functions for the system (5.1)-(5.3). This theorem can be expressed in the form, see eq. (2.148),

$$\int_V dV \left[f_i^{\rm pl} u'_i - f_i^{\rm pl\prime} u_i + E_i^{\rm pl} P'_i - E_i^{\rm pl\prime} P_i - q\varphi^{E'} + q'\varphi^{E} \right] = 0. \tag{5.8}$$

Let us substitute here three variants for the conventional sources $\{f_i^{\rm pl\prime}, E_i^{\rm pl\prime}, \; q', u'_i, P'_i, \varphi^{E'}\}$:

$$\begin{aligned} &\left\{ f_i^{\rm pl\prime}, E_i^{\rm pl\prime}, q', u'_i, P'_i, \varphi^{E'} \right\} = \\ &= \left\{ \delta_{ni}\delta(\mathbf{r}-\mathbf{r}'), 0, 0, G_{ni}^f(\mathbf{r}-\mathbf{r}'), H_{ni}^f(\mathbf{r}-\mathbf{r}'), 0 \right\}, \end{aligned} \tag{5.9}$$

$$\begin{aligned} &\left\{ f_i^{\rm pl\prime}, E_i^{\rm pl\prime}, q', u'_i, P'_i, \varphi^{E'} \right\} = \\ &= \left\{ 0, \delta_{ni}\delta(\mathbf{r}-\mathbf{r}'), 0, G_{ni}^E(\mathbf{r}-\mathbf{r}'), H_{ni}^E(\mathbf{r}-\mathbf{r}'), 0 \right\}, \end{aligned} \tag{5.10}$$

$$\begin{aligned} &\left\{ f_i^{\rm pl\prime}, E_i^{\rm pl\prime}, q', u'_i, P'_i, \varphi^{E'} \right\} = \\ &= \left\{ 0, 0, -\delta(\mathbf{r}-\mathbf{r}'), \phi_i^f \mathbf{r}-\mathbf{r}'), \phi_i^E(\mathbf{r}-\mathbf{r}'), 0 \right\}. \end{aligned} \tag{5.11}$$

Combining (5.8) with (5.9)-(5.11) we obtain, respectively:

$$u_n(\mathbf{r}) = \int_V dV \left[G_{ni}^f(\mathbf{r}-\mathbf{r}') f_i^{\rm pl}(\mathbf{r}') + H_{ni}^f(\mathbf{r}-\mathbf{r}') E_i^{\rm pl}(\mathbf{r}') \right], \tag{5.12}$$

$$P_n(\mathbf{r}) = \int_V dV \left[G_{ni}^E(\mathbf{r}-\mathbf{r}') f_i^{\rm pl}(\mathbf{r}') + H_{ni}^E(\mathbf{r}-\mathbf{r}') E_i^{\rm pl}(\mathbf{r}') \right], \tag{5.13}$$

$$\varphi^E(\mathbf{r}) = \int_V dV \left[\phi_i^f(\mathbf{r}-\mathbf{r}') f_i^{\rm pl}(\mathbf{r}') + \phi_i^E(\mathbf{r}-\mathbf{r}') E_i^{\rm pl}(\mathbf{r}') \right]. \tag{5.14}$$

The Green's functions introduced in eqs.(5.9)-(5.14) satisfy the following system of equations

$$C_{ji} G_{ni}^f(\mathbf{r}) + D_{ji} H_{ni}^f(\mathbf{r}) + \delta_{nj}\delta(\mathbf{r}) = 0, \tag{5.15}$$

$$D_{ji} G_{ni}^f(\mathbf{r}) + B_{ji} H_{ni}^f(\mathbf{r}) - \nabla_j \phi_n^f(\mathbf{r}) = 0, \tag{5.16}$$

$$\nabla_i H^f_{ni}(\mathbf{r}) - \varepsilon_0 \nabla^2 \phi^f_n(\mathbf{r}) = 0, \tag{5.17}$$

$$C_{ji} G^E_{ni}(\mathbf{r}) + D_{ji} H^E_{ni}(\mathbf{r}) = 0, \tag{5.18}$$

$$D_{ji} G^E_{ni}(\mathbf{r}) + B_{ji} H^E_{ni}(\mathbf{r}) - \nabla_j \phi^E_n(\mathbf{r}) + \delta_{nj}\delta(\mathbf{r}) = 0, \tag{5.19}$$

$$\nabla_i H^E_{ni}(\mathbf{r}) - \varepsilon_0 \nabla^2 \phi^E_n(\mathbf{r}) = 0. \tag{5.20}$$

We shall not go here to a bulky procedure of solving this system, but instead just present the resulting expressions for the above Green's functions obtained in [112, 123]:

$$G^f_{ni} = \frac{1}{4\pi}\left\{\nabla_n \nabla_i \left[C_{01} R + C_{11} I_1 - C_{12} I_2\right] + \delta_{ni} \nabla^2 \left[C_{02} R + C_{12} I_2\right]\right\}, \tag{5.21}$$

$$H^f_{ni} = \frac{1}{4\pi}\left\{\nabla_n \nabla_i \left[C_{21} I_1 + C_{22} I_2\right] - C_{22} \delta_{ni} \nabla^2 I_2\right\}, \tag{5.22}$$

$$G^E_{ni} = H^f_{ni} \tag{5.23}$$

$$H^E_{ni} = \frac{1}{4\pi}\left\{\nabla_n \nabla_i \left[\frac{1}{\hat{a}} I_1 - \frac{1}{a} I_2\right] + \frac{1}{a}\delta_{ni} \nabla^2 I_2\right\}, \tag{5.24}$$

$$\phi^f_n = \frac{C_{21}}{4\pi\varepsilon_0} \nabla_n I_1, \tag{5.25}$$

$$\phi^E_n = \frac{1}{4\pi\hat{a}\varepsilon_0} \nabla_n I_1. \tag{5.26}$$

Here the notation is introduced:

$$R = |\mathbf{r}|, \qquad I_\alpha = \frac{\exp\left(-\lambda_\alpha R\right) - 1}{R}, \qquad \alpha = 1, 2, \tag{5.27}$$

$$\lambda_1 \equiv \lambda = \sqrt{\frac{c_{11}\hat{a}}{c_{11} b_{11} - d_{11}^2}}, \qquad \lambda_2 = \sqrt{\frac{c_{44} a}{c_{44}(b_{44} + b_{77}) - d_{44}^2}}, \qquad \hat{a} = a + \varepsilon_0^{-1}, \tag{5.28}$$

$$C_{01} = \frac{c_{44} - c_{11}}{2 c_{11} c_{44}}, \qquad C_{02} = \frac{1}{2 c_{44}}, \tag{5.29}$$

$$C_{11} = \frac{d_{11}^2}{c_{11}^2 \hat{a}}, \quad C_{12} = \frac{d_{44}^2}{c_{44}^2 a}, \quad C_{21} = -\frac{d_{11}}{c_{11}\hat{a}}, \quad C_{22} = \frac{d_{44}}{c_{44} a}. \tag{5.30}$$

We note that the parameter λ_1 coincides with the parameter λ (2.93) in Mindlin's theory of the surface tension [103]. On the other hand, as was mentioned in Ch. 2, the inverse parameter λ^{-1} is a measure of discreetness of the medium and its order of magnitude may be estimated by a lattice parameter. On the other hand, by its definition the parameter λ_2 is of the same order of magnitude as λ_1, i.e. $\lambda_2 \sim \lambda_1 = \lambda \sim b^{-1}$.

(b) Coupled fields of a dislocation loop

Actually, eqs. (5.12)-(5.14) with (5.4)-(5.7) and (5.21)-(5.26) solve the problem of dislocation fields in an infinite Mindlin's dielectric medium. Let us rearrange these expressions to a more explicit form. After substituting (5.4), (5.5) into (5.12)-(5.14) it is convenient to present in analogy with eq. (3.41)

$$G^f_{nl}(\mathbf{r}-\mathbf{r}')\frac{\partial}{\partial x'_i}\delta[(\mathbf{r}-\mathbf{r}^S)\cdot\mathbf{n}^S] =$$
$$= \frac{\partial}{\partial x'_i}\left[G^f_{nl}(\mathbf{r}-\mathbf{r}')\delta[(\mathbf{r}'-\mathbf{r}^S)\cdot\mathbf{n}^S]\right] + G^f_{nl,i}(\mathbf{r}-\mathbf{r}')\delta[(\mathbf{r}'-\mathbf{r}^S))\cdot n^S], \tag{5.31}$$

etc. The first volume integrals by the Gauss theorem will reduce to surface integrals over the infinite surface of the considered unbounded body where the integrands vanish. And the second terms in (5.31) and similar expressions due to the delta function transform to surface integrals over the surface S resting on the dislocation line (Fig. 3.1). As a result, we obtain instead of (5.12)-(5.14):

$$u_n(\mathbf{r}) = b_k\int_S dS_j\left[c_{kjli}G^f_{nl,i}(\mathbf{r}-\mathbf{r}^S) + d_{kjli}H^f_{nl,i}(\mathbf{r}-\mathbf{r}^S)\right], \tag{5.32}$$

$$P_n(\mathbf{r}) = b_k\int_S dS_j\left[c_{kjli}G^E_{nl,i}(\mathbf{r}-\mathbf{r}^S) + d_{kjli}H^E_{nl,i}(\mathbf{r}-\mathbf{r}^S)\right], \tag{5.33}$$

$$\varphi^E(\mathbf{r}) = b_k\int_S dS_j\left[c_{kjli}\phi^f_{l,i}(\mathbf{r}-\mathbf{r}^S) + d_{kjli}\phi^E_{l,i}(\mathbf{r}-\mathbf{r}^S)\right], \tag{5.34}$$

where by definition $dS_l = dSn^S_l$.

Basing on eqs. (5.32)-(5.34), let us also express even more important physical fields of elastic distortions $\beta_{pn} = u_{n,p} - \beta^0_{pn}$, polarization gradient $P_{n,p}$ and electric field $E_p = -\varphi^E_{,p}$:

$$\beta_{pn} = b_k\Big\{\delta_{nk}n_p\delta[(\mathbf{r}-\mathbf{r}^S)\cdot\mathbf{n}^S]+$$
$$+\int_S dS_j\left[c_{kjli}G^f_{nl,ip}(\mathbf{r}-\mathbf{r}^S) + d_{kjli}H^f_{nl,ip}(\mathbf{r}-\mathbf{r}^S)\right]\Big\}, \tag{5.35}$$

$$P_{n,p}(\mathbf{r}) = b_k\int_S dS_j\left[c_{kjli}G^E_{nl,ip}(\mathbf{r}-\mathbf{r}^S) + d_{kjli}H^E_{nl,ip}(\mathbf{r}-\mathbf{r}^S)\right], \tag{5.36}$$

$$E_p(\mathbf{r}) = -b_k\int_S dS_j\left[c_{kjli}\phi^f_{l,ip}(\mathbf{r}-\mathbf{r}^S) + d_{kjli}\phi^E_{l,ip}(\mathbf{r}-\mathbf{r}^S)\right]. \tag{5.37}$$

It can be proved that the above surface integrals (5.32)-(5.37) do not depend on the particular form of the surface S resting on the dislocation line and can be reduced to line integrals along the dislocation loop C (Fig. 3.1). Let us demonstrate this on the example of the elastic distortion field (5.35). It is convenient to start from the identity

$$n_p \delta_{nk} \delta[(\mathbf{r}-\mathbf{r}^S)\cdot\mathbf{n}^S] = \int\limits_S dS_p \delta_{nk}\delta(\mathbf{r}-\mathbf{r}^S) = \\ = -\int\limits_S dS_p \left\{c_{kjli} G^f_{nl,ij}(\mathbf{r}-\mathbf{r}^S) + d_{kjli} H^f_{nl,ij}(\mathbf{r}-\mathbf{r}^S)\right\}, \tag{5.38}$$

where we have made use of eq. (5.15). Now combining eqs. (5.35), (5.38) with (3.60) we arrive at the extension of the Mura representation [150] to the case of Mindlin's isotropic medium:

$$\beta_{pn}(\mathbf{r}) = -b_k \varepsilon_{mjp} \oint\limits_C dl_m \left[c_{kjli} G^f_{nl,i}(\mathbf{r}-\mathbf{r}^C) + d_{kjli} H^f_{nl,i}(\mathbf{r}-\mathbf{r}^C)\right], \tag{5.39}$$

compare with (3.55). Quite similarly

$$P_{n,p}(\mathbf{r}) = -b_k \varepsilon_{mjp} \oint\limits_C dl_m \left[c_{kjli} G^E_{nl,i}(\mathbf{r}-\mathbf{r}^C) + d_{kjli} H^E_{nl,i}(\mathbf{r}-\mathbf{r}^C)\right] \tag{5.40}$$

and

$$E_p(\mathbf{r}) = b_k \varepsilon_{mjp} \oint\limits_C dl_m \left[c_{kjli} \phi^f_{l,i}(\mathbf{r}-\mathbf{r}^C) + d_{kjli} \phi^E_{l,i}(\mathbf{r}-\mathbf{r}^C)\right]. \tag{5.41}$$

Certainly, directly substituting into (5.32)-(5.37) the found expressions for the Green's functions, eqs. (5.21)-(5.26), one can find explicit formulae for the studied coupled fields. We shall present here only expressions for the displacement and the electric potential fields:

$$u_n(\mathbf{r}) = \frac{1}{4\pi}\int\limits_S dS \left\{\left[-\frac{1}{2(1-\nu)}R + \right.\right. \\ + 2\left[(c_{44}C_{11} + d_{44}C_{21})I_1 - (c_{44}C_{12} - d_{44}C_{22})I_2\right]\Big]_{,njq} \bar{\varepsilon}^0_{qj} + \\ + \left[R + 2(c_{44}C_{12} - d_{44}C_{22})I_2\right]_{,ppj} \bar{\varepsilon}^0_{jn} + \\ \left. + \left[\frac{\nu}{2(1-\nu)}R + (c_{12}C_{11} + d_{12}C_{21})I_1\right]_{,npp} \bar{\varepsilon}^0_{kk}\right\}, \tag{5.42}$$

$$\varphi^E(\mathbf{r}) = \frac{1}{4\pi c_{11}\hat{a}\varepsilon_0}\int\limits_S dS \left\{(c_{11}d_{12} - c_{12}d_{11})\, I_{1,pp}\bar{\varepsilon}^0_{kk} + \right. \\ \left. + 2\,(c_{11}d_{44} - c_{44}d_{11})\, I_{1,ki}\bar{\varepsilon}^0_{ik}\right\}, \tag{5.43}$$

where the notation is given by eqs. (5.27)-(5.30) and by

$$R = |\mathbf{r} - \mathbf{r}^S|, \quad \nu = \frac{c_{12}}{2(c_{12} + c_{44})}. \tag{5.44}$$

In fact, eq. (5.42) can be essentially simplified due to the identity

$$c_{44}C_{12} - d_{44}C_{22} = c_{44}\frac{d_{44}^2}{c_{44}^2 a} - d_{44}\frac{d_{44}}{c_{44}a} = 0 \tag{5.45}$$

following from (5.30). Accordingly, instead of (5.42) we obtain

$$\begin{aligned} u_n(\mathbf{r}) = \frac{1}{4\pi}\int\limits_S dS \Big\{ & R_{,ppj}\bar{\varepsilon}^0_{jn} + \frac{1}{2(1-\nu)}\left[\nu R_{,npp}\bar{\varepsilon}^0_{kk} - R_{,njq}\bar{\varepsilon}^0_{qj}\right] \mp \\ & +2(c_{44}C_{11} + d_{44}C_{21})I_{1,njq}\bar{\varepsilon}^0_{qj} + (c_{12}C_{11} + d_{12}C_{21})I_{1,npp}\bar{\varepsilon}^0_{kk}\Big\} \end{aligned} \tag{5.46}$$

5.2 The field of a straight dislocation

Let us consider a straight dislocation line specifying the z-axis of the Cartesian coordinate system $\{x, y, z\}$with the semi-plane $xz, x > 0$ chosen as the surface S resting on the dislocation line (see Fig. 3.3), so that the normal to S is

$$\mathbf{n}^S = (0, 1, 0). \tag{5.47}$$

(a) The edge dislocation

Such a dislocation by definition has the Burgers vector orthogonal to the dislocation line. It is customary to choose for an edge dislocation the surface S so that it would contain both the dislocation line and the Burgers vector. In our case this means that the latter must be directed along the x-axis,

$$\mathbf{b}^{edge} = b(1, 0, 0). \tag{5.48}$$

The corresponding tensor $\bar{\varepsilon}^0_{ij}$ (5.7) with the choice (5.47) and (5.48) clearly has only two non-vanishing components

$$\bar{\varepsilon}^0_{12} = \bar{\varepsilon}^0_{21} = b/2. \tag{5.49}$$

Substituting (5.49) into eq. (5.46) one can easily find for the displacement field

$$u_1(x, y) =$$

$$= \frac{b}{8\pi}\int\limits_0^\infty dx' \int\limits_{-\infty}^\infty dz' \left\{ R_{,pp2} - \frac{1}{1-\nu}R_{,112} + 4(c_{44}C_{11} + d_{44}C_{21})I_{1,112} \right\}, \tag{5.50}$$

$$u_2(x, y) =$$

$$= \frac{b}{8\pi} \int_0^\infty dx' \int_{-\infty}^\infty dz' \left\{ R_{,pp1} - \frac{1}{1-\nu} R_{,221} + 4(c_{44}C_{11} + d_{44}C_{21}) I_{1,221} \right\}, \tag{5.51}$$

$$u_3(x, y) = 0, \tag{5.52}$$

where

$$R = \sqrt{(x' - x)^2 + y^2 + z'^2}. \tag{5.53}$$

Quite similarly, by (5.43), (5.49), one readily obtains the electric potential field

$$\varphi^E(x, y) = \frac{b}{2\pi c_{11} \hat{a} \varepsilon_0} \int_0^\infty dx' \int_{-\infty}^\infty dz' \left(c_{11} d_{44} - c_{44} d_{11} \right) I_{1,12} \tag{5.54}$$

with $I_1(R)$ defined by the same eqs. (5.27) and (5.53).

Formally the reason for an identical vanishing u_3 (5.52) becomes clear from eq. (5.46). Indeed, the first and the last terms in (5.46) must vanish because of the form of the tensor $\bar{\varepsilon}^0_{ij}$ (5.49). And the other terms are zero because $n = 3$ in the given case relates to a differentiation over z of the integrand which by (5.53) is independent of this variable.

After performing the integration in (5.50) and (5.51) with making use of (5.53) and $(5.27)_2$ we arrive at the solutions first derived by Nowacki & Hsieh [112]

$$u_n(x, y) = u_n^0(x, y) + u_n^{\mathrm{M}}(x, y), \quad n = 1,\ 2, \tag{5.55}$$

where $u_n^0(x, y)$ is the classical dislocation field known from the ordinary theory of elasticity and $u_n^{\mathrm{M}}(x, y)$ is the additional field provided by Mindlin's polarization gradient effects. An explicit form for these terms is

$$u_1^0(x, y) = \frac{b}{2\pi} \left(\arctan\frac{y}{x} + \frac{1}{2(1-\nu)} \frac{xy}{r^2} \right), \tag{5.56}$$

$$u_2^0(x, y) = \frac{b}{2\pi} \left(\frac{1-2\nu}{2(1-\nu)} \ln r + \frac{1}{2(1-\nu)} \frac{y^2}{r^2} \right), \tag{5.57}$$

$$u_1^{\mathrm{M}}(x, y) = \frac{b}{\pi} (c_{44}C_{11} + d_{44}C_{21}) J_{,12}, \tag{5.58}$$

$$u_2^{\mathrm{M}}(x, y) = \frac{b}{\pi} (c_{44}C_{11} + d_{44}C_{21}) J_{,22}. \tag{5.59}$$

Here the notation is introduced

$$J(r) = K_0(\lambda r) + \ln r, \qquad r = \sqrt{x^2 + y^2}, \tag{5.60}$$

with $K_0(\lambda r)$ standing for the modified Bessel function of the third kind and $\lambda = \lambda_1$ given by (5.28).

Quite similarly and in the same terms the electric potential field of an edge dislocation is given by

$$\varphi^E(x,y) = \frac{b}{2\pi c_{11}\hat{a}\varepsilon_0}(c_{11}d_{44} - c_{44}d_{11})J_{,2} \tag{5.61}$$

Bearing in mind the identity

$$\frac{d}{dz}K_0(z) = -K_1(z) \tag{5.62}$$

one can replace eqs. (5.58), (5.59) and (5.61) by more explicit relations

$$u_1^{\mathrm{M}}(x,y) = \frac{b}{\pi}(c_{44}C_{11} + d_{44}C_{21})\frac{\partial}{\partial x}\left[\frac{y}{r^2} - \frac{\lambda y}{r}K_1(\lambda r)\right], \tag{5.63}$$

$$u_2^{\mathrm{M}}(x,y) = \frac{b}{\pi}(c_{44}C_{11} + d_{44}C_{21})\frac{\partial}{\partial y}\left[\frac{y}{r^2} - \frac{\lambda y}{r}K_1(\lambda r)\right], \tag{5.64}$$

$$\varphi^E(x,y) = \frac{b}{2\pi c_{11}\hat{a}\varepsilon_0}(c_{11}d_{44} - c_{44}d_{11})\frac{y}{r^2}\left(1 - \lambda r K_1(\lambda r)\right). \tag{5.65}$$

(b) The screw dislocation

For a screw dislocation its Burgers vector is parallel to the dislocation line. In the chosen variant of geometry we have then the Burgers vector:

$$\mathbf{b}^{screw} = b(0,0,1), \tag{5.66}$$

and the tensor $\bar{\varepsilon}_{ij}^0$ (5.7) defined by the components

$$\bar{\varepsilon}_{23}^0 = \bar{\varepsilon}_{32}^0 = b/2. \tag{5.67}$$

It is rather evident that with (5.67) eq. (5.46) gives the only one nonvanishing component,

$$u_1(x,y) = u_2(x,y) = 0, \tag{5.68}$$

$$u_3(x,y) = -\frac{b}{8\pi}\int_0^\infty dx' \int_{-\infty}^\infty dz' R_{,pp2}. \tag{5.69}$$

Thus, for a screw dislocation we arrive at purely classical expression

$$u_3(x,y) = u_3^0(x,y) = \frac{b}{2\pi}\arctan\frac{y}{x} \tag{5.70}$$

without any contribution from the polarization gradient effects.

Substitution of (5.67) into (5.43) leads to

$$\varphi^E(x,y) = \frac{b}{2\pi c_{11}\hat{a}\varepsilon_0} \int\limits_0^\infty dx' \int\limits_{-\infty}^\infty dz' \left(c_{11}d_{44} - c_{44}d_{11}\right) I_{1,23}. \tag{5.71}$$

Taking into account that the function $I_1(R)$ (5.27) with R given by (5.53) is independent of z, we arrive at the statement that fields of straight screw dislocations, in contrast to edge dislocations, do not manifest any electromechanical coupling and do not excite electric fields,

$$\varphi(x,y) = 0, \qquad \mathbf{E}(x,y) = 0. \tag{5.72}$$

This important result, as well as the vanishing correction term u_3^{M}, is also a consequence of the identity (5.45). However for an arbitrary anisotropy we could not expect automatic exclusion of such sort of terms. So in anisotropic bodies screw dislocations should apparently be also sensitive to the polarization gradient effects.

(c) Analysis of the results

Now let us discuss the physical meaning of the found corrections to customary fields of straight edge dislocations. The classical distortion field of a straight dislocation monotonously decreases with distance r from dislocation proportionally to $1/r$ being singular at the dislocation line. Bearing in mind that continuous description of distortions at distances less than a parameter of crystal lattice $a \sim b$ is physically senseless, they usually introduce a cutoff radius $r_0 \sim b$ of the dislocation core, considering all fields only beyond the core at $r > r_0$. Let us find out what sort of dependences characterize the correction terms $u_{n,i}^{\mathrm{M}}$. We start from the displacement field.

Formally, at $\lambda r \ll 1$, when the Bessel function $K_1(\lambda r)$ may be approximated by

$$K_1(\lambda r) \approx \frac{1}{\lambda r} + \frac{\lambda r}{2} \ln \frac{\lambda r}{2}, \tag{5.73}$$

the functions $u_n^{\mathrm{M}}(x,y)$, eqs. (5.63), (5.64), are transformed to

$$\begin{aligned} u_1^{\mathrm{M}}(x,y) &\approx -\frac{b}{2\pi}(c_{44}C_{11} + d_{44}C_{21})\frac{\partial}{\partial x}\left[\lambda^2 y \ln\left(\frac{\lambda r}{2}\right)\right] = \\ &= -\frac{b\lambda^2}{2\pi}(c_{44}C_{11} + d_{44}C_{21})\frac{xy}{r^2}, \end{aligned} \tag{5.74}$$

$$\begin{aligned} u_2^{\mathrm{M}}(x,y) &\approx -\frac{b}{2\pi}(c_{44}C_{11} + d_{44}C_{21})\frac{\partial}{\partial y}\left[\lambda^2 y \ln\left(\frac{\lambda r}{2}\right)\right] = \\ &= -\frac{b\lambda^2}{2\pi}(c_{44}C_{11} + d_{44}C_{21})\left[\frac{y^2}{r^2} + \ln\left(\frac{\lambda r}{2}\right)\right]. \end{aligned} \tag{5.75}$$

Thus, at $\lambda r \ll 1$ the distortions $u^{\mathrm{M}}_{n,i}(x,y)$ are characterized by the same singularity $u^{\mathrm{M}}_{n,i} \propto r^{-1}$ as the classical distortion field. For instance,

$$u^{\mathrm{M}}_{1,2}(x,y) = u^{\mathrm{M}}_{2,1}(x,y) \approx -\frac{b\lambda^2}{2\pi}(c_{44}C_{11} + d_{44}C_{21})\frac{x(x^2-y^2)}{r^4} \tag{5.76}$$

Ordinary they estimate moduli $c_{\alpha\beta}, b_{\alpha\beta}$ and $d_{\alpha\beta}$ so that $cb \sim d^2$. Accordingly, by (5.28), (5.30)

$$\lambda^2(c_{44}C_{11} + d_{44}C_{21}) = \frac{c_{11}\hat{a}}{c_{11}b_{11} - d_{11}^2}\left(\frac{c_{44}d_{11}^2}{c_{11}^2\hat{a}} - \frac{d_{11}d_{44}}{c_{11}a}\right) \sim 1. \tag{5.77}$$

Therefore at the distance of order of the radius of the dislocation core $r \sim r_0 \sim b$ the corrections (5.76) provide contributions comparable with the corresponding components of the classical distortion field of an edge dislocation

$$u^0_{1,2}(x,y) = \frac{b}{2\pi}\left[\frac{x}{r^2} + \frac{1}{2(1-\nu)}\frac{x(x^2-y^2)}{r^4}\right], \tag{5.78}$$

$$u^0_{2,1}(x,y) = \frac{b}{4\pi(1-\nu)}\left[-\frac{2\nu x}{r^2} + \frac{x(x^2-y^2)}{r^4}\right]. \tag{5.79}$$

At the opposite limit $\lambda r \gg 1$ the Bessel function $K_1(\lambda r)$ in eqs. (5.63), (5.64) becomes exponentially small,

$$K_1(\lambda r) \approx \sqrt{\frac{\pi}{2\lambda r}}\exp(-\lambda r), \tag{5.80}$$

and can be omitted. The resulted asymptotic behavior of the terms $u^{\mathrm{M}}_{n,i}$ becomes proportional to r^{-3} as compared with the much more slow decrease $\propto r^{-1}$ of the classical terms (5.78), (5.79). For instance,

$$u^{\mathrm{M}}_{1,2}(x,y) = u^{\mathrm{M}}_{2,1}(x,y) \approx -\frac{2b}{\pi}(c_{44}C_{11} + d_{44}C_{21})\frac{x(x^2-3y^2)}{r^6}. \tag{5.81}$$

Thus, as one could expect, in a complete accordance with the ideology of Mindlin's theory, the appropriate corrections to classical terms become essential only in the regions where the discreteness of the medium turns out to be important, i.e. close to a dislocation core. In this sense the electric coupled field arising on the background of zero classical contribution appears to be of special interest. Let us consider the electric potential of an edge dislocation, eq. (5.65). Bearing in mind the limiting relations (5.73) and (5.80) it is easy to find that

$$\varphi^E(x,y) = \frac{b}{2\pi c_{11}\hat{a}\varepsilon_0}(c_{11}d_{44} - c_{44}d_{11})\begin{cases}\lambda^2 y \ln\left(\dfrac{2}{\lambda r}\right), & \lambda r \ll 1;\\[2mm] \dfrac{y}{r^2}, & \lambda r \gg 1.\end{cases} \tag{5.82}$$

It is clearly seen that both limiting approximations are practically continuous at $\lambda r \sim 1$. The corresponding components of the electric field $E_n = -\varphi^E_{,n}$ are given by

$$E_x(x,y) = E_m \begin{cases} \dfrac{xy}{r^2}, & \lambda r \ll 1, \\ \dfrac{2xy}{\lambda^2 r^4}, & \lambda r \gg 1; \end{cases}, \qquad E_y(x,y) = E_m \begin{cases} \dfrac{y^2}{r^2} - \ln\left(\dfrac{2}{\lambda r}\right), & \lambda r \ll 1, \\ \dfrac{x^2 - y^2}{\lambda^2 r^4}, & \lambda r \gg 1; \end{cases} \tag{5.83}$$

where

$$E_m = \frac{b\lambda^2}{2\pi c_{11} \hat{a} \varepsilon_0}(c_{11}d_{44} - c_{44}d_{11}). \tag{5.84}$$

As follows from (5.83), in contrast to distortion components strongly singular at the limit $r \to 0$, the component E_x is finite at the same limit, and E_y has only logarithmic singularity. Such a radical difference between the distortion and electric field limiting behaviors appears to be rather natural. Indeed, close to the dislocation line distortions loose their physical meaning whereas an electric field remains well defined at any point of a crystal. At large distances from the dislocation both components, E_x and E_y, monotonously decrease $\propto r^{-2}$.

Let us estimate the amplitude E_m (5.84), which corresponds to the level of the electric field in the dislocation core. In accordance with the above considerations one can put into (5.84) $b\lambda \sim 1$. In addition, basing on (5.28) we shall suppose that

$$\hat{a}\varepsilon_0 \sim 1, \quad \lambda d \sim \sqrt{c\hat{a}} \sim \sqrt{c/\varepsilon_0}. \tag{5.85}$$

Then

$$E_m \sim \frac{(b\lambda)(\lambda d)}{2\pi(\hat{a}\varepsilon_0)} \sim \frac{\sqrt{c/\varepsilon_0}}{2\pi}. \tag{5.86}$$

Substituting here the typical magnitude for the elastic modulus $c \sim 10^3$ MPa one obtains

$$E_m \sim 10^8 \mathrm{V/m} = 1 \mathrm{MV/cm}. \tag{5.87}$$

Of course, such a high electric field occurs only very close to the dislocation line, when $\lambda r \sim 1$. Due to the decay of the field $E \propto r^{-2}$, even with the so large field in the dislocation core, at the distances related to the parameter $\lambda r \sim 10^4$, which corresponds only to $r \sim 1\mu$m, the electric field decreases to the level of order of 1 V/m. And at the macroscopic distances $r \sim 1$ cm we shall have a very low electric field $E \sim 10^{-8}$V/m.

Note in passing, that the ordinary piezoelectric effect also provides a rather high electric field close to dislocations which is comparable with the above estimates (5.86), (5.87). However, as we have seen in Ch. 3, this field more slowly decreases with the distance from a dislocation, $E \propto r^{-1}$. Thus, due to the polarization gradient effects even in centrosymmetric dielectrics in the vicinity of dislocations there might exist a very high electric field of the same order as in strong piezoelectrics. This may be very essential for dislocation properties, in particular, for dislocation pinning by charged point defects in ionic crystals, for formation of a charge on dislocations [169, 170] and for such macroscopic phenomena like the Stepanov's effect providing electric polarization of crystals due to a plastic deformation [171, 172]. By the way, it is very well experimentally established that in alkali halides crystals edge dislocations are charged whereas screw dislocation do not carry any electric charge. This observation correlates with the results of this chapter.

Part II

1D and 2D Electro-Elastic Coupled Fields in Media with Surfaces and Interfaces

6

Coupled Fields of Thermal Inclusion in Media with Polarization Gradient

The Mindlin formalism and its extensions have the only trouble - bulky equations of motion and constitutive relations, which provide quite serious technical difficulties in solving particular problems and especially in finding explicit results. Even after the transition from a general anisotropy to centrosymmetric or isotropic media the corresponding equations, as we have seen, remain rather bulky and the number of explicitly solved problems is not large. The first examples of such solutions were given by Mindlin himself for cubic crystals: the description of surface tension [104], of bulk acoustic wave dispersion [103], of coupled fields of a point charge [173], of electromagnetic radiation from a vibrating elastic sphere [174]. Later some of these results have been extended by Chowdhury & Glockner, who found coupled fields of a point charge in a half-space [175] and presented the theory of surface tension for a plate, and cylindrical or spherical shells [176]. It is worthwhile also to mention in this context the theories by Schwartz [177], Askar et al. [178] and Gou [179] who earlier considered the influence of surface curvature on the surface energy and other induced fields of cylindrical and spherical cavities. The other solution in this field has been obtained in our papers [120, 121], where the electro-elastic fields were found in dielectrics with polarization gradient excited by cylindrical or plain thermal inclusions. The content of this chapter is based on the latter publications.

6.1 Plain thermal inclusion in unbounded dielectric medium

(a) Statement of the problem

We consider an infinite plain layer of the thickness $d = 2h$ (Fig. 6.1), part of an isotropic elastic dielectric medium with the same material constants inside and outside the layer. The layer and the outside space are held at constant, but different temperatures. More precisely, we assume the temperature

θ equal θ_0 inside the layer and zero outside it. Such discontinuous temperature field must excite in the medium thermoelastic deformations, which due to the supposed presence of the polarization gradient should be accompanied by coupled electro-elastic fields. This static problem is described by the equations (2.134)-(2.136) taken at $\mathbf{f} = \mathbf{f}^{\mathrm{pl}} = 0, \ddot{u} = 0, q = 0$. As regards to the fourth equation of the system, eq. (2.137), in its stationary form $\kappa\nabla^2\theta + W = 0$, with the prescribed temperature distribution and appropriate choice of heat supply function $W(\mathbf{r})$ it is satisfied identically.

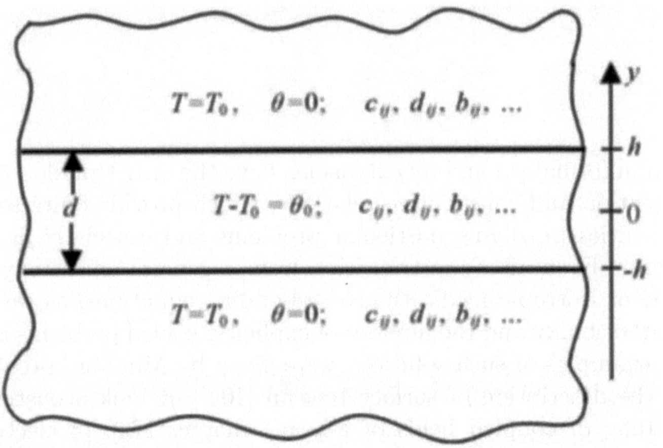

Fig. 6.1. The infinite medium with a plain thermal inclusion

Owing the symmetry of the problem, we shall assume the vector fields of displacement $\mathbf{u}(\mathbf{r})$, polarization $\mathbf{P}(\mathbf{r})$ and electrostatic intensity $\mathbf{E}(\mathbf{r})$ to have only the normal components dependent on only the y-coordinate with the origin chosen in the middle of the layer (Fig. 6.1). Thus, if $\mathbf{n}$ is the unit normal to the layer, the unknown vector fields are looked for in the form $\mathbf{u} = \mathbf{n}u(y)$, $\mathbf{P} = \mathbf{n}P(y)$, $\mathbf{E} = \mathbf{n}E(y)$. Accordingly, we come to the scalar problem with the unknown functions $u(y), P(y)$ and $E(y)$. An additional but known scalar field is the temperature distribution $\theta(y)$, which is a source of all the other fields. This distribution is given by

$$\theta(y) = \theta_0[\mathrm{H}(y+h) - \mathrm{H}(y-h)] = \theta_0 \begin{cases} 0, & y \geq h; \\ 1, & -h \leq y \leq h; \\ 0, & y \leq -h. \end{cases} \tag{6.1}$$

We remind that H(y) in (6.1) is the Heaviside step-function. The solutions for $u(y), P(y)$ and $E(y)$ must be finite at both infinities $y \to \pm\infty$. More precisely, we demand

$$u|_{\pm\infty} \to \pm const, \quad P|_{\pm\infty} \to 0, \quad E|_{\pm\infty} \to 0. \tag{6.2}$$

At both interfaces $y = \pm h$ all three fields must be continuous

$$[u]_{\pm h} = 0, \quad [P]_{\pm h} = 0, \quad [E]_{\pm h} = 0, \tag{6.3}$$

together with the normal projections of mechanical stresses $\mathbf{n} \cdot \sigma \mathbf{n} = \sigma(y)$, electric displacements $\mathbf{D} \cdot \mathbf{n} = D(y) = \varepsilon_0 E(y) + P(y)$ and electric tensor $\mathbf{n} \cdot \mathbf{Q}\mathbf{n} = Q(y)$,

$$[\sigma]_{\pm h} = 0, \quad [D]_{\pm h} = 0, \quad [Q]_{\pm h} = 0. \tag{6.4}$$

In fact, as we shall see, for a given problem the unknown functions $u(y), P(y)$ and $E(y)$ are completely determined by the conditions at infinities (6.2). The above continuities at the interfaces, eqs. (6.3) and (6.4), will be fulfilled automatically.

(a) Equilibrium equations and their solutions

Thus, in the considered case the system of equilibrium equations (2.134)-(2.136) reduces to

$$c\frac{d^2u}{dy^2} + d\frac{d^2P}{dy^2} = \gamma\theta_0[\delta(y+h) - \delta(y-h)], \tag{6.5}$$

$$d\frac{d^2u}{dy^2} + b\frac{d^2P}{dy^2} - aP + E = \eta\theta_0[\delta(y+h) - \delta(y-h)], \tag{6.6}$$

$$\frac{d}{dy}(\varepsilon_0 E + P) = 0, \tag{6.7}$$

where the notation is introduced

$$c = c_{12} + 2c_{44}, \quad d = d_{12} + 2d_{44}, \quad b = b_{12} + 2b_{44}, \tag{6.8}$$

based on eq. (2.100). Eq. (6.7) with making use of conditions $(6.2)_{2,3}$ is trivially integrated providing the relation of the vanishing electric displacement

$$D \equiv \varepsilon_0 E + P = 0. \tag{6.9}$$

With (6.9) eq. (6.6) simplifies to

$$d\frac{d^2u}{dy^2} + b\frac{d^2P}{dy^2} - \hat{a}P = \eta\frac{d\theta}{dy}, \tag{6.10}$$

where

$$\hat{a} = a + \varepsilon_0^{-1}. \tag{6.11}$$

We note that for identically vanishing electric displacement D the continuity condition $(6.4)_2$ is automatically satisfied.

Excluding $u_{,yy}$ from eqs. (6.5), (6.10) we obtain the following differential equation with respect to $P(y)$:

$$\frac{d^2P}{dy^2} - \lambda^2 P = K\theta_0[\delta(y+h) - \delta(y-h)], \tag{6.12}$$

where

$$\lambda = \sqrt{\frac{c\hat{a}}{cb-d^2}}, \quad K = \frac{c\eta - d\gamma}{cb-d^2} = \lambda^2(\alpha_2 - \alpha_1)d/\hat{a}, \tag{6.13}$$

and we have introduced two coefficients of linear expansion α_1 and α_2,

$$\alpha_1 = \gamma/c, \alpha_2 = \eta/d. \tag{6.14}$$

In further considerations we shall suppose that $\alpha_1 \neq \alpha_2$, because in the particular case $\alpha_1 = \alpha_2$, we would have by (6.13) $K = 0$ and by (6.12), (6.9) $-P = 0, E = 0$, i.e. a purely elastic problem, independently of nonvanishing parameter λ. We remind that the parameter λ in (6.13) is the same as in eq. (2.93) and its reality is a consequence of the positive definiteness of the energy, eq. (2.82). Now it is time to mention that the moduli b, d, b_0 and η describe a space dispersion of the medium. According to Ch. 2 (see the consideration below eq. (2.29)),

$$d, b_0, \eta \propto l, \quad b \propto l^2, \tag{6.15}$$

where the length parameter l is expected to be of order of the lattice discreteness measure. Accordingly, we shall assume that $\lambda^{-1} \sim l$. It is clear that in the limit $l \to 0$ all the above moduli (and λ^{-1}) are vanishing and we return to the ordinary formulation of the Voigt-Toupin theory of a dispersionless piezoelectricity.

The solution of eq. (6.12) satisfying the conditions at infinities $(6.2)_2$ can be easily found:

$$P(y) = \frac{K\theta_0}{\lambda}\left[e^{-\lambda h}\text{sh}\lambda y - \text{sh}\lambda(y+h)H(-y-h) - \text{sh}\lambda(y-h)H(y-h)\right] =$$
$$= \frac{K\theta_0}{\lambda}\begin{cases} e^{-\lambda y}\text{sh}\lambda h, & y \geq h; \\ e^{-\lambda h}\text{sh}\lambda y, & -h \leq y \leq h; \\ -e^{\lambda y}\text{sh}\lambda h, & y \leq -h. \end{cases} \tag{6.16}$$

The function $P(y)$ (6.16) is odd, $P(-y) = -P(y)$, which reflects the antisymmetric source function in eq. (6.12).

Now, let us substitute into eq. (6.5) the derivative $P_{,yy}$ expressed from (6.12) with the function $P(y)$ defined by (6.16). This leads to the equation for the displacement field $u(y)$:

$$\frac{d^2u}{dy^2} = K_1\theta_0[\delta(y+h) - \delta(y-h)] - \lambda_1^2 P, \tag{6.17}$$

where

$$K_1 = \frac{b\gamma - d\eta}{cb - d^2}, \quad \lambda_1^2 = \frac{d}{c}\lambda^2 = \frac{d\hat{a}}{cb - d^2}. \tag{6.18}$$

A displacement field $u(y)$ is defined up to arbitrary constant. It is convenient to choose this constant so that $u(y)$ would be anti-symmetric like $P(y)$. This was the reason for a choice of the conditions at infinities in the form of eq. $(6.2)_1$. The corresponding solution of eq. (6.17) fitting $(6.2)_1$ is given by

$$u(y) = \alpha_1\theta_0 \begin{cases} [h - (k/\lambda)e^{-\lambda y}\text{sh}\lambda h], & y \geq h; \\ [y - (k/\lambda)e^{-\lambda h}\text{sh}\lambda y], & -h \leq y \leq h; \\ [-h + (k/\lambda)e^{\lambda y}\text{sh}\lambda h], & y \leq -h. \end{cases} \tag{6.19}$$

Here k is a dimensionless parameter of order of unity defined by

$$k = \frac{Kd}{\gamma} = \frac{\alpha_2/\alpha_1 - 1}{bc/d^2 - 1}. \tag{6.20}$$

It is easily checked that the found fields $u(y)$ and $P(y)$ are continuous at $y = \pm h$, which together with (6.9) fit all the three conditions (6.3). On the other hand, their derivatives defining the only non-vanishing components of distortions ($\beta = u_{,y}$) and polarization gradient ($\pi = P,y$) are discontinuous:

$$\beta(y) = \frac{du}{dy} = \alpha_1\theta_0 \begin{cases} ke^{-\lambda y}\text{sh}\lambda h, & y > h; \\ [1 - ke^{-\lambda h}\text{ch}\lambda y], & -h < y < h; \\ ke^{\lambda y}\text{sh}\lambda h, & y < -h; \end{cases} \tag{6.21}$$

$$\pi(y) = \frac{dP}{dy} = K\theta_0 \begin{cases} -e^{-\lambda y}\text{sh}\lambda h, & y > h; \\ e^{-\lambda h}\text{ch }\lambda y, & -h < y < h; \\ -e^{\lambda y}\text{sh }\lambda h, & y < -h. \end{cases} \tag{6.22}$$

Fortunately, such behavior at the interfaces $y = \pm h$ of distortions and polarization gradient does not contradict any physical conditions. In accordance with eq. (6.4) the functions, which cannot have any discontinuities, are the normal projections of electric displacements, $D(y)$, mechanical stresses, $\sigma(y)$, and electric tensor, $Q(y)$. As we have seen, the first one is continuous due to the identity (6.9). However, the other two must be verified. Basing on eqs. (2.130), (2.132) one has

$$\sigma(y) = c\beta(y) + d\pi(y) - \gamma\theta(y), \tag{6.23}$$

$$Q(y) = d\beta(y) + b\pi(y) - \eta\theta(y) + b_0. \tag{6.24}$$

Substituting here eqs. (6.1), (6.21) and (6.22) we obtain

$$\sigma(y) = 0, \tag{6.25}$$

$$Q(y) = b_0 + d(\alpha_1 - \alpha_2)\theta_0 \begin{cases} e^{-\lambda y}\mathrm{sh}\lambda h, & y \geq h; \\ [1 - e^{-\lambda h}\mathrm{ch}\lambda y], & -h \leq y \leq h; \\ e^{\lambda y}\mathrm{sh}\lambda h, & y \leq -h. \end{cases} \tag{6.26}$$

An identically vanishing stress (6.25) is rather expectable result for the problem without other sources apart from the temperature distribution (6.1) and with no fixing boundaries at infinities, eq. $(6.2)_1$. Anyway, this provides a trivial continuity of $\sigma(y)$. One can check that the function $Q(y)$ is also continuous at $y = \pm h$. Thus, the found solutions are mathematically correct.

(b) Comments on the results

The considered problem in a classical statement, i.e. for a medium without Mindlin's couplings, has an absolutely trivial solution. Of course, there should be no electric fields, i.e. $E = 0$, $P = 0$, $D = 0$. Mechanical displacements outside the layer must be homogeneous and distortions must be distinct of zero only inside of the layer:

$$u(y) = \alpha_1\theta_0 \begin{cases} h, & y \geq h; \\ y, & -h \leq y \leq h; \\ -h, & y \leq -h; \end{cases} \quad \beta(y) = \alpha_1\theta_0 \begin{cases} 0, & y > h; \\ 1, & -h < y < h; \\ 0, & y < -h. \end{cases} \tag{6.27}$$

And the corresponding mechanical stress must vanish everywhere:

$$\sigma(y) = c\beta(y) - \gamma\theta(y) = 0, \tag{6.28}$$

which is evident both physically and mathematically from (6.1), $(6.27)_2$ and $(6.14)_1$. As we have seen, vanishing $D(y)$ and $\sigma(y)$ fields retain even in a Mindlin's medium. However, the electric field with account for Mindlin's couplings does not vanish anymore,

$$E(y) = -\frac{K\theta_0}{\varepsilon_0\lambda} \begin{cases} e^{-\lambda y}\mathrm{sh}\ \lambda h, & y \geq h; \\ e^{-\lambda h}\mathrm{sh}\ \lambda y, & -h \leq y \leq h; \\ -e^{\lambda y}\mathrm{sh}\ \lambda h, & y \leq -h. \end{cases} \tag{6.29}$$

And a behavior of the fields $u(y)$ (6.19) and $\beta(y)$ (6.21) replacing the classical distributions (6.27) acquires new features. For instance, the medium outside the layer ($|y| > h$) is now not free of distortions: starting at the interfaces from the magnitudes comparable with the classical level $\alpha_1\theta_0$ (see (6.27)) the function $\beta(y)$ quickly decrease proportionally to $\exp(-\lambda|y - h|)$. The same factor characterizes the electric field (6.29) close to the interfaces. Taking into account that $\lambda^{-1} \sim l$ is of order of several lattice parameters, the obtained corrections are not negligible only in very narrow zones and could hardly be important for strain fields. However, the electric field arises on the background of zero with the amplitude at $y \approx \pm h$, which can be rather high, as will be shown. Such field might create non-trivial physical consequences even in microscopically narrow regions. We shall return to this aspect of the problem at the end of the chapter in Conclusions.

6.2 Thermal inclusion at the surface of a half-infinite dielectric

(a) Equations and boundary conditions

Let us now consider the analogous but less symmetric problem: the half-infinite dielectric medium with a heated layer of the thickness h at the surface (Fig. 6.2). The temperature distribution in this case is determined by

$$\theta(y) = \theta_0 \mathrm{H}(h - y) = \theta_0 \begin{cases} 0, & y \geq h; \\ 1, & 0 \leq y \leq h. \end{cases} \tag{6.30}$$

The equations of equilibrium are very similar to (6.5)-(6.7):

$$c\frac{d^2u}{dy^2} + d\frac{d^2P}{dy^2} = -\gamma\theta_0\delta(y - h), \tag{6.31}$$

$$d\frac{d^2u}{dy^2} + b\frac{d^2P}{dy^2} - aP + E = -\eta\theta_0\delta(y - h), \tag{6.32}$$

$$\frac{d}{dy}(\varepsilon_0 E + P) = 0. \tag{6.33}$$

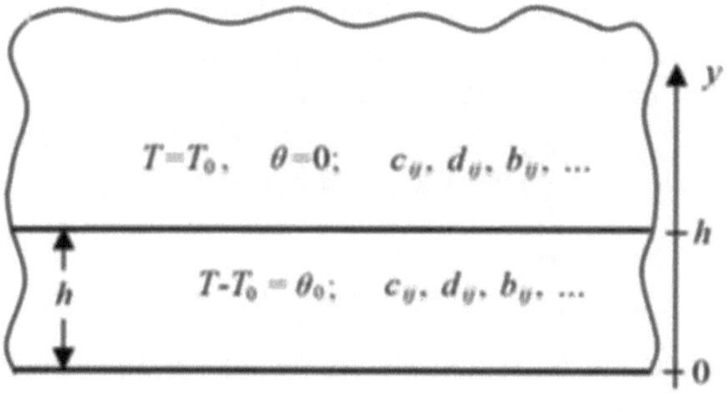

Fig. 6.2. The half-infinite medium with a thermal inclusion at the surface

Choosing boundary conditions at the surface $y = 0$, we shall suppose that the surface is free of any sources. Mechanically, this means the vanishing tractions,

$$\sigma(0) = 0. \tag{6.34}$$

In fact, it is rather evident that in the given case for the same reasons as before the thermoelastic stress $\sigma(y)$ must vanish. So, the condition (6.34) will hardly help us in solving the problem. On the other hand, in the absence of electrical sources at the surface the problem becomes also insensitive to the ordinary electric conditions at $y = 0$. Indeed, for a considered one-dimensional problem the surface is automatically equipotential and could be supposed to have a zero potential $\varphi(0) = 0$, which would be satisfied identically, without providing any additional condition for unknown functions. The same situation occurs for the condition of a continuity of the normal components of electric displacements in the medium ($D = P+\varepsilon_0 E$) and in the adjoined vacuum (D_v), because both identically vanish at any y. In fact, the only non-trivial electric condition at the surface is provided by the standard Mindlin's restriction for the electric tensor,

$$Q(0) = 0. \tag{6.35}$$

By (6.24) this is equivalent to the equation

$$d\frac{du}{dy}\bigg|_{y=0} + b\frac{dP}{dy}\bigg|_{y=0} = \eta\theta_0 - b_0. \tag{6.36}$$

The conditions at infinity will be also very essential. They are similar to (6.2),

$$u|_\infty \to 0, \quad P|_\infty \to 0, \quad E|_\infty \to 0. \tag{6.37}$$

And at the interface we expect, as before, to obtain automatically the continuity

$$[u]_h = 0, \quad [P]_h = 0, \quad [E]_h = 0. \tag{6.38}$$

With the above conditions at infinity the solution of eq. (6.33) again reduces to the identity (6.9), which indeed makes the requirement of continuity $D(0) = D_v(0)$ to be trivial: 0=0. Then the left-hand side of (6.32) becomes identical with that of eq. (6.10). Combining this result with eq. (6.31), we arrive at transformed eqs. (6.12) and (6.17),

$$\frac{d^2P}{dy^2} - \lambda^2 P = -K\theta_0\delta(y-h), \tag{6.39}$$

$$\frac{d^2u}{dy^2} = -K_1\theta_0\delta(y-h) - \lambda_1^2 P. \tag{6.40}$$

Here the parameters λ, K, λ_1 and K_1 are defined by the same eqs. (6.13) and (6.18). In contrast to the previous consideration, where the function $P(y)$ was found independently of $u(y)$, now the above system, eqs. (6.39), (6.40), cannot be solved in succession, because boundary condition (6.36) combines the both unknown functions.

(b) Solutions and analysis

The general solution of eq. (6.39) is

$$P(y) = Ae^{\lambda y} + Be^{-\lambda y} - \frac{K\theta_0}{\lambda}\mathrm{sh}[\lambda(y-h)]\mathrm{H}(y-h), \tag{6.41}$$

where A and B are constants which are supposed to be found from the boundary conditions. One of these constants (A) can be found from eq. $(6.37)_2$:

$$A = \frac{K\theta_0}{2\lambda}e^{-\lambda h}. \tag{6.42}$$

After integrating eq. (6.40) and differentiating (6.41) it is easy to find

$$\frac{du}{dy} = \frac{\lambda_1^2}{\lambda}\left[Be^{-\lambda y} + \frac{K\theta_0}{2\lambda}e^{-\lambda|y-h|}\mathrm{sgn}(y-h)\right] + \frac{\gamma\theta_0}{c}\mathrm{H}(h-y), \tag{6.43}$$

$$\frac{dP}{dy} = -\lambda\left[Be^{-\lambda y} + \frac{K\theta_0}{2\lambda}e^{-\lambda|y-h|}\mathrm{sgn}(y-h)\right], \tag{6.44}$$

where $\mathrm{sgn}(y-h)$ means a sign of the argument. At the surface $y=0$ eqs. (6.42)-(6.44) give

$$\left.\frac{du}{dy}\right|_{y=0} = \frac{\gamma\theta_0}{c} - \frac{\lambda_1^2}{\lambda}(A-B), \quad \left.\frac{dP}{dy}\right|_{y=0} = \lambda(A-B). \tag{6.45}$$

With (6.45) the boundary condition (6.36) reduces to the equation with respect to $A-B$. The solution is

$$B = A - \frac{K\theta_0}{\lambda} + \frac{\lambda b_0}{\hat{a}}. \tag{6.46}$$

With found derivatives (6.43), (6.44) and known temperature distribution (6.30) we can calculate the stress $\sigma(y)$ (6.23) and the electric tensor $Q(y)$ (6.24):

$$\sigma(y) = 0, \tag{6.47}$$

$$Q(y) = b_0\left(1-e^{-\lambda y}\right) +$$
$$+d(\alpha_2-\alpha_1)\theta_0\left\{e^{-\lambda y} + \frac{1}{2}\left(e^{-\lambda|y-h|} - e^{-\lambda(y+h)}\right) - \right. \tag{6.48}$$
$$\left. - e^{-\lambda(y-h)}\mathrm{H}(y-h) - \mathrm{H}(h-y)\right\}.$$

As we expected, the body is free of stresses. On the other hand, the function $Q(y)$ vanishes only at the surface, in accordance with (6.35), being continuous at $y=h$.

Now let us return to the fields $u(y)$ and $P(y)$. After the substitution of the found constants A and B, (6.42) and (6.46), into eqs. (6.41), (6.43) and the further integration of eq. (6.43) with taking into account the condition $(6.37)_1$ we arrive at the final solutions for $u(y)$ and $P(y)$:

$$u(y) = -\alpha_1\theta_0(h-y)\mathrm{H}(h-y) - \frac{\lambda d}{\hat{a}c}[b_0 - d(\alpha_2-\alpha_1)\theta_0]e^{-\lambda y} - \frac{\lambda d^2(\alpha_2-\alpha_1)\theta_0}{\hat{a}c}\begin{cases} e^{-\lambda y}\mathrm{ch}\lambda h, & y \geq h; \\ e^{-\lambda h}\mathrm{ch}\lambda y, & y \leq h; \end{cases} \tag{6.49}$$

$$P(y) = \frac{\lambda}{\hat{a}}\left\{\left[b_0 - d(\alpha_2-\alpha_1)\theta_0\left(1-\tfrac{1}{2}e^{-\lambda h}\right)\right]e^{-\lambda y} + \tfrac{1}{2}d(\alpha_2-\alpha_1)\theta_0 e^{-\lambda|y-h|}\right\} \tag{6.50}$$

In view of the proved identity (6.9), the corresponding electric field distribution $E(y) = -\varepsilon_0^{-1}P(y)$ again might be essential only in the very narrow zones, this time close to the surface ($y = 0$) and to the interface ($y = h$) :

$$\begin{aligned} E\big|_{y\approx 0} &\approx -\frac{\lambda}{\varepsilon_0\hat{a}}\left[b_0 - d(\alpha_2-\alpha_1)\theta_0\right]e^{-\lambda y}, \\ E\big|_{y\approx h} &\approx -\frac{\lambda d(\alpha_2-\alpha_1)\theta_0}{2\varepsilon_0\hat{a}}e^{-\lambda|y-h|}. \end{aligned} \tag{6.51}$$

It is natural that the magnitude E_m of this field at the interface is determined by the same physical parameters as in the above infinite medium and could be rather high ($E_m \sim 10^5$ V/m, see eq. (6.130) in Conclusions). On the other hand, at the surface we have additional source for electric effects - a surface tension, determined by the Mindlin's modulus b_0 (see Sec. 2.3 $(a)2^0$). That is why the first expression in (6.51) for $E|_{y=0}$ has more complex structure than the formula for $E|_{y=h}$.

6.3 Plain thermal inclusion in the middle of a dielectric plate

(a) A plate with faces free of tension

Now we consider the dielectric plate, shown in Fig. 6.3, with a symmetric position of the thermal inclusion in the middle of the plate, which represents some extension of the problem solved in Sec. 6.1. Indeed, the temperature distribution (6.1) and the corresponding equilibrium equations (6.5)-(6.7) remain valid for the considered case of a plate. The continuity conditions (6.3), (6.4) at the interfaces $y = \pm h$ also can serve as good criteria of correct calculations. However, the boundary conditions should be formulated from the beginning. For the same reason, as in Sec.6.2, again the only non-trivial electric boundary requirement is given by the Mindlin's condition of vanishing the normal components of the electric tensor **Q** at the surfaces.

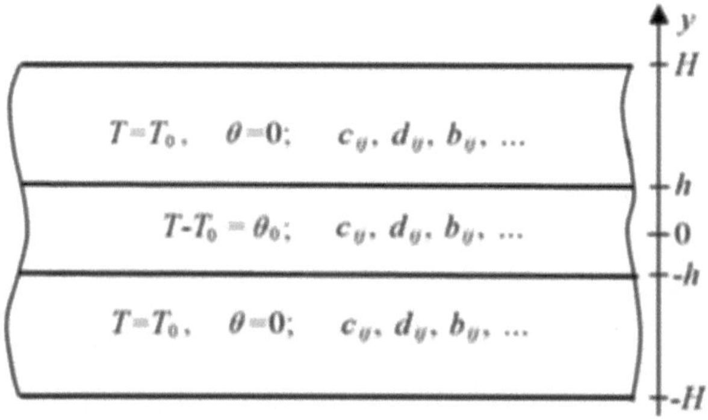

Fig. 6.3. The dielectric plate with a thermal inclusion in the middle

In our terms this means

$$Q(\pm H) = 0. \tag{6.52}$$

In this sub-section for mechanical boundary conditions we choose the free surfaces,

$$\sigma(\pm H) = 0. \tag{6.53}$$

It is obvious that eqs. (6.12) and (6.17) keep their form for the plate. We are looking for a solution of eq. (6.12) in the form

$$P(y) = A\mathrm{sh}\lambda y - \frac{K\theta_0}{\lambda}[\mathrm{sh}\lambda(y+h)\mathrm{H}(-y-h) + \mathrm{sh}\lambda(y-h)\mathrm{H}(y-h)], \tag{6.54}$$

where A is the unknown constant. The corresponding polarization gradient is equal

$$\frac{dP}{dy} = A\lambda\mathrm{ch}\lambda y - K\theta_0[\mathrm{ch}\lambda(y+h)\mathrm{H}(-y-h) + \mathrm{ch}\lambda(y-h)\mathrm{H}(y-h)]. \tag{6.55}$$

The integration of eq. (6.17) gives the distortion

$$\begin{aligned}\frac{du}{dy} &= B - A\lambda\frac{d}{c}[\mathrm{ch}\lambda y - 1] - \frac{\gamma\theta_0}{c}[\mathrm{H}(-y-h) + \mathrm{H}(y-h)] + \\ &+ \frac{d}{c}K\theta_0[\mathrm{ch}\lambda(y+h)\mathrm{H}(-y-h) + \mathrm{ch}\lambda(y-h)\mathrm{H}(y-h)]\end{aligned} \tag{6.56}$$

with the additional unknown constant B. The both functions (6.55) and (6.56) are even and take equal values at the surfaces $y = \pm H$,

$$\left.\frac{du}{dy}\right|_{\pm H} = B - A\lambda\frac{d}{c}[\mathrm{ch}\lambda H - 1] - \frac{\gamma\theta_0}{c} + \frac{d}{c}K\theta_0\mathrm{ch}\lambda(H-h), \tag{6.57}$$

$$\left.\frac{dP}{dy}\right|_{\pm H} = A\lambda\mathrm{ch}\lambda H - K\theta_0\mathrm{ch}\lambda(H-h). \tag{6.58}$$

Substitution of (6.1), (6.57) and (6.58) into (6.23), (6.24) with(6.52), (6.53) leads to the linear system of two equations determining A and B :

$$d\lambda A + cB = \gamma\theta_0, \tag{6.59}$$

$$\left(\frac{\hat{a}}{\lambda d}\mathrm{ch}\lambda H + \frac{\lambda d}{c}\right) A + B = \alpha_1\theta_0 + (\alpha_2 - \alpha_1)\theta_0 \mathrm{ch}\lambda(H-h) - \frac{b_0}{d}. \tag{6.60}$$

The solution is

$$A = -\frac{\lambda d}{\hat{a}} \frac{[b_0 + d(\alpha_1 - \alpha_2)\theta_0 \mathrm{ch}\lambda(H-h)]}{\mathrm{ch}\lambda H}, \quad B = \frac{\gamma\theta_0}{c} - \frac{\lambda d}{c} A. \tag{6.61}$$

With found A and B eqs. (6.55), (6.56) and (6.1) allow us to derive the functions $\sigma(y)$ and $Q(y)$, defined by (6.23), (6.24):

$$\sigma(y) = -\gamma\theta_0 + c\left(B + \frac{\lambda d}{c} A\right) = 0, \tag{6.62}$$

$$\begin{gathered} Q(y) = d(\alpha_1 - \alpha_2)\theta_0\{1 - \mathrm{ch}\lambda(H-h) + \\ +[\mathrm{ch}\lambda(y+h) - 1]\mathrm{H}(-y-h) + [\mathrm{ch}\lambda(y-h) - 1]\mathrm{H}(y-h)\} - \\ -[b_0 + d(\alpha_1 - \alpha_2)\theta_0 \mathrm{ch}\lambda(H-h)](\mathrm{ch}\lambda y/\mathrm{ch}\lambda H - 1). \end{gathered} \tag{6.63}$$

The customary identical zero for the stresses is a direct consequence of the free surfaces boundary conditions. As to the function $Q(y)$, it is continuous at the interfaces $y = \pm h$, which is the only feature important for us regarding this function.

The distribution of displacements $u(y)$ is obtained by integrating the distortion field (6.56) with A and B taken from (6.61):

$$\begin{gathered} u(y) = \alpha_1\theta_0[y - (y+h)\mathrm{H}(-y-h) - (y-h)\mathrm{H}(y-h)] + \\ +\frac{d}{c}\left\{\left(\frac{\lambda b_0}{\hat{a}} - \frac{K\theta_0}{\lambda}\mathrm{ch}\lambda(H-h)\right)\frac{\mathrm{sh}\lambda y}{\mathrm{ch}\lambda H} + \right. \\ \left. + \frac{K\theta_0}{\lambda}[\mathrm{sh}\lambda(y+h)\mathrm{H}(-y-h) + \mathrm{sh}\lambda(y-h)\mathrm{H}(y-h)]\right\} \end{gathered} \tag{6.64}$$

This function is chosen to be even and its continuity at $y = \pm h$ is easily verified.

The corresponding continuous distribution of polarization $P(y)$ is found by combining eqs. (6.54) and $(6.61)_1$:

$$\begin{gathered} P(y) = -\left\{\left(\frac{\lambda b_0}{\hat{a}} - \frac{K\theta_0}{\lambda}\mathrm{ch}\lambda(H-h)\right)\frac{\mathrm{sh}\lambda y}{\mathrm{ch}\lambda H} + \right. \\ \left. + \frac{K\theta_0}{\lambda}[\mathrm{sh}\lambda(y+h)\mathrm{H}(-y-h) + \mathrm{sh}\lambda(y-h)\mathrm{H}(y-h)]\right\} \end{gathered} \tag{6.65}$$

And the electric field is again given by the identity (6.9),

$$E(y) = -\varepsilon_0^{-1} P(y). \tag{6.66}$$

As before, the both fields $P(y)$ and $E(y)$ are most interesting characteristics of studied coupled fields, because close to the surfaces and to the interfaces their magnitudes could be rather high. For instance, the electric field in these narrow zones can be represented in the form

$$E\big|_{y\approx\pm H} \approx \pm\frac{\lambda b_0}{\varepsilon_0 \hat{a}} e^{-\lambda(H-|y|)}, \quad E\big|_{y\approx\pm h} \approx \pm\frac{\lambda d(\alpha_1-\alpha_2)\theta_0}{2\varepsilon_0 \hat{a}} e^{-\lambda\left|h-|y|\right|}. \tag{6.67}$$

In contrast to the case of the heated layer at the surface, this time the pre-surface electric field is completely determined by a surface tension (compare expressions $(6.67)_1$ and $(6.51)_1$).

(b) A plate with clamped surfaces

In this sub-section we shall for the first time consider the situation when thermal expansion is not free, being limited by clamped surfaces. Under such boundary condition one cannot expect anymore the vanishing thermoelastic stresses. Thus, we shall solve the same problem, as in Sec. 6.3(a), with the requirement (6.53) replaced by

$$u(\pm H) = 0. \tag{6.68}$$

Eqs. (6.54)-(6.58) remain valid, however we should add to them the relations

$$\begin{aligned} u(y) = {} & \left(B + A\lambda\frac{d}{c}\right) y - \alpha_1\theta_0[(y+h)\mathrm{H}(-y-h) + (y-h)\mathrm{H}(y-h)] + \\ & + \frac{d}{c}\left\{-A\mathrm{sh}\lambda y + \frac{K\theta_0}{\lambda}[\mathrm{sh}\lambda(y+h)\mathrm{H}(-y-h) + \mathrm{sh}\lambda(y-h)\mathrm{H}(y-h)]\right\} \end{aligned} \tag{6.69}$$

and

$$u(\pm H) = \pm\left\{\frac{d}{c}(\lambda H - \mathrm{sh}\lambda y)A + HB - \alpha_1\theta_0(H-h) + \frac{d}{c}\frac{K\theta_0}{\lambda}\mathrm{sh}\lambda(H-h)\right\}. \tag{6.70}$$

Thus the system determining A and B is formed by the combination of eqs. (6.68), (6.70) with (6.60). The solution of the system is given by

$$A = \frac{\lambda}{\hat{a}}\left\{\frac{\dfrac{hd\alpha_1\theta_0}{H} - b_0 + d(\alpha_2-\alpha_1)\theta_0\left(ch[\lambda(H-h)] + \dfrac{\Delta sh[\lambda(H-h)]}{\lambda H}\right)}{\mathrm{ch}\lambda H + \dfrac{\Delta\mathrm{sh}\lambda H}{\lambda H}}\right\} \tag{6.71}$$

$$B = \alpha_1\theta_0\left(1-\frac{h}{H}\right) - \frac{\lambda d}{c}\left(1-\frac{\mathrm{sh}\lambda H}{\lambda H}\right)A - \frac{\Delta(\alpha_2-\alpha_1)\theta_0}{\lambda H}\mathrm{sh}\lambda(H-h). \tag{6.72}$$

Here the notation is introduced

$$\Delta = \frac{d^2}{cb - d^2}. \tag{6.73}$$

As we know, the magnitude $\lambda^{-1} \sim l$ is of order of several parameters of a lattice, therefore for any macroscopic H, h and $H - h$ the parameters $\lambda H, \lambda h$ and $\lambda(H - h)$ must be very large. So we may neglect in the above bulky formulae (6.71) and (6.72) by exponentially small terms proportional to $\exp(-2\lambda H)$, $\exp[-2\lambda(H-h)]$ and $\exp(-\lambda H)/\lambda H$. This leads to more compact expressions

$$A \approx \frac{\lambda d}{\hat{a}} \left\{ 2\left(\alpha_1\theta_0 \frac{h}{H} - \frac{b_0}{d}\right) e^{-\lambda H} + (\alpha_2 - \alpha_1)\theta_0 e^{-\lambda h} \right\}, \tag{6.74}$$

$$B \approx -\frac{\lambda d}{c} A + \alpha_1\theta_0 \left(1 - \frac{h}{H}\right) + \frac{\Delta}{2\lambda H}\left(\alpha_1\theta_0 \frac{h}{H} - \frac{b_0}{d}\right). \tag{6.75}$$

With (6.74), (6.75) eq. (6.69) gives

$$\begin{aligned} u(y) \approx \alpha_1\theta_0 &\left\{ \left(1 - \frac{h}{H}\right) y - (y+h)\mathrm{H}(-y-h) - (y-h)\mathrm{H}(y-h) \right\} + \\ &+ \frac{\Delta}{2\lambda H}\left(\alpha_1\theta_0 \frac{h}{H} - \frac{b_0}{d}\right) y + \frac{\Delta}{\lambda}\{(\alpha_2 - \alpha_1)\theta_0[\mathrm{sh}\lambda(y+h)\mathrm{H}(-y-h) + \\ &+ \mathrm{sh}\lambda(y-h)\mathrm{H}(y-h) - e^{-\lambda h}\mathrm{sh}\lambda y] - 2\left(\alpha_1\theta_0 \frac{h}{H} - \frac{b_0}{d}\right) e^{-\lambda H}\mathrm{sh}\lambda y \right\}. \end{aligned} \tag{6.76}$$

Similarly, one obtains from (6.54)

$$\begin{aligned} P(y) \approx \frac{2\lambda d}{\hat{a}} &\left(\alpha_1\theta_0 \frac{h}{H} - \frac{b_0}{d}\right) e^{-\lambda H}\mathrm{sh}\lambda y + \\ + \frac{\lambda d}{\hat{a}}(\alpha_2 - \alpha_1)\theta_0 &\left[e^{-\lambda h}\mathrm{sh}\lambda y - \mathrm{sh}\lambda(y+h)\mathrm{H}(-y-h) - \mathrm{sh}\lambda(y-h)\mathrm{H}(y-h)\right]. \end{aligned} \tag{6.77}$$

These fields, $u(y)$ and $P(y)$, are clearly continuous everywhere including the interfaces $y = \pm h$, in contrast to the corresponding fields of distortions $\beta(y) = u_{,y}(y)$ and polarization gradient $\pi(y) = P_{,y}(y)$:

$$\begin{aligned} \frac{du}{dy} \approx \alpha_1\theta_0 &\left\{ \left(1 - \frac{h}{H}\right) - \mathrm{H}(-y-h) - \mathrm{H}(y-h)] \right\} + \\ &+ \Delta\left(\alpha_1\theta_0 \frac{h}{H} - \frac{b_0}{d}\right)\left[\frac{1}{2\lambda H} - 2e^{-\lambda H}\mathrm{ch}\lambda y\right] + \\ + \Delta(\alpha_2 - \alpha_1)\theta_0 &\left\{ -e^{-\lambda h}\mathrm{ch}\lambda y + \mathrm{ch}\lambda(y+h)\mathrm{H}(-y-h) + \mathrm{ch}\lambda(y-h)H(y-h) \right\}, \end{aligned} \tag{6.78}$$

$$\begin{aligned} \frac{dP}{dy} \approx \frac{\Delta c}{d} &\{(\alpha_2 - \alpha_1)\theta_0\left[e^{-\lambda h}\mathrm{ch}\lambda y - \mathrm{ch}\lambda(y+h)\mathrm{H}(-y-h) - \right. \\ &\left. - \mathrm{ch}\lambda(y-h)\mathrm{H}(y-h)\right] + 2\left(\alpha_1\theta_0 \frac{h}{H} - \frac{b_0}{d}\right) e^{-\lambda H}\mathrm{ch}\lambda y \right\}. \end{aligned} \tag{6.79}$$

They are discontinuous at $y = \pm h$. On the other hand, the linear superposition (6.20), (6.21) of these fields with the third discontinuous field of temperature distribution (6.1) form continuous fields $\sigma(y)$ and $Q(y)$. We shall present only the stress field, which does not vanish this time. The result is easily found by combining eqs. $(6.62)_1$ and (6.75):

$$\sigma(y) \approx -\gamma\theta_0 \frac{h}{H} + \frac{\Delta}{2\lambda H}\left(\gamma\theta_0 \frac{h}{H} - \frac{cb_0}{d}\right). \tag{6.80}$$

The second term here represents a Mindlin's correction to the result of classical theory given by the first term. Here we must note that a homogeneous character of the obtained field $\sigma(y)$ is an exact result, which has nothing to do with our neglect by exponentially small terms. Indeed, as it follows from eqs. (6.20), (6.55), (6.56) and (6.1),

$$\sigma(y) = cB + \lambda dA - \gamma\theta_0 = const. \tag{6.81}$$

Of course, the above small corrections to thermoelastic stresses are much less interesting than Mindlin's electric coupled field (6.66). As before it might be essential only in the vicinity of surfaces or interfaces, where again there are narrow zones of high electric field:

$$\begin{aligned} E_{y\approx\pm H} &\approx \pm\frac{\lambda}{\varepsilon_0\hat{a}}\left(b_0 - d\alpha_1\theta_0\frac{h}{H}\right)e^{-\lambda(H-|y|)} \\ E_{y\approx\pm h} &\approx \pm\frac{\lambda d}{2\varepsilon_0\hat{a}}\left(\alpha_1-\alpha_2\right)\theta_0 e^{-\lambda|h-|y||}. \end{aligned} \tag{6.82}$$

It is very natural that the fields at the interfaces $(6.82)_2$ and $(6.67)_2$ coincide. Indeed, such very localized dependencies could hardly "feel" boundary conditions at the surfaces, especially after our approximation relating to the transition from (6.71) to (6.74).

6.4 Plain thermal inclusion at the surface of a dielectric plate

(a) A plate with one face free and the other face clamped

In this section we shall extend the problem solved in Sec. 6.2 for a half-infinite dielectric medium, Fig. 6.2, by introducing into consideration the second surface, which earlier was at infinity (Fig. 6.4). The temperature distribution (6.30) and the equilibrium equations (6.31)-(6.33) and (6.39), (6.40) retain the same. We also inherit the boundary conditions (6.34) and (6.35). At the new surface $y = H$ we shall put

$$u(H) = 0, \qquad Q(H) = 0. \tag{6.83}$$

The first of this conditions is just a reformulation of $(6.37)_1$. It makes more definite the function $u(y)$, which is principally defined to an arbitrary constant. In fact, the requirement $(6.83)_1$ is trivially compatible with the condition $\sigma(H) = 0$, because, as will be shown, the demand (6.34) automatically provides the identity $\sigma(y) = 0$. That is why we shall not separately consider the boundary problem related to the case of both faces being free. On the other hand, the problem of the plate with both clamped faces, as we have just seen, eliminates identical vanishing of stresses and needs independent consideration (see the next subsection).

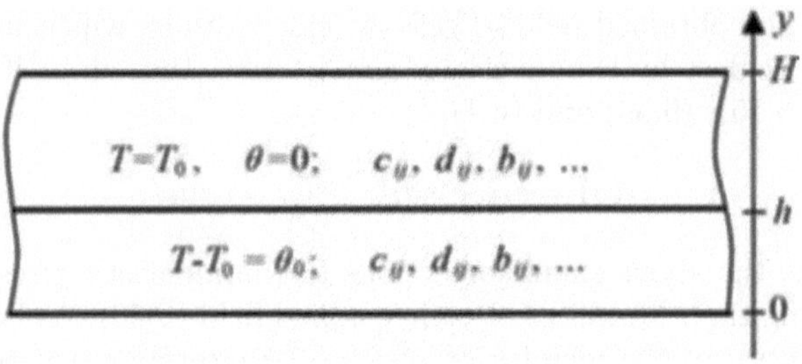

Fig. 6.4. The dielectric plate with a thermal inclusion at the surface

Thus, we can again look for a solution in the form (6.41). After substitution of this function into eq. (6.40) and integration of the latter one finds the distortion

$$\frac{du}{dy} = C - \alpha_1\theta_0 \mathrm{H}(y-h) - \frac{\lambda d}{c}\left[Ae^{\lambda y} - Be^{-\lambda y} - \frac{K\theta_0}{\lambda}\mathrm{ch}\lambda(y-h)\mathrm{H}(y-h)\right], \tag{6.84}$$

where C is the third unknown constant. Differentiation of (6.41) gives additionally

$$\frac{dP}{dy} = \lambda\left[Ae^{\lambda y} - Be^{-\lambda y} - \frac{K\theta_0}{\lambda}\mathrm{ch}\lambda(y-h)\mathrm{H}(y-h)\right]. \tag{6.85}$$

We note that identically

$$\frac{du}{dy} + \frac{d}{c}\frac{dP}{dy} = C - \alpha_1\theta_0\mathrm{H}(y-h). \tag{6.86}$$

With this observation and eq. (6.30) the stress $\sigma(y)$ (6.23) must be identically equal to a constant,

$$\sigma(y) = cC - \gamma\theta_0. \tag{6.87}$$

This general conclusion is very similar to the above analogous statement (6.81). So, if any face of the plate is free of tension, i.e. if $\sigma = 0$ at any of surfaces, as supposed to occur for the face $y = 0$, eq. (6.34), then everywhere

$$\sigma(y) = 0 \tag{6.88}$$

and therefore

$$C = \alpha_1 \theta_0. \tag{6.89}$$

With the known C and the boundary condition $(6.83)_1$ we can transform eq. (6.86) into

$$u(y) = \alpha_1 \theta_0 (y - h)\mathrm{H}(h - y) - \frac{d}{c}\left[P(y) - P(H)\right]. \tag{6.90}$$

Thus, now for evaluating the fields $u(y)$ and $P(y)$ we need to find the two unknown constants A and B. In view of (6.24), (6.30), the boundary conditions (6.35) and $(6.83)_2$ take the form

$$d\frac{du}{dy}\bigg|_{y=0} + b\frac{dP}{dy}\bigg|_{y=0} = \eta\theta_0 - b_0, \tag{6.91}$$

$$d\frac{du}{dy}\bigg|_{y=H} + b\frac{dP}{dy}\bigg|_{y=H} = -b_0. \tag{6.92}$$

Substituting here eqs. (6.84) and (6.85) we obtain the system

$$A - B = \frac{K\theta_0}{\lambda} - \frac{\lambda b_0}{\hat{a}}, \tag{6.93}$$

$$Ae^{\lambda H} - Be^{-\lambda H} = \frac{K\theta_0}{\lambda}\mathrm{ch}\lambda(H - h) - \frac{\lambda b_0}{\hat{a}}, \tag{6.94}$$

with the solution

$$A = \frac{K\theta_0}{2\lambda}\frac{\mathrm{ch}\lambda(H - h) - e^{-\lambda H}}{\mathrm{sh}\lambda H} - \frac{\lambda b_0}{2\hat{a}}\frac{1 - e^{-\lambda H}}{\mathrm{sh}\lambda H}, \tag{6.95}$$

$$B = \frac{K\theta_0}{2\lambda}\frac{\mathrm{ch}\lambda(H - h) - e^{\lambda H}}{\mathrm{sh}\lambda H} - \frac{\lambda b_0}{2\hat{a}}\frac{1 - e^{\lambda H}}{\mathrm{sh}\lambda H}. \tag{6.96}$$

Substituting the found constants A and B into (6.41) one obtains the polarization distribution

$$P(y) = \frac{K\theta_0}{2\lambda}\left[\frac{\mathrm{ch}\lambda(H - h)\mathrm{ch}\lambda y - \mathrm{ch}\lambda(H - y)}{\mathrm{sh}\lambda H} - \mathrm{sh}\lambda(y - h)H(y - h)\right] - \frac{\lambda b_0}{\hat{a}}\frac{\mathrm{ch}\lambda y - \mathrm{ch}\lambda(H - y)}{\mathrm{sh}\lambda H}, \tag{6.97}$$

which together with eq. (6.90) determines also a continuous displacement field. One can check that the corresponding field $Q(y)$ of the electric tensor is also continuous. We shall not present here its explicit form. As before, we are more interested in the electric field $E(y) = -\varepsilon_0^{-1}P(y)$ close to the surfaces

$y = 0, y = H$ and the interface $y = h$, which can be easily found by some manipulating with eq. (6.97)

$$E|_{y\approx 0} \approx -\frac{\lambda}{\varepsilon_0 \hat{a}} \left[b_0 - d(\alpha_2 - \alpha_1)\theta_0\right] e^{-\lambda y},$$
$$E|_{y\approx h} \approx -\frac{\lambda d(\alpha_2 - \alpha_1)\theta_0}{2\varepsilon_0 \hat{a}} e^{-\lambda|y-h|}, \tag{6.98}$$
$$E|_{y\approx H} \approx \frac{\lambda b_0}{\varepsilon_0 \hat{a}} e^{-\lambda(H-y)},$$

where the first two localized fields naturally coincide with those found in Sec. 6.2 for similar zones of the half-infinite medium, eq. (6.51). The field at the face H in our accuracy is completely determined by Mindlin's surface tension.

(b) A plate with clamped faces

Below we shall solve for the same plate shown in Fig. 6.4 the boundary problem

$$u(0) = 0, \quad Q(0) = 0, \quad u(H) = 0, \quad Q(H) = 0. \tag{6.99}$$

We can start from eqs. (6.41), (6.84)-(6.86). Integration of eq. (6.86) with making use of the boundary conditions $(6.99)_{1,3}$ leads to the relation

$$u(y) = -\frac{d}{c}\left[P(y) - P(0)\right] + Cy - \alpha_1\theta_0(y-h)\mathrm{H}(y-h), \tag{6.100}$$

where

$$C = \alpha_1\theta_0\left(1 - \frac{h}{H}\right) + \frac{d}{cH}\left[P(H) - P(0)\right]. \tag{6.101}$$

Combining eqs. (6.84)-(6.86) with the boundary conditions (6.91), (6.92) we have

$$\left.\frac{\hat{a}}{\lambda^2}\frac{dP}{dy}\right|_{y=0} = d\left(\alpha_2\theta_0 - C\right) - b_0, \tag{6.102}$$

$$\left.\frac{\hat{a}}{\lambda^2}\frac{dP}{dy}\right|_{y=H} = d\left(\alpha_1\theta_0 - C\right) - b_0. \tag{6.103}$$

Substituting here the corresponding derivatives of $P(y)$ from (6.85) one obtains the system similar to (6.93), (6.94):

$$Ae^{\lambda H} - Be^{-\lambda H} = \kappa_1 + \frac{K\theta_0}{\lambda}\mathrm{ch}\lambda(H-h), \tag{6.104}$$

$$A - B = \kappa_2, \tag{6.105}$$

where the coefficients $\kappa_i (i = 1, 2)$ are defined by

$$\kappa_i = \frac{\lambda}{\hat{a}}\left[d\left(\alpha_i\theta_0 - C\right) - b_0\right]. \tag{6.106}$$

The solution of the above system is given by

$$
\begin{aligned}
A &= \frac{\kappa_1 - \kappa_2 e^{-\lambda H} + (K\theta_0/\lambda)\mathrm{cth}\lambda(H-h)}{2\mathrm{sh}\lambda H}, \\
B &= \frac{\kappa_1 - \kappa_2 e^{\lambda H} + (K\theta_0/\lambda)\mathrm{cth}\lambda(H-h)}{2\mathrm{sh}\lambda H}.
\end{aligned}
\tag{6.107}
$$

Now we can also find the unknown constant C. Combining (6.107) with (6.41) we arrive at the equation

$$
\begin{aligned}
&P(H) - P(0) = \\
-C\frac{2\lambda d}{\hat{a}} th\tfrac{1}{2}\lambda H &+ \frac{\lambda}{\hat{a}}\left[d(\alpha_1+\alpha_2)\theta_0 - 2b_0\right]\mathrm{th}\frac{1}{2}\lambda H + \\
&+ \frac{K\theta_0}{\lambda}\frac{\mathrm{sh}\lambda(h-\frac{1}{2}H)}{\mathrm{ch}\frac{1}{2}\lambda H},
\end{aligned}
\tag{6.108}
$$

which forms with eq. (6.101) the necessary system allowing us to exclude the difference $P(H) - P(0)$ and to evaluate C. In fact, for our purposes there is no need to present a bulky exact solution. Omitting the small exponents and the terms of order of $(\lambda H)^{-2}$ one obtains

$$
\begin{aligned}
C &\approx \alpha_1\theta_0\left(1-\frac{h}{H}\right) + \frac{\Delta}{\lambda H}\left\{\left[\alpha_2 - \alpha_1\left(1-\frac{2h}{H}\right)\right]\theta_0 - \frac{2b_0}{d}\right\} \approx \\
&\approx \alpha_1\theta_0\left(1-\frac{h}{H}\right).
\end{aligned}
\tag{6.109}
$$

Thus, with the found C the parameters $\kappa_{1,2}$ become definite together with the fields $u(y)$ (6.100) and $P(y)$ (6.41). For instance, the explicit form of the latter is

$$
\begin{aligned}
&P(y) = \\
&= \frac{\kappa_1\mathrm{ch}\lambda y - \kappa_2\mathrm{ch}\lambda(H-y)}{\mathrm{sh}\lambda H} + \\
&+ \frac{(K\theta_0/\lambda)\left[\mathrm{ch}\lambda y\,\mathrm{ch}\lambda(H-h) - \mathrm{sh}\lambda H\,\mathrm{sh}\lambda(y-h)\mathrm{H}(y-h)\right]}{\mathrm{sh}\lambda H}.
\end{aligned}
\tag{6.110}
$$

The found polarization distribution as well as the accompanied electric field $E(y) = -\varepsilon_0^{-1}P(y)$ are again concentrated basically close to the surfaces and the interface:

$$
\begin{aligned}
E\big|_{y\approx 0} \approx \frac{\kappa_2}{\varepsilon_0}e^{-\lambda y}, \quad E\big|_{y\approx h} &\approx -\frac{\lambda d(\alpha_2-\alpha_1)\theta_0}{2\varepsilon_0\hat{a}}e^{-\lambda|y-h|} \\
E\big|_{y\approx H} &\approx -\frac{\kappa_1}{\varepsilon_0}e^{-\lambda|H-y|},
\end{aligned}
\tag{6.111}
$$

where $\kappa_{1,2}$ are given by eq. (6.106). We note that the localized field $(6.111)_{2,3}$ are completely identical with $(6.82)_{2,1}$, respectively, and eq. $(6.111)_1$ at $H \to \infty$ naturally coincides with $(6.51)_1$.

6.5 Cylindrical thermal inclusion in an infinite dielectric medium

(a) Equations and boundary conditions

In this section we return to the above mentioned problem first solved by J.P. Nowacki & Trimarco [120]. Consider a cylindrical thermal inclusion of the radius r_0 in an infinite dielectric medium. As before, the inclusion and the surrounding space are supposed to have the same material constants and to differ from each other only by temperature:

$$\theta(r) = \theta_0[1 - \mathrm{H}(r - r_0)] = \begin{cases} \theta_0, \ 0 \leq r \leq r_0; \\ 0, \quad r > r_0. \end{cases} \tag{6.112}$$

In (6.112) r is counting from the axis of the cylinder in the plane orthogonal to this axis. Owing to the cylindrical symmetry of the problem we can assume the vector fields $\mathbf{u}$, $\mathbf{P}$ and $\mathbf{E}$ to have only radial components and to be dependent only on the radius r :

$$\mathbf{u}(\mathbf{r}) = u(\mathbf{r})\mathbf{n}_r, \quad \mathbf{P}(\mathbf{r}) = P(\mathbf{r})\mathbf{n}_r, \quad E(\mathbf{r}) = E(\mathbf{r})\mathbf{n}_r, \tag{6.113}$$

where the unit vector $\mathbf{n}_r = \mathbf{r}/r$ is introduced. The scalar functions $u(r), P(r)$ and $E(r)$ must be continuous at the interface $r = r_0$ and vanish at infinity:

$$[u]_{r_0} = 0, \quad [P]_{r_0}, = 0 \quad [E]_{r_0} = 0. \tag{6.114}$$

$$u|_\infty \to 0, \quad P|_\infty \to 0, \quad E|_\infty \to 0. \tag{6.115}$$

With these conditions eq. (2.136) provides as before the relation (6.9) of zero electric induction,

$$P(r) + \varepsilon_0 E(r) = 0. \tag{6.116}$$

Let us substitute (6.113) into the system (2.134), (2.135) with taking into account that in the cylindrical coordinates $\{r, \varphi, z\}$ with the orthogonal basis $\{\mathbf{n}_r, \mathbf{n}_\varphi, \mathbf{n}_z\}$ one has

$$\nabla = \mathbf{n}_r \frac{\partial}{\partial r} + \mathbf{n}_\varphi \frac{\partial}{r\partial\varphi} + \mathbf{n}_z \frac{\partial}{\partial z}, \quad \frac{\partial \mathbf{n}_r}{\partial r} = 0, \quad \frac{\partial \mathbf{n}_r}{\partial \varphi} = \mathbf{n}_\varphi, \quad \frac{\partial \mathbf{n}_\varphi}{\partial \varphi} = -\mathbf{n}_r, \tag{6.117}$$

and correspondingly (say for u)

$$\mathbf{n}_r \left[\nabla^2(u\mathbf{n}_r)\right] = \nabla^2 u - \frac{u}{r^2} = \frac{d}{dr}\left(\frac{1}{r}\frac{d}{dr}(ru)\right) = \mathbf{n}_r \nabla\nabla \cdot (u\mathbf{n}_r). \tag{6.118}$$

As a result, after some transformations with making use of (6.116) we arrive at the system

$$\frac{d}{dr}\left\{\frac{1}{r}\frac{d}{dr}\left[r\left(cu + dP\right)\right]\right\} = \gamma\frac{d\theta}{dr}, \tag{6.119}$$

$$\frac{d}{dr}\left\{\frac{1}{r}\frac{d}{dr}\left[r\left(du + bP\right)\right]\right\} - \hat{a}P = \eta\frac{d\theta}{dr}. \tag{6.120}$$

(b) Solutions and analysis

Excluding u from eqs. (6.119) and (6.120) one obtains

$$\frac{d^2P}{dr^2} + \frac{1}{r}\frac{dP}{dr} - \left(\lambda^2 + \frac{1}{r^2}\right) P = -K\theta_0\delta(r - r_0). \tag{6.121}$$

The obtained equation with respect to $P(r)$ can be solved straightforwardly. It is very well known that the homogeneous part of eq. (6.121) is solved by the fundamental system of linearly independent modified Bessel functions $I_1(\lambda r)$ and $K_1(\lambda r)$. One can check that inhomogeneous eq. (6.121) combined with conditions (6.114), (6.115) are solved by the combination

$$P(r) = r_0 K\theta_0 \begin{cases} \mathrm{I}_1(\lambda r)\mathrm{K}_1(\lambda r_0), \; 0 \le r \le r_0; \\ \mathrm{I}_1(\lambda r_0)\mathrm{K}_1(\lambda r), \quad r \ge r_0. \end{cases} \tag{6.122}$$

Substituting (6.122) into (6.121) one should notice that the first derivative is discontinuous at $r = r_0$: .

$$\left[\frac{dP}{dr}\right]_{r_0} = -K\theta_0 = -\theta_0\lambda^2(\alpha_2 - \alpha_1)d/\hat{a}. \tag{6.123}$$

With a known field $P(r)$ we can easily find the fields $E(r)$ and $u(r)$, the first by eq. (6.116) and the second by the integration of eq. (6.119), which gives

$$u(r) + \frac{d}{c}P(r) = \frac{\alpha_1\theta_0}{2} \begin{cases} r, & 0 \le r \le r_0; \\ r_0^2/r, & r \ge r_0. \end{cases} \tag{6.124}$$

The functions $E(r)$ and $u(r)$ are continuous as well as $P(r)$, however their derivatives $E'(r)$ and $u'(r)$ are discontinuous at r_0 and by (6.19), (6.123) and (6.124) with making use of eqs. (6.14) and (6.20) their jumps at the interface are

$$\left[\frac{dE}{dr}\right]_{r_0} = \frac{K\theta_0}{\varepsilon_0} = \theta_0\lambda^2(\alpha_2 - \alpha_1)d/\varepsilon_0\hat{a}, \quad \left[\frac{du}{dr}\right]_{r_0} = \alpha_1\theta_0\left[k - 1\right], \tag{6.125}$$

where k is defined by (6.20). One can check that the stress field in this case does not vanish. However its component σ_{rr} is continuous at the interface, as it must be. On the other hand, the other nonvanishing components $\sigma_{\varphi\varphi}$ and σ_{zz} have their jumps at $r = r_0$:

$$[\sigma_{\varphi\varphi}]_{r_0} = [\sigma_{zz}]_{r_0} = \gamma\theta_0\left[1 - \frac{c_{12}}{c_{11}} + \frac{k}{d}\left(\frac{c_{12}}{c_{11}}d_{11} - d_{12}\right)\right]. \tag{6.126}$$

One of the most important features of the found solutions is again related to high electric fields in the narrow zone from both sides of the interface.

Indeed, for macroscopic and even for mesoscopic radius r_0 of the inclusion and for $r \sim r_0$, as we know from the above considerations, $\lambda r_0, \lambda r \gg 1$. For such large arguments the involved into (6.122) Bessel functions can be estimated by the asymptotics

$$I_1(z) \approx \frac{e^z}{\sqrt{2\pi z}}, \quad K_1(z) \approx \sqrt{\frac{\pi}{2z}} e^{-z}. \tag{6.127}$$

Accordingly, close to r_0 by (6.122), (6.116) the field $E(r)$ is

$$E(r) \approx -\frac{K\theta_0}{2\varepsilon_0 \lambda} e^{-\lambda|r-r_0|}, \tag{6.128}$$

which with $(6.13)_2$ naturally coincides with eqs. $(6.51)_2$, $(6.98)_2$ and $(6.111)_2$ if to replace in the latter h by r_0.

We stress that the dimensionless parameter k (6.20) in eqs. (6.125) and (6.126) is generally of order of unity. Hence the contributions of Mindlin's terms into the above jumps are of the same order as contributions of classical terms. However the importance of non-classical contributions to strains and stresses is limited only by the narrow layer around the interface, in the same way as in (6.128). Nevertheless in some microscopic phenomena the found additional localized fields can play an important and unexpected role which is customarily ignored by the classical continual mechanics.

6.6 Conclusions

Thus, the theory of polarization gradient applied to a series of one-dimensional problems related to isotropic dielectrics predicts that a plain or cylindrical thermal inclusion must excite, apart from the ordinary thermoelastic strains, the additional electro-mechanical fields localized close to surfaces and interfaces. It is interesting that with "switching off" the Mindlin's coupling effects ($d \to 0, b \to 0, b_0 \to 0, \eta \to 0$) the new fields disappear not due to vanishing of their amplitudes, but by means of decreasing of zones, where they act. Let us look at this problem for the example of an electric component arising on the background of zero. Indeed, after such switching off the polarization gradient coupling effects, the amplitudes $E_m \sim K\theta_0/\lambda\varepsilon_0$ of such fields do not disappear. Basing on (6.13), (6.15) and supposing that $|bc - d^2| \sim d^2$, the magnitude of E_m at the interface can be estimated as

$$E_m \sim \alpha_1\theta_0/(c\varepsilon_0)^{1/2}. \tag{6.129}$$

By the way, this is not at all a low field. At $\alpha_1\theta_0 \sim 10^{-4}, c \sim 10^3$ MPa one obtains the estimate

$$E_m \sim 10^5 \text{ V/m}. \tag{6.130}$$

So, close to the interface where the temperature jump occurs, the field E can be expressed as

$$E \sim E_m \exp[-\lambda|\Delta r|] \tag{6.131}$$

with Δr denoting the distence from the interface. Thus, the formal disappearance of the field happens because the pre-interface zone of the field penetration $|\Delta r| \sim 1/\lambda \sim l$ for $l \to 0$ becomes infinitely narrow. As we already mentioned, in realistic conditions of ordinary dielectrics the parameter $\lambda^{-1} \sim l$ should be expected of order of several lattice parameters and the material constants d, b and η must be very small. This means that in ordinary dielectrics the considered thermal inclusion should excite rather high electric fields (6.130) in the very narrow zones around surfaces or interfaces. It appears that the accomplished analysis and the made prediction might be useful in some applications.

7

Green's Functions for Piezoelectric Strip with a General Line Source

In this chapter we are going to derive the static Green function describing the 2D coupled fields in arbitrary piezoelectric strip excited by a general line defect parallel to the surfaces and consisting of the four line sources introduced in Sec. 3.3 (c): the line of forces $\mathbf{f}$, the line of charge q, the straight dislocation with the Burgers vector $\mathbf{b}$, and its electrostatic analog of the strength $\Delta\varphi$. As we have seen in Ch. 3, the latter line defect is completely determined by the discontinuity $\Delta\varphi$ in the electrical potential across the arbitrary plane cut ending on the line, which in our case determines the coinciding positions of all the mentioned sources. The first studies in this field were accomplished for a particular case of straight dislocations in isotropic elastic plates [55-59]. These results were later extended for anisotropic [53] and possibly inhomogeneous [54] purely elastic (nonpiezoelectric) infinite strips. The further extension of the theory for the case of arbitrary piezoelectric strips has been developed in our paper [28]. The content of Chapter 7 is based on this paper.

7.1 Statement of the problem and basic equations

(a) Some preliminary remarks

The aim of this chapter is a derivation of the electro-elastic fields excited in a piezoelectric plate by the general line defect defined above. Of course, such an object, as the general line defect first introduced by Lothe & Barnett [63], is only conventionally a "defect", because among its four constituent line sources there is only one lattice defect - the dislocation. So, perhaps, it would be more correct to talk about the general line source rather than about a defect. Anyway, independently of the terminology introduced and accepted before us, below we are going to find the physical fields excited by such a combined source. In our analysis it will be convenient to follow the method developed in [54] for purely elastic anisotropic continuously layered media, specifying the problem to a homogeneous medium and simultaneously extending it for a piezoelectric case. Such an extension turns out to be not quite straightforward because of the electrical boundary conditions, which are determined by a continuity of the normal component of electric displacement on the interfaces between the strip and the adjoined isotropic dielectric media. Our consideration will be based on the Stroh-like approach presented in details in Chapter 3 (Sec. 3.3 (c)) where the problem was already formulated for the same general line defect in unbounded medium.

(b) Basic equations

Consider the infinite piezoelectric plate of thickness d in the coordinate system specified by the following conditions: the xz-plane is parallel to the surfaces with the axis z collinear to the general defect which intersects the plane xy at the point x', y' (see Fig. 7.1). The material of the plate is supposed to be of unrestricted anisotropy admitting piezoelectricity, and is characterized by the elastic moduli tensor c_{ijkl}, the piezoelectric constants $eikl$ and the electric permittivity ε_{ij}. These tensors form coefficients in the constitutive equations relating the stress σ_{ij} and the electric displacement D_i to the elastic strain β_{ij} and the electric field E_i :

$$\sigma_{ij} = c_{ijkl}\beta_{kl} - e_{lij}E_l, \tag{7.1}$$

$$D_j = e_{jlk}\beta_{kl} + \varepsilon_{jl}E_l. \tag{7.2}$$

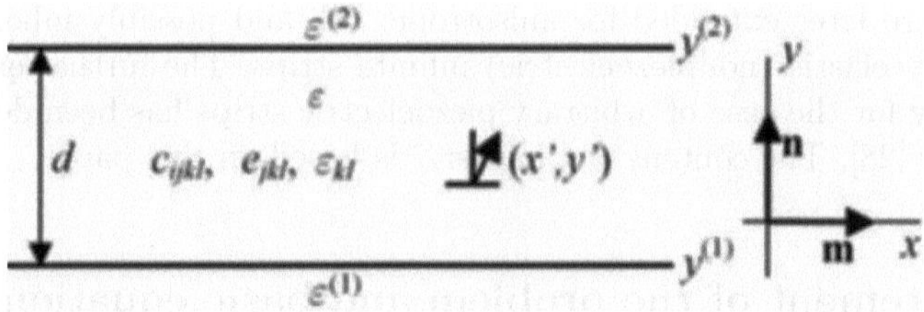

Fig. 7.1. Geometry of the problem and choice of the coordinate system.

We remind that by eq. (3.12) in presence of a dislocation and an electrostatic line defect the elastic distortion β_{kl} and the electric field E_l are not equal to gradients of the displacement vector $\mathbf{u}$ and the electric potential φ, respectively, as usual. In this case one should write instead:

$$u_{k,l} = \beta_{kl} + \beta^0_{kl}, \tag{7.3}$$

$$-\varphi_{,l} = E_l - E^0_l, \tag{7.4}$$

with

$$\boldsymbol{\beta}^0 = -\mathbf{b} \otimes \mathbf{n}\mathrm{H}(x - x')\delta(y - y'), \quad \mathbf{E}^0 = -\Delta\varphi\mathbf{n}\mathrm{H}(x - x')\delta(y - y'). \tag{7.5}$$

Here $\mathbf{n}$ is the unit vector along the y-axis (Fig. 7.1). The corresponding equilibrium equations have the form

$$\sigma_{ij,j} = -f_i \delta(x - x')\delta(y - y'), \tag{7.6}$$

$$D_{j,j} = q\delta(x - x')\delta(y - y'). \tag{7.7}$$

As is shown in Ch. 3, Sec. 3.3 (c), the considered 2D problem can be reduced to one eight-dimensional equation

$$(\mathbf{I}\partial_y - \mathbf{N}\partial_x)\boldsymbol{\eta}(x, y) = -\mathbf{g}\mathrm{H}(x - x')\delta(y - y'), \tag{7.8}$$

where $\mathbf{N}$ is the 8×8 constant matrix, eq. (3.138), depending only on the orientation of the frame $\{\mathbf{m}, \mathbf{n}\}$ and on the material constants c_{iJKl} (see eqs. (3.17) and (3.51)), $\mathbf{g}$ is the strength of the general line defect,

$$\mathbf{g} = [\mathbf{b}, \Delta\varphi, -\mathbf{f}, q]^t, \tag{7.9}$$

and $\boldsymbol{\eta}(x, y)$ is the unknown eight-component vector function

$$\boldsymbol{\eta}(x, y) = [\mathbf{U}(x, y), \mathbf{V}(x, y)]^t. \tag{7.10}$$

Its four-dimensional block-vectors

$$\mathbf{U}(x, y) = \begin{pmatrix} \mathbf{u}(x, y) \\ \varphi(x, y) \end{pmatrix}, \quad \mathbf{V}(x, y) = \begin{pmatrix} \mathbf{J}(x, y) \\ D(x, y) \end{pmatrix} \tag{7.11}$$

represent an extended "displacement" $\mathbf{U}$ and a generalized stress function $\mathbf{V}$ defined by

$$(\boldsymbol{\sigma} + \boldsymbol{\sigma}^0)_{ix} = J_{i,y}, \quad (\boldsymbol{\sigma} + \boldsymbol{\sigma}^0)_{iy} = -J_{i,x}; \tag{7.12}$$

$$(\mathbf{D} + \mathbf{D}^0)_x = D_{,y}, \quad (\mathbf{D} + \mathbf{D}^0)_y = -D_{,x}, \tag{7.13}$$

where

$$\boldsymbol{\sigma}^0 = \mathbf{f} \otimes \mathbf{m}\mathrm{H}(x - x')\delta(y - y'), \quad \mathbf{D}^0 = -q\mathbf{m}\mathrm{H}(x - x')\delta(y - y'). \tag{7.14}$$

Certainly, with definitions (7.14) one can put in eqs. (7.12), (7.13) $\sigma^0_{iy} = 0$ and $D^0_y = 0$.

Note that eq.(7.8) is equivalent to the system of eight differential equations of the first order, in contrast to the standard system of four differential equations of the second order given by a combination of eqs. (7.6), (7.7) with (7.1)-(7.5). As a result, the right-hand side of (7.8) contains the Heaviside function $\mathrm{H}(x - x')$, instead of the delta-function $\delta(x - x')$ in (7.6), (7.7).

(c) General solution and some additional properties of the eigenvectors $\boldsymbol{\xi}_\alpha$

It is convenient to look for a solution of eq. (7.8) in the form of a Fourier integral,

$$\boldsymbol{\eta}(x,y) = \int_{-\infty}^{\infty} dk \exp(ikx)\boldsymbol{\eta}(k,y). \tag{7.15}$$

As is shown in Sec. 3.3 (c), this leads to the expression

$$\begin{aligned}\boldsymbol{\eta}(x,y) = \frac{1}{2\pi i}\sum_{\alpha=1}^{8}\int_{-\infty}^{\infty}\frac{dk}{k}\exp\left\{ik[x-x'+p_\alpha(y-y')]\right\}\times \\ \times\left[b_\alpha(k)-(\boldsymbol{\xi}_\alpha\cdot\mathbf{Tg})\,\mathrm{H}(y-y')\right],\end{aligned} \tag{7.16}$$

where $\boldsymbol{\xi}_a$ and p_α $(\alpha = 1,\ldots,8)$ are the eigenvectors and eigenvalues of the matrix $\mathbf{N}$, and b_α are eight unknown functions of k which are supposed to be found from the boundary conditions.

In accordance with a block structure (7.10), (7.11) of the formalism let us denote:

$$\boldsymbol{\xi}_\alpha = \begin{pmatrix}\mathbf{U}_\alpha\\ \mathbf{V}_\alpha\end{pmatrix}, \quad \mathbf{U}_\alpha = \begin{pmatrix}\mathbf{A}_\alpha\\ \varphi_\alpha\end{pmatrix} \quad \mathbf{V}_\alpha = \begin{pmatrix}\mathbf{L}_\alpha\\ D_\alpha\end{pmatrix}. \tag{7.17}$$

In these terms the completeness equation (3.147) is transformed to quite a series of elementary relations, which will be useful in further considerations:

$$\sum_{\alpha=1}^{8}\mathbf{U}_\alpha\otimes\mathbf{U}_\alpha=\mathbf{0},\quad \sum_{\alpha=1}^{8}\mathbf{V}_\alpha\otimes\mathbf{V}_\alpha=\mathbf{0},\quad \sum_{\alpha=1}^{8}\mathbf{U}_\alpha\otimes\mathbf{V}_\alpha=\mathbf{I}; \tag{7.18}$$

$$\sum_{\alpha=1}^{8}\mathbf{A}_\alpha\otimes\mathbf{A}_\alpha=\mathbf{0},\quad \sum_{\alpha=1}^{8}\mathbf{L}_\alpha\otimes\mathbf{L}_\alpha=\mathbf{0},\quad \sum_{\alpha=1}^{8}\mathbf{A}_\alpha\otimes\mathbf{L}_\alpha=\mathbf{I}; \tag{7.19}$$

$$\sum_{\alpha=1}^{8}\varphi_\alpha\mathbf{A}_\alpha=\mathbf{0},\quad \sum_{\alpha=1}^{8}\varphi_\alpha\mathbf{L}_\alpha=\mathbf{0},\quad \sum_{\alpha=1}^{8}D_\alpha\mathbf{A}_\alpha=\mathbf{0},\quad \sum_{\alpha=1}^{8}D_\alpha\mathbf{L}_\alpha=0; \tag{7.20}$$

$$\sum_{\alpha=1}^{8}\varphi_\alpha\varphi_\alpha=\mathbf{0},\quad \sum_{\alpha=1}^{8}D_\alpha D_\alpha=\mathbf{0},\quad \sum_{\alpha=1}^{8}\varphi_\alpha D_\alpha=\mathbf{0}. \tag{7.21}$$

For our further needs let us also introduce the three other representations of eigenvectors. The first of them is the so called F-representation, defined by Lothe & Barnett [63, 64]:

$$\boldsymbol{\xi}_{F\alpha} = \begin{pmatrix}\mathbf{U}_{F\alpha}\\ \mathbf{V}_{F\alpha}\end{pmatrix}, \quad \mathbf{U}_{F\alpha} = \begin{pmatrix}\mathbf{A}_\alpha\\ D_\alpha\end{pmatrix}, \quad \mathbf{V}_{F\alpha} = \begin{pmatrix}\mathbf{L}_\alpha\\ \varphi_\alpha\end{pmatrix}. \tag{7.22}$$

By definition, the vectors $\boldsymbol{\xi}_{F\alpha}$ are eigenvectors of the matrix $\mathbf{N}_F$, which is obtained from $\mathbf{N}$ by interchanging the 4-th and the 8-th row, and by the subsequent interchange of the 4-th and the 8-th column. The eigenvectors $\boldsymbol{\xi}_{F\alpha}$ of the matrix $\mathbf{N}_F$ clearly concern the same eigenvalues p_α and satisfy the relations of orthogonality and completeness similar to eqs. (3.146), (3.147) and (7.18).

The other two, the so-called boundary representations, there were first introduced in our paper [28]. As we shall see, they will work for the cases of the boundary conditions related to a continuity of normal components of electric displacements on the plate surfaces when they are adjoined to some dielectric media (Fig. 7.1). Now we have to consider some more complex combinations of the same values $\mathbf{A}_\alpha, \mathbf{L}_\alpha, \varphi_\alpha$, and D_α :

$$\boldsymbol{\xi}_\alpha^{(\mathrm{j})\pm} = \begin{pmatrix} \mathbf{U}_\alpha^{(\mathrm{j})\pm} \\ \mathbf{V}_\alpha^{(\mathrm{j})\pm} \end{pmatrix}, \quad \mathbf{U}_\alpha^{(\mathrm{j})\pm} = \begin{pmatrix} \mathbf{A}_\alpha \\ \varphi_\alpha \pm D_\alpha / i\varepsilon^{(\mathrm{j})} \end{pmatrix}, \quad \mathbf{V}_\alpha^{(\mathrm{j})\pm} = \begin{pmatrix} \mathbf{L}_\alpha \\ D_\alpha \pm i\varepsilon^{(\mathrm{j})} \varphi_\alpha \end{pmatrix}. \tag{7.23}$$

Finally the boundary F-representation is given by

$$\boldsymbol{\xi}_{F\alpha}^{(\mathrm{j})\pm} = \begin{pmatrix} \mathbf{U}_{F\alpha}^{(\mathrm{j})\pm} \\ \mathbf{V}_{F\alpha}^{(\mathrm{j})\pm} \end{pmatrix}, \quad \mathbf{U}_{F\alpha}^{(\mathrm{j})\pm} = \begin{pmatrix} \mathbf{A}_\alpha \\ D_\alpha \pm i\varepsilon^{(\mathrm{j})} \varphi_\alpha \end{pmatrix}, \quad \mathbf{V}_\alpha^{(\mathrm{j})\pm} = \begin{pmatrix} \mathbf{L}_\alpha \\ \varphi_\alpha \pm D_\alpha / i\varepsilon^{(\mathrm{j})} \end{pmatrix}. \tag{7.24}$$

The 4-vectors introduced in (7.23) and (7.24) have several important properties, which directly follow from eqs. (7.19)-(7.21):

$$\sum_{\alpha=1}^{8} \mathbf{U}_\alpha^{(\mathrm{j})+} \otimes \mathbf{U}_\alpha^{(\mathrm{j})-} = \sum_{\alpha=1}^{8} \mathbf{V}_\alpha^{(\mathrm{j})+} \otimes \mathbf{V}_\alpha^{(\mathrm{j})-} = \sum_{\alpha=1}^{8} \mathbf{U}_{F\alpha}^{(\mathrm{j})+} \otimes \mathbf{U}_{F\alpha}^{(\mathrm{j})-} = \sum_{\alpha=1}^{8} \mathbf{V}_{F\alpha}^{(\mathrm{j})+} \otimes \mathbf{V}_{F\alpha}^{(\mathrm{j})-} = 0, \tag{7.25}$$

$$\sum_{\alpha=1}^{8} \mathbf{U}_\alpha^{(\mathrm{j})+} \otimes \mathbf{U}_{F\alpha}^{(\mathrm{j})-} = \sum_{\alpha=1}^{8} \mathbf{V}_\alpha^{(\mathrm{j})+} \otimes \mathbf{V}_{F\alpha}^{(\mathrm{j})-} = \sum_{\alpha=1}^{8} \mathbf{U}_{F\alpha}^{(\mathrm{j})+} \otimes \mathbf{U}_\alpha^{(\mathrm{j})-} = \sum_{\alpha=1}^{8} \mathbf{V}_{F\alpha}^{(\mathrm{j})+} \otimes \mathbf{V}_\alpha^{(\mathrm{j})-} = 0. \tag{7.26}$$

Some of these relations will be used below for the calculation of the coefficients b_α in eq. (7.15).

7.2 Boundary conditions and their block representations

(a) Four types of boundary conditions in terms of two representations

The unknown Fourier amplitude $\boldsymbol{\eta}(k,y)$ in (7.15), (7.16) can be easily expressed in terms of physical quantities directly involved in boundary conditions. Indeed, in view of (7.10)-(7.14), for $y \neq y'$ it is defined by the following combination of components

$$\boldsymbol{\eta}(k,y) = [\mathbf{U}(k,y), \mathbf{V}(k,y)]^t = [\mathbf{u}(k,y), \varphi(k,y), \mathbf{J}(k,y), D(k,y)]^t, \qquad (7.27)$$

where

$$\mathbf{J}(k,y) = \frac{i}{k}\mathbf{n}\boldsymbol{\sigma}(k,y), \quad D(k,y) = \frac{i}{k}D_n(k,y), \qquad (7.28)$$

with $D_n = D_y = \mathbf{n}\cdot\mathbf{D}$. On each of the boundaries $y = y^{(1)}$ and $y = y^{(2)}$ we will distinguish free ($\mathbf{n}\boldsymbol{\sigma} = \mathbf{0}$) or clamped ($\mathbf{u} = \mathbf{0}$) states of the surface. On the other hand, the same boundaries can be electrically open $(D_n = 0)^1$, closed ($\varphi = 0$) or to be adjoined to a dielectric medium (in particular, to vacuum), when the condition of a continuity of D_n on the boundary must be satisfied.

In principle, it is very advantageous to deal in further development with a 4-component block form of boundary conditions. For instance, in terms of the above formalism, the conditions of the clamped metallized surface $y = y^{(j)}$ ($\mathbf{u} = \mathbf{0}, \varphi = 0$) are described by the singlee four-dimensional equation $\mathbf{U}(k, y^{(j)}) = \mathbf{0}$. And the conditions of the mechanically free, electrically open boundary $y = y^{(j)}$($\mathbf{n}\boldsymbol{\sigma} = \mathbf{0}, D_n = 0$) are reduced to the equation $\mathbf{V}(k, y^{(j)}) = \mathbf{0}$. In order to describe on equal footing also the block pairs $(\mathbf{u}, D_n)$ and $(\mathbf{n}\boldsymbol{\sigma}, \varphi)$, it is convenient to introduce additionally the F-representation of the $\boldsymbol{\eta}$-function, associated with (7.22):

$$\boldsymbol{\eta}_F(k,y) = [\mathbf{U}_F(k,y), \mathbf{V}_F(k,y)]^t = [\mathbf{u}(k,y), D(k,y), \mathbf{J}(k,y), \varphi(k,y)]^t. \quad (7.29)$$

Obviously, the corresponding function $\boldsymbol{\eta}_F(x,y)$ obeys the same equation (7.8) if we replace there $\mathbf{g}$ (7.9) by

$$\mathbf{g}_F = [\mathbf{b}, q, -\mathbf{f}, \Delta\varphi]^t \qquad (7.30)$$

and the matrix $\mathbf{N}$ by $\mathbf{N}_F$. Accordingly, the mechanically clamped electrically open surface is described by the condition $\mathbf{U}_F(k, y^{(j)}) = \mathbf{0}$. And the boundary condition of a free metallized surface is given by the equation $\mathbf{V}_F(k, y^{(j)}) = \mathbf{0}$. Here we would like to stress that the F-representation is just another form

[1] This terminology was introduced by Ingebrigtsen [180] and was widely used in further literature (see, e.g., [62-64]). The condition $D_n = 0$ formally corresponds to an adjoined medium with a zero permittivity which can be understood as a limiting situation of very high electric permittivity in the medium itself compared to that in the adjoined medium. On the other hand, $D_n = 0$ is a condition of a vanishing charge density at the surface.

of the same solution: the function $\eta_F(k, y)$ differs from $\eta(k, y)$ only by the order in the list of the same physical fields. However, such rearrangement has proved to be a very efficient instrument in the theory of surface waves in semi-infinite piezoelectric media [63, 64] due to allowance for an analysis of all the mentioned series of boundary conditions in the framework of a unified formalism.

(b) Universal boundary conditions

The both representations will be applied in further analysis. However, with two surfaces, the number of combinations of different boundary conditions appears to be too large. For instance, if we considere all physically different combinations of the above two mechanical conditions ($\mathbf{u} = \mathbf{0}$ and $\mathbf{n}\boldsymbol{\sigma} = \mathbf{0}$) and three electrical conditions ($\varphi = 0,\ D_n = 0$ plus the adjoined vacuum) on each of the two surfaces, we would arrive at 21 boundary problems. That is why we shall consider instead only one universal electrical condition concerning the plate bounded by the two isotropic dielectric media with the permittivities $\varepsilon^{(1)}$ and $\varepsilon^{(2)}$. Then the above three variants of the electrical boundary condition ($\varphi = 0, D_n = 0$ or the adjoined vacuum) at $y = y^{(j)}$ are obtained by the choice for $\varepsilon^{(j)} : \varepsilon^{(j)} = \infty, \varepsilon^{(j)} = 0$ or $\varepsilon^{(j)} = \varepsilon_0$, respectively.

Let us derive an explicit form for such a more general boundary condition. In the adjoined isotropic media, the potential $\varphi(x, y)$ is a harmonic function. So its Fourier image $\varphi(k, y)$ must obey the equation

$$\frac{\partial^2}{\partial y^2}\varphi(k, y) - k^2\varphi(k, y) = 0 \tag{7.31}$$

with the solution

$$\begin{aligned} \varphi(k, y \le y^{(1)}) &= \varphi_1 \exp[|k|(y - y^{(1)})], \\ \varphi(k, y \ge y^{(2)}) &= \varphi_2 \exp[-|k|(y - y^{(2)})]. \end{aligned} \tag{7.32}$$

Thus

$$\begin{aligned} D_n(k, y \le y^{(1)}) = -\varepsilon^{(1)}\partial_y\varphi(k, y \le y^{(1)}) &= -\varepsilon^{(1)}|k|\varphi(k, y \le y^{(1)}), \\ D_n(k, y \ge y^{(2)}) &= \varepsilon^{(2)}|k|\varphi(k, y \ge y^{(2)}). \end{aligned} \tag{7.33}$$

Taking into account that at the electrical interfaces $y = y^{(j)}$ both the potential φ and the normal component of the electric displacement $D_n = D_y = \mathbf{n} \cdot \mathbf{D}$ are continuous, the resulting boundary conditions, internal for the plate, take the form (see also [64]):

$$\begin{aligned} \mathbf{n} \cdot \mathbf{D}(k, y^{(1)}) + \varepsilon^{(1)}|k|\varphi(k, y^{(1)}) &= 0, \\ \mathbf{n} \cdot \mathbf{D}(k, y^{(2)}) - \varepsilon^{(2)}|k|\varphi(k, y^{(2)}) &= 0, \end{aligned} \tag{7.34}$$

or in terms of the components D of decomposition (7.27),

$$\begin{aligned} D(k,y^{(1)}) + i\varepsilon^{(1)}\mathrm{sgn}(k)\varphi(k,y^{(1)}) &= 0, \\ D(k,y^{(2)}) - i\varepsilon^{(2)}\mathrm{sgn}(k)\varphi(k,y^{(2)}) &= 0. \end{aligned} \tag{7.35}$$

For further convenience of formulating boundary conditions in a block form, let us introduce the two boundary representations of the solution by means of some rearrangements of its both possible forms, $\boldsymbol{\eta}(k,y)$, eq. (7.27), and $\boldsymbol{\eta}_F(k,y)$, eq. (7.29), by defining on each of the surfaces a pair of new 8-vectors:

$$\boldsymbol{\eta}^{\pm}(k,y^{(j)}) = [\mathbf{U}^{\pm}(k,y^{(j)}), \mathbf{V}^{\pm}(k,y^{(j)})]^t, \quad j=1,2, \tag{7.36}$$

with

$$\mathbf{U}^{\pm}(k,y^{(j)}) = [\mathbf{u}(k,y^{(j)}), \varphi(k,y^{(j)}) \pm D(k,y^{(j)})/i\varepsilon^{(j)}]^t, \tag{7.37}$$

$$\mathbf{V}^{\pm}(k,y^{(j)}) = [\mathbf{J}(k,y^{(j)}), D(k,y^{(j)}) \pm i\varepsilon^{(j)}\varphi(k,y^{(j)})]^t, \tag{7.38}$$

and

$$\boldsymbol{\eta}_F^{\pm}(k,y^{(j)}) = [\mathbf{U}_F^{\pm}(k,y^{(j)}), \mathbf{V}_F^{\pm}(k,y^{(j)})]^t, \quad j=1,2, \tag{7.39}$$

with

$$\mathbf{U}_F^{\pm}(k,y^{(j)}) = [\mathbf{u}(k,y^{(j)}), D(k,y^{(j)}) \pm i\varepsilon^{(j)}\varphi(k,y^{(j)})]^t, \tag{7.40}$$

$$\mathbf{V}_F^{\pm}(k,y^{(j)}) = [\mathbf{J}(k,y^{(j)}), \varphi(k,y^{(j)}) \pm D(k,y^{(j)})/i\varepsilon^{(j)}]^t. \tag{7.41}$$

In eqs. (7.36)-(7.41) the choice of a sign is determined in the same way as in (7.35), both by the number j of the surface ($y^{(1)}$ or $y^{(2)}$) and by the sign of k :

$$\pm = (-1)^{j+1}\mathrm{sgn}(k). \tag{7.42}$$

In these terms we shall have to consider on each boundary only two variants of boundary conditions: the free surface described by any of two equivalent equations,

$$\mathbf{V}^{\pm}(k,y^{(j)}) = \mathbf{0} \quad \text{or} \quad \mathbf{V}_F^{\pm}(k,y^{(j)}) = \mathbf{0}, \tag{7.43}$$

and the clamped surface which also has an alternative description:

$$\mathbf{U}^{\pm}(k,y^{(j)}) = \mathbf{0} \quad \text{or} \quad \mathbf{U}_F^{\pm}(k,y^{(j)}) = \mathbf{0}. \tag{7.44}$$

Note that the equations in the above pairs become non-equivalent in the limiting cases $\varepsilon^{(j)} \to 0$ and $\varepsilon^{(j)} \to \infty$. In particular, the condition of the metallized free surface is described by the second equation of the pair (7.43), which after passing to the limit $\varepsilon^{(j)} \to \infty$ becomes just

$$\mathbf{V}_F(k,y^{(j)}) = \mathbf{0}. \tag{7.45}$$

For the metallized clamped surface one has from the first equation of the pair (7.44) in the same limit:

$$\mathbf{U}(k,y^{(j)}) = \mathbf{0}. \tag{7.46}$$

Quite similarly, the condition of a free electrically open surface is given by eq. $(7.43)_1$ at $\varepsilon^{(j)} = 0$, and $(7.44)_2$ taken at $\varepsilon^{(j)} = 0$ describes a clamped electrically open surface.

7.3 Green's function for the plate with free surfaces

(a) Three integral representations of solution

In this case the boundary conditions are

$$\mathbf{V}^{\pm}(k, y^{(1)}) = \mathbf{0}, \qquad \mathbf{V}^{\mp}(k, y^{(2)}) = \mathbf{0}, \tag{7.47}$$

where in accordance with (7.42) the upper or lower sign is chosen for $k > 0$ or $k < 0$, respectively. In particular, for the case $k > 0$ in terms of eq. (7.16), conditions (7.47) reduce to

$$\begin{gathered}\sum_{\alpha=1}^{8} \mathbf{V}_{\alpha}^{(1)+} \exp\left[ikp_{\alpha}\left(y^{(1)} - y'\right)\right] b_{\alpha} = \mathbf{0},\\ \sum_{\alpha=1}^{8} \mathbf{V}_{\alpha}^{(2)-} \exp\left[ikp_{\alpha}\left(y^{(2)} - y'\right)\right] \{b_{\alpha} - (\boldsymbol{\xi}_{\alpha} \cdot \mathbf{Tg})\} = \mathbf{0}.\end{gathered} \tag{7.48}$$

In view of $(7.25)_2$, the first equation of this system is satisfied identically by the substitution

$$b_{\alpha}(k > 0) = \mathbf{V}_{\alpha}^{(1)-} \cdot \mathbf{G}(k > 0) \exp\left\{-ikp_{\alpha}\left(y^{(1)} - y'\right)\right\}. \tag{7.49}$$

With (7.49), eq. $(7.48)_2$ gives

$$\begin{aligned}\mathbf{G}(k > 0) = \sum_{\beta=1}^{8} &\left\{\sum_{\gamma=1}^{8} \mathbf{V}_{\gamma}^{(2)-} \otimes \mathbf{V}_{\gamma}^{(1)-} \exp\left[ikp_{\gamma}d\right]\right\}^{-1} \times\\ &\times \mathbf{V}_{\beta}^{(2)-} (\boldsymbol{\xi}_{\beta} \cdot \mathbf{Tg}) \exp\left[ikp_{\beta}\left(y^{(2)} - y'\right)\right],\end{aligned} \tag{7.50}$$

where $d = y^{(2)} - y^{(1)}$. Quite similarly one can find the expression for $b_{\alpha}(k < 0)$ and $\mathbf{G}(k < 0)$, which are obtained from (7.49), (7.50) by the replacements $\mathbf{V}_{\gamma,\beta}^{(1,2)-} \to \mathbf{V}_{\gamma,\beta}^{(1,2)+}$. Let us introduce the 8×8 matrix with components

$$m_{\alpha\beta}(\mathrm{k}) = \begin{cases} m_{\alpha\beta}^{(+)}(k), \; k < 0,\\ m_{\alpha\beta}^{(-)}(k), \; k > 0,\end{cases} \tag{7.51}$$

where

$$m_{\alpha\beta}^{(\pm)}(k) = \mathbf{V}_{\alpha}^{(1)\pm} \cdot \left\{\sum_{\gamma=1}^{8} \mathbf{V}_{\gamma}^{(2)\pm} \otimes \mathbf{V}_{\gamma}^{(1)\pm} \exp\left[ik\left(p_{\gamma} - p_{\beta}\right)d\right]\right\}^{-1} \mathbf{V}_{\beta}^{(2)\pm}. \tag{7.52}$$

With this notation for any sign of k the coefficients b_{α} in (7.16) are written as

$$b_\alpha(k) = \sum_{\beta=1}^{8} m_{\alpha\beta}(k)\left(\boldsymbol{\xi}_\beta \cdot \mathbf{Tg}\right)\exp\left[ik\left(p_\beta - p_\alpha\right)\left(y^{(1)} - y'\right)\right], \tag{7.53}$$

and the general solution (7.16) takes the form

$$\begin{aligned}\boldsymbol{\eta}(x,y) = \frac{1}{2\pi i}\sum_{\alpha,\beta=1}^{8}\int_{-\infty}^{\infty}\frac{dk}{k}\boldsymbol{\xi}_\alpha\left(\boldsymbol{\xi}_\beta \cdot \mathbf{Tg}\right)\exp\{ik\left(x - x' + \right.\\ \left. + p_\alpha(y - y^{(1)}) - p_\beta(y' - y^{(1)})\right)\}\left[m_{\alpha\beta}(k) - \delta_{\alpha\beta}\mathrm{H}(y - y')\right].\end{aligned} \tag{7.54}$$

This is the integral representation of the Green function for a plate with free surfaces (we remind that here condition (3.141) is implied). It is easily verified that the solution (7.54) with (7.51), (7.52) satisfies both eq. (7.8) and boundary conditions (7.47). Eq. (7.54) describes the electro-elastic fields for any permittivities of adjoining media, including the limiting situations when one or both boundaries become electrically open: $\varepsilon^{(\mathrm{j})} \to 0$. In the latter case one should just respectively replace in (7.52) the corresponding vectors $\mathbf{V}_\delta^{(\mathrm{j})\pm}$ by $\mathbf{V}_\delta$ (see (7.23) at $\varepsilon^{(\mathrm{j})} = 0$).

However, as was expected, this representation does not work for electrically closed (metallized) boundaries. In accordance with the above considerations, the corresponding limit $\varepsilon^{(\mathrm{j})} \to \infty$ for one or both permittivities is admitted by the F-representation which is obtained from eqs. (7.54), (7.51), (7.52) by the replacements of all vectors $\mathbf{V}_\delta^{(\mathrm{j})\pm}(\delta = \alpha, \gamma, \beta)$ by $\mathbf{V}_{F\delta}^{(\mathrm{j})\pm}$, respectively. For any finite permittivities $\varepsilon^{(\mathrm{j})}$ both representations are completely equivalent. But only the F-representation gives a clear limit at $\varepsilon^{(\mathrm{j})} \to \infty$ relating to the transformation of the corresponding vector $\mathbf{V}_{F\delta}^{(\mathrm{j})\pm}$ into $\mathbf{V}_{F\delta}$.

Note that due to properties (7.26), we could put into substitution (7.49) the vector $\mathbf{V}_{F\alpha}^{(1)-}$ instead of $\mathbf{V}_\alpha^{(1)-}$. This would lead us again to eq. (7.54) but with the alternative expression for $m_{\alpha\beta}^{(\pm)}(k)$ which differs from (7.52) by the replacement $\mathbf{V}_{\alpha,\gamma}^{(1)\pm} \to \mathbf{V}_{F\alpha,F\gamma}^{(1)\pm}$. The resulting solution gives the third representation for $\boldsymbol{\eta}(x,y)$ admitting this time the different limits: $\varepsilon^{(1)} \to \infty, \varepsilon^{(2)} \to 0$.

(b) Analysis of convergence of the Fourier integral

Now we must prove the convergence of the integral in eq. (7.54). This will require an analysis of the asymptotic behavior of the integrand at $k \to \pm\infty$. Note that the dependence of the integrand on k is completely determined by the exponents explicitly presented in (7.54), (7.51), (7.52) and by k in the denominator of (7.54). In order to make our further notation more systematic, let us introduce the matrix

$$\mathbf{W}_k^{\pm} = \sum_{\alpha=1}^{8} \boldsymbol{\xi}_\alpha^{(2)\pm} \otimes \mathbf{T}\boldsymbol{\xi}_\alpha^{(1)\pm} \exp(ikdp_\alpha) =$$

$$= \begin{pmatrix} \sum_{\alpha=1}^{8} \mathbf{U}_\alpha^{(2)\pm} \otimes \mathbf{V}_\alpha^{(1)\pm} \exp(ikdp_\alpha) & \sum_{\alpha=1}^{8} \mathbf{U}_\alpha^{(2)\pm} \otimes \mathbf{U}_\alpha^{(1)\pm} \exp(ikdp_\alpha) \\ \sum_{\alpha=1}^{8} \mathbf{V}_\alpha^{(2)\pm} \otimes \mathbf{V}_\alpha^{(1)\pm} \exp(ikdp_\alpha) & \sum_{\alpha=1}^{8} \mathbf{V}_\alpha^{(2)\pm} \otimes \mathbf{U}_\alpha^{(1)\pm} \exp(ikdp_\alpha) \end{pmatrix} \equiv$$

$$\equiv \begin{pmatrix} \left[\mathbf{W}_k^{\pm}\right]_{11} & \left[\mathbf{W}_k^{\pm}\right]_{12} \\ \left[\mathbf{W}_k^{\pm}\right]_{21} & \left[\mathbf{W}_k^{\pm}\right]_{22} \end{pmatrix}. \tag{7.55}$$

In these terms the matrix $m_{\alpha\beta}^{(\pm)}(k)$ (7.52) takes the form

$$\mathbf{m}_{\alpha\beta}^{(\pm)}(k) = \mathbf{V}_\alpha^{(1)\pm} \cdot \left\{\left[\mathbf{W}_k^{\pm}\right]_{21}\right\}^{-1} \mathbf{V}_\beta^{(2)\pm} \exp(ikdp_\beta) \tag{7.56}$$

In view of the rule of numbering accepted in Ch. 3, eqs. (3.144), (3.145),

$$\boldsymbol{\xi}_{\alpha+4} = \boldsymbol{\xi}_\alpha^*, \quad p_{\alpha+4} = p_\alpha^*, \quad \operatorname{Im} p_\alpha > 0, \quad \alpha = 1, \ldots, 4, \tag{7.57}$$

the 4×4 matrix $\left[\mathbf{W}_k^{\pm}\right]_{21}$ can be presented in the form

$$\left[\mathbf{W}_k^{\pm}\right]_{21} = \left[\mathbf{w}_k^{\pm}\right]_{21} + \left\{\left[\mathbf{w}_{-k}^{\pm}\right]_{21}\right\}^*, \tag{7.58}$$

where

$$\left[\mathbf{w}_k^{\pm}\right]_{21} = \sum_{\alpha=1}^{4} \mathbf{V}_\alpha^{(2)\pm} \otimes \mathbf{V}_\alpha^{(1)\pm} \exp(ikdp_\alpha). \tag{7.59}$$

It is easily checked that the sum over α in (7.59) is equivalent to the matrix product

$$\left[\mathbf{w}_k^{\pm}\right]_{21} = \mathbf{V}^{(2)\pm}\mathbf{d}^{(k)}(\mathbf{V}^{(1)\pm})^t, \tag{7.60}$$

where $\mathbf{V}^{(2)\pm}$ is the matrix with the columns $\mathbf{V}_\alpha^{(2)\pm}$:

$$\mathbf{V}^{(2)\pm} = \left[\mathbf{V}_1^{(2)\pm}, \mathbf{V}_2^{(2)\pm}, \mathbf{V}_3^{(2)\pm}, \mathbf{V}_4^{(2)\pm}\right];$$

similarly, the matrix $(\mathbf{V}^{(1)\pm})^t$ is built of the rows $\mathbf{V}_\alpha^{(1)\pm}$; and $\mathbf{d}^{(k)}$ is the diagonal 4×4 matrix with the elements

$$d_{\alpha\beta}^{(k)} = \delta_{\alpha\beta} \exp[ikdp_\alpha]. \tag{7.61}$$

In accordance with rule (7.57), all the terms in (7.59) are characterized by $\operatorname{Im} p_\alpha > 0$. Therefore the first term in the right-hand side of (7.58) is exponentially large at $k \to -\infty$ and exponentially small at $k \to \infty$; the second term, *vice versa*, is small at $k \to -\infty$ and large at $k \to \infty$. In other words,

$$\text{at } k \to -\infty: \quad \{[\mathbf{w}_k]_{21}\}^{-1} \to 0 \quad \text{and} \quad [\mathbf{w}_{-k}]_{21}^* \to 0; \tag{7.62}$$

$$\text{at } k \to \infty : \quad [\mathbf{w}_k]_{21} \to 0 \quad \text{and} \quad \{[\mathbf{w}_{-k}]^*_{21}\}^{-1} \to 0; \tag{7.63}$$

(for brevity here and below we omit the superscript $\pm$). Accordingly, at $k \to -\infty$

$$\begin{aligned} \{[\mathbf{W}_k]_{21}\}^{-1} &= \{[\mathbf{w}_k]_{21} + [\mathbf{w}_{-k}]^*_{21}\}^{-1} = \\ &= (\mathbf{I} + \{[\mathbf{w}_k]_{21}\}^{-1}[\mathbf{w}_{-k}]^*_{21})^{-1}\{[\mathbf{w}_k]_{21}\}^{-1} \approx \\ &\approx \{[\mathbf{w}_k]_{21}\}^{-1} - \{[\mathbf{w}_k]_{21}\}^{-1}[\mathbf{w}_{-k}]^*_{21}\{[\mathbf{w}_k]_{21}\}^{-1}, \end{aligned} \tag{7.64}$$

and at $k \to \infty$

$$\{[\mathbf{W}_k]_{21}\}^{-1} \approx \{[\mathbf{w}_{-k}]^*_{21}\}^{-1} - \{[\mathbf{w}_{-k}]^*_{21}\}^{-1}[\mathbf{w}_k]_{21}\{[\mathbf{w}_{-k}]^*_{21}\}^{-1}. \tag{7.65}$$

Thus, for any sign of k, at large k the matrix $\{[\mathbf{W}_k]_{21}\}^{-1}$ is exponentially decaying,

$$\{[\mathbf{W}_k]_{21}\}^{-1} \sim \exp[-|k| d \max \, \mathrm{Im}\, p_\gamma]. \tag{7.66}$$

The convergence of the integral in eq. (7.54) is determined by the function

$$\begin{aligned} &\mathbf{V}^{(1)\pm}_\alpha \cdot \left\{[\mathbf{W}^\pm_k]_{21}\right\}^{-1} \mathbf{V}^{(2)\pm}_\beta \exp\left(ik[p_\alpha(y - y^{(1)}) + p_\beta(y^{(2)} - y')]\right) - \\ &\qquad -\delta_{\alpha\beta}\mathrm{H}(y - y') \exp[ikp_\alpha(y - y')]. \end{aligned} \tag{7.67}$$

With the rule (7.57) and the estimate (7.66), it is evident that only the terms corresponding to $\alpha, \beta = 1, \ldots, 4$ at $k \to -\infty$ and to $\alpha, \beta = 5, \ldots, 8$ at $k \to \infty$ should be checked in (7.67). Any other combinations of α, β in (7.67) provide a guaranteed exponential decay at $|k| \to \infty$. Let us prove that exponentially large terms for "dangerous" numbers α, β in (7.67) compensate each other at $|k| \to \infty$. For the proof we have to find the asymptotic form of the matrix $m_{\alpha\beta}(k)$. Substituting the approximate form (7.64) into eq. (7.56) with making use of definition (7.51), (7.52) and representation (7.60), one obtains for the region $k \to -\infty$:

$$\begin{aligned} m_{\alpha\beta}(k) &= \left[\left(\mathbf{V}^{(1)+}\right)^t \left\{\left[\mathbf{W}^+_k\right]_{21}\right\}^{-1} \mathbf{V}^{(2)+}\right]_{\alpha\beta} \exp\left(ikdp_\beta\right) \approx \\ &\approx \delta_{\alpha\beta} - \left\{\left(\mathbf{V}^{(2)+}\right)^{-1} \left[\mathbf{w}^+_{-k}\right]^*_{21} \left[\left(\mathbf{V}^{(1)+}\right)^t\right]^{-1}\right\}_{\alpha\beta} \exp\left(-ikdp_\beta\right) \end{aligned} \tag{7.68}$$

where $\alpha, \beta = 1, \ldots, 4$. And quite similarly, for the range $k \to \infty$ and the numbers $\alpha, \beta = 5, \ldots, 8$:

$$m_{\alpha\beta}(k) \approx \delta_{\alpha\beta} - \left\{\left(\mathbf{V}^{(2)-}\right)^{-1} \left[\mathbf{w}^-_k\right]_{21} \left[\left(\mathbf{V}^{(1)-}\right)^t\right]^{-1}\right\}_{\alpha\beta} \exp\left(-ikdp_\beta\right). \tag{7.69}$$

Now it can be easily verified that the expected compensation of "bad" terms indeed occurs and the integrand in (7.54) with (7.68), (7.69) takes quite a safe convergent form. Note in addition that due to our initial choice of the integration path along $k = \mathrm{Re}k - i\varepsilon$ (see eq. (3.141) and Fig. 3.4) the pole at $k = 0$ does not disturb the convergence as well.

(c) Solutions for physical quantities

The found solution, eq. (7.54), determines the four "potential" functions $\boldsymbol{\eta}(x,y) = [\mathbf{u}(x,y), \varphi(x,y), \mathbf{J}(x,y), D(x,y)]$, which allow us to find the corresponding physical quantities: the elastic distortion tensor $\boldsymbol{\beta}(x,y)$, eq. (7.3), the electric field $\mathbf{E}(x,y)$, eq. (7.4), the stress tensor $\boldsymbol{\sigma}(x,y)$ and the electric displacement $\mathbf{D} = \mathbf{D}' + \mathbf{D}^0$, eqs. (7.11)-(7.13):

$$\begin{aligned} \begin{pmatrix} \boldsymbol{\beta} \\ -\mathbf{E} \end{pmatrix} &= \begin{pmatrix} \dfrac{\partial \mathbf{u}}{\partial x} \otimes \mathbf{m} + \dfrac{\partial \mathbf{u}}{\partial y} \otimes \mathbf{n} \\ \dfrac{\partial \varphi}{\partial x}\mathbf{m} + \dfrac{\partial \varphi}{\partial y}\mathbf{n} \end{pmatrix} - \begin{pmatrix} \boldsymbol{\beta}^0 \\ \mathbf{E}^0 \end{pmatrix}, \\ \begin{pmatrix} \boldsymbol{\sigma} \\ \mathbf{D} \end{pmatrix} &= \begin{pmatrix} \dfrac{\partial \mathbf{J}}{\partial y} \otimes \mathbf{m} - \dfrac{\partial \mathbf{J}}{\partial x} \otimes \mathbf{n} \\ \dfrac{\partial D}{\partial y}\mathbf{m} - \dfrac{\partial D}{\partial x}\mathbf{n} \end{pmatrix} - \begin{pmatrix} \boldsymbol{\sigma}^0 \\ \mathbf{D}^0 \end{pmatrix}, \end{aligned} \tag{7.70}$$

where the quantities $\boldsymbol{\beta}^0, \mathbf{E}^0, \boldsymbol{\sigma}^0$ and $\mathbf{D}^0$ are defined by eqs. (7.5) and (7.14). In further calculations it is convenient to choose an observation point beyond the cut-surface, i.e. to suppose that $y \neq y'$. Then putting in (7.70) $\boldsymbol{\beta}^0 = \mathbf{E}^0 = \boldsymbol{\sigma}^0 = \mathbf{D}^0 = 0$ and combining (7.70) with (7.54) one obtains:

$$\begin{pmatrix} \boldsymbol{\beta} \\ \mathbf{E} \end{pmatrix} = \sum_{\alpha,\beta=1}^{8} \begin{pmatrix} \mathbf{A}_\alpha \\ -\varphi_\alpha \end{pmatrix} \otimes (\mathbf{m} + \mathrm{p}_\alpha \mathbf{n}) \left(\boldsymbol{\xi}_\beta \cdot \mathbf{Tg}\right) I_{\alpha\beta}(x,y|x',y'), \tag{7.71}$$

$$\begin{pmatrix} \boldsymbol{\sigma} \\ \mathbf{D} \end{pmatrix} = \sum_{\alpha,\beta=1}^{8} \begin{pmatrix} \mathbf{L}_\alpha \\ D_\alpha \end{pmatrix} \otimes (p_\alpha \mathbf{m} - \mathbf{n}) \left(\boldsymbol{\xi}_\beta \cdot \mathbf{Tg}\right) I_{\alpha\beta}(x,y|x',y'), \tag{7.72}$$

with

$$\begin{aligned} &I_{\alpha\beta}(x,y|x',y') = \\ &\frac{1}{2\pi}\int_{-\infty}^{\infty} dk \exp\left\{ik[x - x' + p_\alpha(y - y^{(1)}) - p_\beta(y' - y^{(1)})]\right\} \times \\ &\times \left[m_{\alpha\beta}(k) - \delta_{\alpha\beta}\mathrm{H}(y - y')\right] \end{aligned} \tag{7.73}$$

This integral differs from the Fourier integral in (7.54) only by the absence of k in the denominator. So the convergence of $I_{\alpha\beta}$ is guaranteed by the above consideration. The convolution $\boldsymbol{\xi}_\beta \cdot \mathbf{Tg}$ in (7.71), (7.72) is determined by the strength (7.9) of the four supposed line sources:

$$\boldsymbol{\xi}_\beta \cdot \mathbf{Tg} = \mathbf{L}_\beta \cdot \mathbf{b} + D_\beta \Delta\varphi - \mathbf{A}_\beta \cdot \mathbf{f} + \varphi_\beta q. \tag{7.74}$$

Of course, all these sources are independent and anyone of them can vanish. For instance, if a dislocation is the only source, one should put into (7.71), (7.72) $\boldsymbol{\xi}_\beta \cdot \mathbf{Tg} = \mathbf{L}_\beta \cdot \mathbf{b}$.

We consider solutions (7.71), (7.72) as Green's functions because they allow finding the electro-elastic fields excited by any 2D distributions of the above sources. For example, the electric field $\mathbf{E}(x, y)$, produced in the plate by the distribution of the line forces $\mathbf{f}(x, y)$, is given by

$$\mathbf{E}(x,y) = \sum_{\alpha,\beta=1}^{8} \varphi_\alpha(\mathbf{m} + p_\alpha \mathbf{n}) \int_{-\infty}^{\infty} dx' \int_{y^{(1)}}^{y^{(2)}} dy' \left(\mathbf{A}_\beta \cdot \mathbf{f}(x', y')\right) I_{\alpha\beta}(x, y|x', y'). \tag{7.75}$$

The direct calculation shows that the function $m_{\alpha\beta}(k)$ (7.51), (7.52) has a pole at $k = 0$, with the following limiting behavior from both sides:

$$\lim_{k \to -0} \left[k m_{\alpha\beta}^{(+)}(k) \right] = \lim_{k \to +0} \left[k m_{\alpha\beta}^{(-)}(k) \right] = m_{\alpha\beta}^{(0)}/id, \tag{7.76}$$

where the notation is introduced

$$m_{\alpha\beta}^{(0)} = \mathbf{L}_\alpha \cdot \left\{ \sum_{\gamma=1}^{8} p_\gamma \mathbf{L}_\gamma \otimes \mathbf{L}_\gamma \right\}^{-1} \mathbf{L}_\beta. \tag{7.77}$$

Thus, at $kd \ll 1$ in the main approximation one has:

$$m_{\alpha\beta}(k) \approx m_{\alpha\beta}^{(0)}/ikd. \tag{7.78}$$

The obtained result enables us to find the asymptotic value of the integral $I_{\alpha\beta}$ (7.73) at $|x - x'| \to \infty$. Indeed, in this limit due to the oscillatory term $\exp[ik(x - x')]$ and the pole in (7.78), the main contribution to the integral is given by very small k ($kd \to 0$). Substituting (7.78) into (7.73) we obtain the integral similar to the Fourier image of the Heaviside function,

$$H(x - x') = \frac{1}{2\pi i} \int_{-\infty}^{\infty} \frac{dk}{k} \exp\left[ik(x - x')\right], \tag{7.79}$$

eq. (3.140). The evaluation of the integral can be done also quite similarly by the deformation of the integration path (3.141) into one of the two contours shown in Fig. 3.4 (see the considerations after eq. (3.141)). The result is

$$I_{\alpha\beta} = (m_{\alpha\beta}^{(0)}/d) H(x - x'). \tag{7.80}$$

With (7.80) one can find the physical fields (7.71), (7.72) at infinity $|x - x'| \to \infty$. In view of (7.74) and (7.19)-(7.21), the only source which can provide non-vanishing fields at infinity, is the line of the force $\mathbf{f}$:

$$\sum_{\beta=1}^{8} m_{\alpha\beta}^{(0)} (\boldsymbol{\xi}_\beta \cdot \mathbf{Tg}) = -\mathbf{L}_\alpha \cdot \left\{ \sum_{\gamma=1}^{8} p_\gamma \mathbf{L}_\gamma \otimes \mathbf{L}_\gamma \right\}^{-1} \mathbf{f}/d. \tag{7.81}$$

On the other hand, due to properties (7.19)-(7.21), even with eq. (7.81),

$$\mathbf{E} \cdot \mathbf{m} = \mathbf{D} \cdot \mathbf{n} = \boldsymbol{\sigma}\mathbf{n} = 0. \tag{7.82}$$

However, the elastic distortion $\boldsymbol{\beta}$ and the components $\mathbf{E} \cdot \mathbf{n}, \mathbf{D} \cdot \mathbf{m}$, and $\boldsymbol{\sigma}\mathbf{m}$ do not vanish at $x \to \infty$. For instance,

$$\boldsymbol{\sigma}^{\infty}\mathbf{m} = \mathbf{f}/d, \tag{7.83}$$

$$\mathbf{D}^{\infty} \cdot \mathbf{m} = -\sum_{\alpha=1}^{8} p_\alpha D_\alpha \left\{ \mathbf{L}_\alpha \cdot \left[\sum_{\beta=1}^{8} p_\beta \mathbf{L}_\beta \otimes \mathbf{L}_\beta \right]^{-1} \mathbf{f} \right\} / d \tag{7.84}$$

Such behavior of fields not vanishing at infinity $x \to \infty$ is caused by the applied non-equilibrated force $\mathbf{f}$. Note that such specific features are well known for Green's functions, which are rather mathematical than physical objects. On the other hand, an application of the found Green's function to the plate loaded by the equilibrated force distribution $\mathbf{f}(x, y)$ satisfying the condition

$$\int_{-\infty}^{\infty} dx' \int_{y^{(1)}}^{y^{(2)}} dy' \mathbf{f}(x', y') = \mathbf{0} \tag{7.85}$$

will automatically exclude the above troubles at infinity (see eq. (7.75)).

7.4 Green's functions for plates with other boundary conditions

(a) Green's function for the plate with clamped surfaces

In this section we consider the solution (7.16) for the case when both surfaces of the plate are clamped, i.e. are blocked against any displacements,

$$\mathbf{u}(x, y^{(1)}) = \mathbf{u}(x, y^{(2)}) = \mathbf{0}, \tag{7.86}$$

and the electrical boundary conditions, as before, are determined by the continuity of the normal component of electric displacement D_n. In other words, instead of eqs. (7.47), one should write

$$\mathbf{U}^{\pm}(k, y^{(1)}) = \mathbf{0}, \qquad \mathbf{U}^{\mp}(k, y^{(2)}) = \mathbf{0} \tag{7.87}$$

with the same rule of sign choice as in (7.47). The procedure of obtaining the unknown coefficients b_α in (7.16) is very similar to that in the previous section. So we will not repeat it but just present the final result which has the form of eqs. (7.54), (7.71)-(7.73) if the matrices $m_{\alpha\beta}(k), m^{\pm}_{\alpha\beta}(k)$, given by eqs. (7.51), (7.52), are replaced by $r_{\alpha\beta}(k), r^{\pm}_{\alpha\beta}(k)$, respectively, with

$$r_{\alpha\beta}^{(\pm)}(k) = \mathbf{U}_{\alpha}^{(1)\pm} \cdot \left\{ \left[\mathbf{W}_k^{\pm} \right]_{12} \right\}^{-1} \mathbf{U}_{\beta}^{(2)\pm} \exp\left(ikdp_{\beta} \right). \tag{7.88}$$

Again the limiting value of the $I_{\alpha\beta}$ integral at $|x - x'| \to \infty$ is given by eq. (7.80) after the replacement there of the matrix $m_{\alpha\beta}^{(0)}$ (7.77) by

$$r_{\alpha\beta}^{(0)} = \mathbf{A}_{\alpha} \cdot \left\{ \sum_{\gamma=1}^{8} p_{\gamma} \mathbf{A}_{\gamma} \otimes \mathbf{A}_{\gamma} \right\}^{-1} \mathbf{A}_{\beta}. \tag{7.89}$$

The plate with clamped boundaries is equilibrated by definition. That is why this time the force **f** is not involved in the relation analogous to (7.81). Instead, one has

$$\sum_{\beta=1}^{8} r_{\alpha\beta}^{(0)} (\boldsymbol{\xi}_{\beta} \cdot \mathbf{Tg}) = -\mathbf{A}_{\alpha} \cdot \left\{ \sum_{\gamma=1}^{8} p_{\gamma} \mathbf{A}_{\gamma} \otimes \mathbf{A}_{\gamma} \right\}^{-1} \mathbf{b}/d. \tag{7.90}$$

Thus, in the considered case we again arrive at non-vanishing physical fields at the infinity $x \to \infty$. These fields are: the y-component of the elastic distortion,

$$\boldsymbol{\beta}^{(\infty)} \mathbf{n} = \mathbf{b}/d, \tag{7.91}$$

the x-component of the electric displacement,

$$\mathbf{D}^{\infty} \cdot \mathbf{m} = \sum_{\alpha=1}^{8} p_{\alpha} D_{\alpha} \left\{ \mathbf{A}_{\alpha} \cdot \left[\sum_{\beta=1}^{8} p_{\beta} \mathbf{A}_{\beta} \otimes \mathbf{A}_{\beta} \right]^{-1} \mathbf{b} \right\} / d \tag{7.92}$$

and the two components of the stress field,

$$\boldsymbol{\sigma}^{\infty} \cdot \mathbf{m} = \sum_{\alpha=1}^{8} p_{\alpha} \mathbf{L}_{\alpha} \left\{ \mathbf{A}_{\alpha} \cdot \left[\sum_{\beta=1}^{8} p_{\beta} \mathbf{A}_{\beta} \otimes \mathbf{A}_{\beta} \right]^{-1} \mathbf{b} \right\} / d, \tag{7.93}$$

$$\boldsymbol{\sigma}^{\infty} \cdot \mathbf{n} = -\left[\sum_{\beta=1}^{8} p_{\beta} \mathbf{A}_{\beta} \otimes \mathbf{A}_{\beta} \right]^{-1} \mathbf{b}/d. \tag{7.94}$$

This demonstrates a specific property of dislocation field in the plate with clamped boundaries which was not noticed earlier. Formally, a dislocation is defined by a plastic displacement **b** at the cut surface ending on the dislocation line. With clamped boundaries, the cut surface cannot meet the boundaries where any displacements are forbidden. Therefore in the considered case, the only possible cut surface is a half-plane parallel to the boundaries (see eq. (7.5)). The elastic distortion (7.91) and the other fields non-vanishing at infinity $x \to \infty$ arise as an elastic response to the plastic displacement **b** which must be compensated by the elastic strain $\sim b/d$ at any distance from the dislocation, in order to provide on the boundaries a zero total displacement. On the other hand, one should remember that the order of magnitude of the parameter b/d with $b \sim 10^{-8}$ cm is not negligible only for very thin plates.

(b) The case when one surface is free and the other is clamped

The last combination of boundary conditions, which we are going to consider, is given by

$$\mathbf{U}^{\pm}(k, y^{(1)}) = \mathbf{0}, \qquad \mathbf{V}^{\mp}(k, y^{(2)}) = \mathbf{0}. \tag{7.95}$$

The development similar to that in the previous two sections leads again to the Green function in the form (7.54), (7.71)-(7.73), with $m_{\alpha\beta}(k), m^{\pm}_{\alpha\beta}(k)$ this time replaced by $s_{\alpha\beta}(k), s^{\pm}_{\alpha\beta}(k)$, with

$$s^{(\pm)}_{\alpha\beta}(k) = \mathbf{U}^{(1)\pm}_{\alpha} \cdot \left\{ \left[\mathbf{W}^{\mp}_{k} \right]_{22} \right\}^{-1} \mathbf{V}^{(2)\pm}_{\beta} \exp\left(ikdp_{\beta} \right). \tag{7.96}$$

As before, the final Fourier integral is well convergent. However, in contrast to the previous two cases, the matrix function $s_{\alpha\beta}(k)$ does not contain any pole at $k = 0$, which secures quite an ordinary behavior of physical fields without any specific features at infinity. Indeed, they are definitely not expected for a given boundary problem.

7.5 Conclusions

Thus we have found the Green functions describing the electro-elastic fields created by internal electro-mechanical 2D sources in the piezoelectric plate placed between two dielectric spaces. The problem is solved for the plate surfaces characterized by the three types of mechanical boundary conditions: (i) both surfaces are free; (ii) both surfaces are clamped; and (iii) one surface is free and the other is clamped. The results are expressed in terms of eigenvectors and eigenvalues of the extended Stroh matrix which are supposed to be known. For the general case of arbitrary anisotropy, the obtained Green's functions contain one integration corresponding to a back transition from the Fourier k-space to the real x-space. In accordance with our analysis, these integrals are well convergent being characterized by integrands exponentially decaying at both infinities. As usual, such integrals cannot be calculated analytically for unrestricted anisotropy. Nevertheless, we have managed to study their behavior at a very large distance from the defect as compared with the plate thickness d. The physical fields relating to the boundary problems (i) and (ii) have proved to contain terms not vanishing at infinity. In the first case (i) such behavior is caused by the non-equilibrated load $\mathbf{f}$ along the line (see eq. (7.6)). Of course, this does not lead to any physical troubles. It is evident that application of the corresponding Green's function to any equilibrated problem would automatically provide vanishing fields at infinity. However, in the second case (ii) relating to clamped boundaries, the non-vanishing terms at infinity, as was explained above, represent specific features of the dislocation field under the given conditions. In particular, this makes the dislocations to be quite dangerous sources of abnormally high internal stresses in mesoscopically thin interlayers between the rigid media.

8

Green's Functions for Piezoelectric Strip with Line Surface Sources

In this chapter the formalism developed in Ch. 7 will be used for a determination of electro-elastic fields in a piezoelectric strip with line sources situated at its surfaces. There will be considered lines of forces applied to free surfaces, steps of displacements at clamped surfaces (dislocations at rigid interfaces), lines of electric charges and steps of electric potential discontinuities. In the cases when the surface does not contain any electrical sources it will be supposed to be adjoined to an isotropic dielectric medium. The found Green's functions can be used for determination of coupled electro-elastic fields in the strip excited by any one-dimensional surface distributions of forces, displacements, electric charges or potential. The material of the chapter is based on our publication [29]. Before this paper the appropriate Green's functions were already known [54] for arbitrary anisotropic and even transversely inhomogeneous (graded or stratified) strips but without piezoelectric coupling.

8.1 Basic equations and boundary conditions

(a) Statement of the problem

We shall consider the same infinite piezoelectric plate as in the previous chapter. It has the thickness d and is made of material characterized by the elastic moduli tensor c_{ijkl}, the piezoelectric constants e_{jkl} and the permittivity ε_{kl}. The coordinate system is supposed to be chosen so that its plane xz is parallel to the surfaces of the plate which are specified by the coordinates $y^{(1)}$ and $y^{(2)}$ (Fig. 7.1). The problem is to find 2D distributions of electro-elastic fields in the xy plane excited by different combinations of line sources situated on the surfaces of the plate and parallel to the z-axis.

As before, the considered coupled physical fields will be described in the Stroh-like eight-dimensional formalism with the unknown function $\boldsymbol{\eta}(x, y)$, defined by

$$\boldsymbol{\eta}(x,y)=\begin{pmatrix}\mathbf{U}(x,y)\\ \mathbf{V}(x,y)\end{pmatrix},\quad \mathbf{U}(x,y)=\begin{pmatrix}\mathbf{u}(x,y)\\ \varphi(x,y)\end{pmatrix},\quad \mathbf{V}(x,y)=\begin{pmatrix}\mathbf{J}(x,y)\\ D(x,y)\end{pmatrix},\tag{8.1}$$

where the components of the stress function 4-vector $\mathbf{V}(x,y)$ this time are related to the stress tensor $\boldsymbol{\sigma}(x,y)$ and the electric displacement vector $\mathbf{D}(x,y)$ in the more simple manner

$$\boldsymbol{\sigma}\mathbf{m}=\frac{\partial \mathbf{J}}{\partial y},\quad \boldsymbol{\sigma}\mathbf{n}=-\frac{\partial \mathbf{J}}{\partial x},\quad \mathbf{D}\cdot\mathbf{m}=\frac{\partial D}{\partial y},\quad \mathbf{D}\cdot\mathbf{n}=-\frac{\partial D}{\partial x}.\tag{8.2}$$

The equation of motion in these terms is given by a homogeneous version of (7.8):

$$(\mathbf{I}\partial_y-\mathbf{N}\partial_x)\boldsymbol{\eta}(x,y)=\mathbf{0},\tag{8.3}$$

with 8×8 matrix $\mathbf{N}$ defined by eq. (3.138). The solution is again looked for in the form of the Fourier integral

$$\boldsymbol{\eta}(x,y)=\int\limits_{-\infty}^{\infty} dk\exp(ikx)\boldsymbol{\eta}(k,y),\tag{8.4}$$

where the Fourier amplitude $\boldsymbol{\eta}(k,y)$ satisfies the equation

$$(\mathbf{I}\partial_y-ik\mathbf{N})\boldsymbol{\eta}(k,y)=\mathbf{0},\tag{8.5}$$

which can be immediately solved in terms of eigenvectors $\boldsymbol{\xi}_\alpha$ and eigenvalues p_α of the matrix $\mathbf{N}$ (supposed to be nondegenerate in a nonsemisimple sense [181]):

$$\boldsymbol{\eta}(k,y)=\sum_{\alpha=1}^{8} c_\alpha\boldsymbol{\xi}_\alpha\exp[ikp_\alpha y],\tag{8.6}$$

where the coefficients c_α are supposed to be found from boundary conditions.

(b) Boundary conditions

As we have seen in Ch. 7, the Fourier amplitude

$$\boldsymbol{\eta}(k,y)=[\mathbf{U}(k,y),\mathbf{V}(k,y)]^t=[\mathbf{u}(k,y),\varphi(k,y),\mathbf{J}(k,y),D(k,y)]^t,\tag{8.7}$$

is directly related to the physical quantities involved in boundary conditions. In particular,

$$\mathbf{J}(k,y)=\frac{i}{k}\mathbf{n}\boldsymbol{\sigma}(k,y),\quad D(k,y)=\frac{i}{k}\mathbf{n}\cdot\mathbf{D}(k,y),\tag{8.8}$$

and all four components in the right-hand side of (8.7) can be easily expressed in terms of given surface distributions of displacements, potential, forces and charges. For example, let on the surfaces of the plate $y=y^{(1)}$ and $y=y^{(2)}$ we have, respectively,

$$\mathbf{u}(x, y^{(1)}) = \mathbf{b}^{(1)}\mathrm{H}(x - x^{(1)}), \mathbf{u}(x, y^{(2)}) = \mathbf{b}^{(2)}\mathrm{H}(x - x^{(2)}), \tag{8.9}$$

and/or

$$\varphi(x, y^{(1)}) = \Delta\varphi^{(1)}\mathrm{H}(x - x^{(1)}), \quad \varphi(x, y^{(2)}) = \Delta\varphi^{(2)}\mathrm{H}(x - x^{(2)}), \tag{8.10}$$

and/or

$$\mathbf{n}\boldsymbol{\sigma}(x, y^{(1)}) = -\mathbf{F}^{(1)}\delta(x - x^{(1)}), \quad \mathbf{n}\boldsymbol{\sigma}(x, y^{(2)}) = \mathbf{F}^{(2)}\delta(x - x^{(2)}), \tag{8.11}$$

and/or

$$\mathbf{n} \cdot \mathbf{D}(x, y^{(1)}) = q^{(1)}\delta(x - x^{(1)}), \quad \mathbf{n} \cdot \mathbf{D}(x, y^{(2)}) = -q^{(2)}\delta(x - x^{(2)}). \tag{8.12}$$

After the Fourier transformations in eqs. (8.9)-(8.12) with making use of eq. (8.8) and taking into account the integral representations

$$\begin{aligned} \mathrm{H}(x - x') &= \frac{1}{2\pi i}\int_{-\infty}^{\infty} \frac{dk}{k} \exp[ik(x - x')], \\ \delta(x - x') &= \frac{1}{2\pi}\int_{-\infty}^{\infty} dk \exp[ik(x - x')] \end{aligned} \tag{8.13}$$

the corresponding boundary conditions for the 8-vector $\boldsymbol{\eta}$ at $y = y^{(j)}$ $(j = 1, 2)$ can be summed in the k- and x-space, respectively, in the following combined form:

$$\begin{aligned} \boldsymbol{\eta}(k, y^{(j)}) &= [\mathbf{u}(k, y^{(j)}), \varphi(k, y^{(j)}), \mathbf{J}(k, y^{(j)}), D(k, y^{(j)})]^t \\ &= [\mathbf{b}^{(j)}, \Delta\varphi^{(j)}, \pm\mathbf{F}^{(j)}, \mp q^{(j)}]^t \exp(-ikx^{(j)})/2\pi ik, \end{aligned} \tag{8.14}$$

$$\begin{aligned} \boldsymbol{\eta}(x, y^{(j)}) = [\mathbf{u}(x, y^{(j)}), \varphi(x, y^{(j)}), \mathbf{J}(x, y^{(j)}), D(x, y^{(j)})]^t = \\ [\mathbf{b}^{(j)}, \Delta\varphi^{(j)}, \pm\mathbf{F}^{(j)}, \mp q^{(j)}]^t\mathrm{H}(x - x^{(j)}), \end{aligned} \tag{8.15}$$

where the upper sign corresponds to $j = 1$.

In $(8.13)_1$ and at all further integrations over k for the convergence of corresponding integrals we shall imply that:

$$k = \mathrm{Re}k - i\varepsilon, \quad \varepsilon \to +0, \tag{8.16}$$

as it used to be done in Chs. 3 and 7 (see eq. (3.141) and Fig. 3.4).

Of course, at formulating physically meaningful boundary conditions one should not set on the same surface simultaneously displacements and loads or potential and charges. Unfortunately, this limitation excludes the possibility of derivation of one universal Green's function describing any possible realization of the boundary problem. On the contrary, as we shall see, the number of physically different boundary problems requiring for their solution separate

Green's functions is rather large. This number additionally increases if to take into account the boundary problems related to the situations when the electrical conditions are determined not by electric charge or electric potential distributions on the surface but its adjoining to a dielectric subspace. As was shown in Ch. 7, in the latter case the electrical boundary condition is given by

$$D(k, y^{(j)}) \pm i\varepsilon^{(j)}\varphi(k, y^{(j)}) \text{ sgn } (k) = 0. \tag{8.17}$$

Thus we shall distinguish boundary problems setting on each of the surfaces one of the blocks:

$$[\mathbf{u}(k, y^{(j)}), \varphi(k, y^{(j)})]^t = [\mathbf{b}^{(j)}, \Delta\varphi^{(j)}]^t \exp(-ikx^{(j)})/2\pi ik, \tag{8.18}$$

$$[\mathbf{u}(k, y^{(j)}), D(k, y^{(j)})]^t = [\mathbf{b}^{(j)}, \mp q^{(j)}]^t \exp(-ikx^{(j)})/2\pi ik, \tag{8.19}$$

$$\begin{gathered}[\mathbf{u}(k, y^{(j)}), \varphi(k, y^{(j)}) \pm \text{ sgn } (k) D(k, y^{(j)})/i\varepsilon^{(j)}]^t \\ = [\mathbf{b}^{(j)}, 0]^t \exp(-ikx^{(j)})/2\pi ik,\end{gathered} \tag{8.20}$$

$$[\mathbf{J}(k, y^{(j)}), D(k, y^{(j)})]^t = [\pm\mathbf{F}^{(j)}, \mp q^{(j)}]^t \exp(-ikx^{(j)})/2\pi ik, \tag{8.21}$$

$$[\mathbf{J}(k, y^{(j)}), \varphi(k, y^{(j)})]^t = [\pm\mathbf{F}^{(j)}, \Delta\varphi^{(j)}]^t \exp(-ikx^{(j)})/2\pi ik, \tag{8.22}$$

$$\begin{gathered}[\mathbf{J}(k, y^{(j)}), D(k, y^{(j)}) \pm \text{ sgn } (k) i\varepsilon^{(j)}\varphi(k, y^{(j)})]^t = \\ = [\pm\mathbf{F}^{(j)}, 0]^t \exp(-ikx^{(j)})/2\pi ik,\end{gathered} \tag{8.23}$$

where again the upper sign corresponds to $j = 1$.

The block form (8.1) of initially introduced unknown 8-vector $\boldsymbol{\eta}$ is directly applicable for solving the boundary problems (8.18) and (8.21). For the boundary problems (8.19), (8.22) the F-representation of the unknown 8-vector $\boldsymbol{\eta}_F$ introduced in Ch. 7 appears to be more adequate. It is obtained by a simple rearrangement of the order of components in the initial 8-vector $\boldsymbol{\eta} : \varphi \leftrightarrow D$. As a result, the Fourier amplitude $\boldsymbol{\eta}_F(k, y)$ is decomposed as

$$\begin{gathered}\boldsymbol{\eta}_F(k, y) = [\mathbf{U}_F(k, y), \mathbf{V}_F(k, y)]^t, \quad \mathbf{U}_F(k, y) = [\mathbf{u}(k, y), D(k, y)]^t, \\ \mathbf{V}_F(k, y) = [\mathbf{J}(k, y), \varphi(k, y)]^t.\end{gathered} \tag{8.24}$$

Finally for the boundary problems (8.20) and (8.23) it is convenient to introduce the so-called boundary representations:

$$\begin{gathered}\mathbf{U}^{\pm}(k, y^{(j)}) = [\mathbf{u}(k, y^{(j)}), \varphi(k, y^{(j)}) \pm D(k, y^{(j)})/i\varepsilon^{(j)}]^t, \\ \mathbf{V}^{\pm}(k, y^{(j)}) = [\mathbf{J}(k, y^{(j)}), D(k, y^{(j)}) \pm i\varepsilon^{(j)}\varphi(k, y^{(j)})]^t;\end{gathered} \tag{8.25}$$

and

$$\begin{gathered}\mathbf{U}_F^{\pm}(k, y^{(j)}) = [\mathbf{u}(k, y^{(j)}), D(k, y^{(j)}) \pm i\varepsilon^{(j)}\varphi(k, y^{(j)})]^t, \\ \mathbf{V}_F^{\pm}(k, y^{(j)}) = [\mathbf{J}(k, y^{(j)}), \varphi(k, y^{(j)}) \pm D(k, y^{(j)})/i\varepsilon^{(j)}]^t,\end{gathered} \tag{8.26}$$

where the choice of sign is determined by the rule

$$\pm = (-1)^{j+1} \operatorname{sgn}(k). \tag{8.27}$$

It is evident that for non-vanishing and finite permittivities $\varepsilon^{(1)}$ and/or $\varepsilon^{(2)}$ the representations (8.25) and (8.26) are absolutely equivalent. However only the representation (8.25) allows one to describe the boundary problem at $y = y^{(j)}$ relating to a metallized surface ($\varepsilon^{(j)} \to \infty$) with prescribed displacements, or the condition of a vanishing electric charge density ($\varepsilon^{(j)} \to 0$) on a surface with prescribed loads. And only the representation (8.26) admits the contrary limits for $\varepsilon^{(j)}$. On the other hand, in accordance with eq. (13),

$$\begin{aligned} \mathbf{U}(k, y^{(j)}) &= \sum_{\alpha=1}^{8} c_\alpha \mathbf{U}_\alpha \exp\left(ikp_\alpha y^{(j)}\right), \\ \mathbf{V}(k, y^{(j)}) &= \sum_{\alpha=1}^{8} c_\alpha \mathbf{V}_\alpha \exp\left(ikp_\alpha y^{(j)}\right); \end{aligned} \tag{8.28}$$

$$\begin{aligned} \mathbf{U}_F(k, y^{(j)}) &= \sum_{\alpha=1}^{8} c_\alpha \mathbf{U}_{F\alpha} \exp\left(ikp_\alpha y^{(j)}\right), \\ \mathbf{V}_F(k, y^{(j)}) &= \sum_{\alpha=1}^{8} c_\alpha \mathbf{V}_{F\alpha} \exp\left(ikp_\alpha y^{(j)}\right); \end{aligned} \tag{8.29}$$

$$\begin{aligned} \mathbf{U}^{\pm}(k, y^{(j)}) &= \sum_{\alpha=1}^{8} c_\alpha \mathbf{U}_\alpha^{(j)} \pm \exp\left(ikp_\alpha y^{(j)}\right), \\ \mathbf{V}^{\pm}(k, y^{(j)}) &= \sum_{\alpha=1}^{8} c_\alpha \mathbf{V}_\alpha^{(j)} \pm \exp\left(ikp_\alpha y^{(j)}\right); \end{aligned} \tag{8.30}$$

$$\begin{aligned} \mathbf{U}_F^{\pm}(k, y^{(j)}) &= \sum_{\alpha=1}^{8} c_\alpha \mathbf{U}_{F\alpha}^{(j)} \pm \exp\left(ikp_\alpha y^{(j)}\right), \\ \mathbf{V}_F^{\pm}(k, y^{(j)}) &= \sum_{\alpha=1}^{8} c_\alpha \mathbf{V}_{F\alpha}^{(j)\pm} \exp\left(ikp_\alpha y^{(j)}\right). \end{aligned} \tag{8.31}$$

Here the block 4-vectors $\mathbf{U}_\alpha$, $\mathbf{V}_\alpha$, $\mathbf{U}_{F\alpha}$, $\mathbf{V}_{F\alpha}$, $\mathbf{U}_\alpha^{(j)\pm}$, $\mathbf{V}_\alpha^{(j)\pm}$, $\mathbf{U}_{F\alpha}^{(j)\pm}$ and $\mathbf{V}_{F\alpha}^{(j)\pm}$ are defined by eqs. (7.17), (7.22)-(7.24).

8.2 Strips with prescribed loads at both surfaces

In this section we shall derive the Green's functions related to six boundary problems united by the same mechanical conditions at the surfaces - the prescribed loads, eqs. (8.11). There will be considered a series of non-equivalent combinations of boundary conditions given by eqs. (8.21)-(8.23) and schematically shown in Fig. 8.1. The loads will be supposed to be equilibrated, i.e. $\mathbf{F}^{(1)} + \mathbf{F}^{(2)} = \mathbf{0}$.

(a) Surfaces adjoined to dielectric media

Let us start from the case of an absence of any surface electrical sources. The strip is supposed to be situated between two isotropic dielectric media with permittivities $\varepsilon^{(1)}$ (at $y < y^{(1)}$) and $\varepsilon^{(2)}$ (at $y > y^{(2)}$) (Fig. 8.1(*a*)). The corresponding boundary conditions are obtained by combining the relations (8.23), $(8.25)_2$ and $(8.30)_2$ taken at both $j = 1$ and $j = 2$. With the rule (8.27) at $k > 0$ one has

$$\sum_{\alpha=1}^{8} c_\alpha \mathbf{V}_\alpha^{(1)+} \exp\left(ikp_\alpha y^{(1)}\right) = \left[\mathbf{F}^{(1)},\ 0\right]^t \exp\left[-ikx^{(1)}\right]/2\pi ik \tag{8.32}$$

$$\sum_{\alpha=1}^{8} c_\alpha \mathbf{V}_\alpha^{(2)-} \exp\left(ikp_\alpha y^{(2)}\right) = -\left[\mathbf{F}^{(2)},\ 0\right]^t \exp\left[-ikx^{(2)}\right]/2\pi ik \tag{8.33}$$

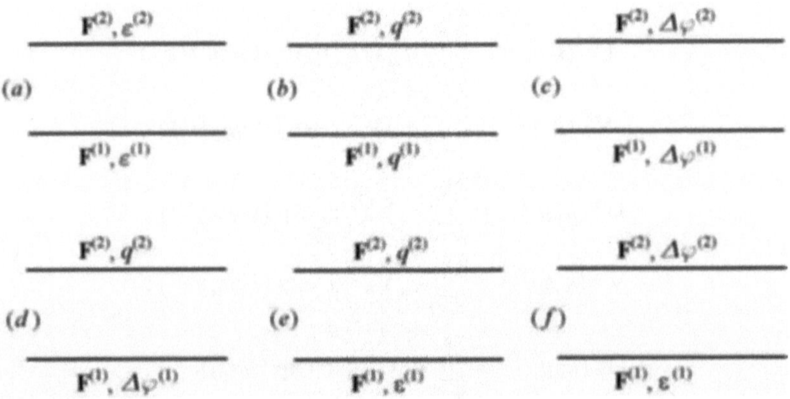

Fig. 8.1. Six non-equivalent boundary problems for a piezoelectric strip loaded at both surfaces.

It is convenient to reduce finding the eight unknown coefficients c_α in the system (8.32), (8.33) to the looking for the two unknown 4-vectors $\mathbf{c}^{(1)}$ and $\mathbf{c}^{(2)}$ by means of the substitution

$$\begin{aligned} c_\alpha = &\Big\{\mathbf{c}^{(1)} \cdot \mathbf{V}_\alpha^{(1)-} \exp[-ik(x^{(2)} + p_\alpha y^{(1)})] \\ &+\mathbf{c}^{(2)} \cdot \mathbf{V}_\alpha^{(2)+} \exp[-ik(x^{(1)} + p_\alpha y^{(2)})]\Big\}/2\pi ik. \end{aligned} \tag{8.34}$$

Putting (8.34) into eqs.(8.32) and (8.33) with making use of the relation $(7.25)_2$ one obtains

$$\sum_{\gamma=1}^{8} \mathbf{V}_\gamma^{(2)-} \otimes \mathbf{V}_\gamma^{(1)-} \exp(ikp_\gamma d)\, \mathbf{c}^{(1)} = -\boldsymbol{\Phi}_0^{(2)}, \tag{8.35}$$

$$\sum_{\gamma=1}^{8} \mathbf{V}_{\gamma}^{(1)+} \otimes \mathbf{V}_{\gamma}^{(2)+} \exp\left(-ikp_{\gamma}d\right) \mathbf{c}^{(2)} = \boldsymbol{\Phi}_0^{(1)}, \tag{8.36}$$

with

$$\boldsymbol{\Phi}_0^{(j)} = [\mathbf{F}^{(j)}, 0]^t. \tag{8.37}$$

After a straightforward exclusion of $\mathbf{c}^{(1)}$ and $\mathbf{c}^{(2)}$ from eqs. (8.34)-(8.36) it is easy to find

$$\begin{aligned} c_{\alpha}(k>0) = \Big\{ & a_{\alpha}^{(21)+}(k) \exp\left[-ik\left(x^{(1)} + p_{\alpha}y^{(1)}\right)\right] - \\ & - a_{\alpha}^{(12)-}(k) \exp\left[-ik\left(x^{(2)} + p_{\alpha}y^{(2)}\right)\right] \Big\} / 2\pi ik, \end{aligned} \tag{8.38}$$

$$\begin{aligned} c_{\alpha}(k<0) = \Big\{ & a_{\alpha}^{(21)-}(k) \exp\left[-ik\left(x^{(1)} + p_{\alpha}y^{(1)}\right)\right] - \\ & - a_{\alpha}^{(12)+}(k) \exp\left[-ik\left(x^{(2)} + p_{\alpha}y^{(2)}\right)\right] \Big\} / 2\pi ik, \end{aligned} \tag{8.39}$$

where

$$a_{\alpha}^{(ij)\pm}(k) = \mathbf{V}_{\alpha}^{(i)\pm} \cdot \left\{ \sum_{\gamma=1}^{8} \mathbf{V}_{\gamma}^{(j)\pm} \otimes \mathbf{V}_{\gamma}^{(i)\pm} \exp\left[ik(p_{\gamma} - p_{\alpha})(y^{(j)} - y^{(i)})\right] \right\}^{-1} \boldsymbol{\Phi}_0^{(j)}. \tag{8.40}$$

With the known coefficients $c_{\alpha}(k)$ the resulting Green's function is completely determined by eqs. (8.4) and (8.6):

$$\begin{aligned} \boldsymbol{\eta}(x,y) = \frac{1}{2\pi i} \sum_{\alpha=1}^{8} \int_{-\infty}^{\infty} \frac{dk}{k} \boldsymbol{\xi}_{\alpha} \Big(& a_{\alpha}^{(21)}(k) \exp\left\{ik[x - x^{(1)} + p_{\alpha}(y - y^{(1)})]\right\} - \\ & - a_{\alpha}^{(12)}(k) \exp\left\{ik[x - x^{(2)} + p_{\alpha}(y - y^{(2)})]\right\} \Big), \end{aligned} \tag{8.41}$$

where

$$a_{\alpha}^{(21)}(k) = \begin{cases} a_{\alpha}^{(21)-}(k), & k<0, \\ a_{\alpha}^{(21)+}(k), & k>0; \end{cases} \quad a_{\alpha}^{(12)}(k) = \begin{cases} a_{\alpha}^{(12)+}(k), & k<0, \\ a_{\alpha}^{(12)-}(k), & k>0. \end{cases} \tag{8.42}$$

As was mentioned above, the used representation provides a complete description of electro-elastic fields in the considered plate for any realization of initial boundary problem except of the situations when one or both surfaces are metallized (i.e. when $\varphi(k, y^{(j)}) = 0$ for $j = 1$ or/and $j = 2$ and simultaneously $\varepsilon^{(j)} \to \infty$). In such cases the F-representation (8.24) is more adequate. The transformation of the solution (8.41) to the F-representation is given by the trivial replacing in (8.40) the all $\mathbf{V}_{\alpha,\gamma}^{(j)\pm}$ by $\mathbf{V}_{F\alpha,F\gamma}^{(j)\pm}$ $(j = 1, 2)$, respectively. Now one can go to the limit $\varepsilon^{(j)} \to \infty$ at $j = 1$ or/and $j = 2$ which leads to the respective simplification $\mathbf{V}_{F\alpha,F\gamma}^{(j)\pm} \to \mathbf{V}_{F\alpha,F\gamma}$ in the renormalized eq. (8.40) (of course, if both surfaces are metallized, the superscripts $\pm$ in eq. (8.40) should be omitted and eq. (8.42) becomes an identity).

An analysis of convergency of the integral in eq. (8.41) can be done quite similarly to the consideration in Ch. 7. The conclusion is the same: the integrand is exponentially small at infinities and the convergency of the integral is guaranteed. Moreover, such integrals can be analytically taken when the eigenvalues $p_\alpha(\alpha = 1, ..., 8)$ are purely imaginary. This may occur if the material of the plate is characterized by a symmetry plane parallel to the surfaces or by a 2-fold symmetry axis perpendicular to them (unfortunately, the above condition for a correct calculation of analogous integrals was missed in [54]). Piezoelectric extension of the consideration [54] meets the new trouble: the functions $a_\alpha^{(ij)}(k)$ (8.42) are not analytical on the line $\mathrm{Re}k = 0$ in the complex plane k. However, this problem disappears in the particular cases $\varepsilon^{(j)} \to 0$ and $\varepsilon^{(j)} \to \infty$. Below we consider the latter case of metallized surfaces.

The result of integration is expressed as a sum of residues at $k = 0$ and at the poles k_n determined by equations

$$\begin{aligned} \left| \sum_{\gamma=1}^{8} \mathbf{V}_{F\gamma} \otimes \mathbf{V}_{F\gamma} \exp\left(-ikp_\gamma d\right) \right| &= 0, \\ \left| \sum_{\gamma=1}^{8} \mathbf{V}_{F\gamma} \otimes \mathbf{V}_{F\gamma} \exp\left(ikp_\gamma d\right) \right| &= 0. \end{aligned} \tag{8.43}$$

It is easily checked that these equations determine the same set of complex roots. Taking into account the known fact that the eigenvectors $\boldsymbol{\xi}_{F\gamma}$ and the eigenvalues p_γ exist in complex conjugate pairs, see (7.57), it is clear that eqs. (8.43) have the structure

$$F_1(k) + iF_2(k) = 0, \quad F_1(k) - iF_2(k) = 0, \tag{8.44}$$

where F_1 and F_2 are real functions. Therefore i) both equations (8.43) have the same sets of roots, and ii) these roots k_n must form complex conjugate pairs. We shall denote the roots with positive imaginary parts as k_n^+, and their conjugate counterparts as k_n^-.

With this observation the result of integration can be expressed in the form

$$\boldsymbol{\eta}(x, y) = \boldsymbol{\eta}^{(1)}(x, y) - \boldsymbol{\eta}^{(2)}(x, y), \tag{8.45}$$

where

$$\boldsymbol{\eta}^{(j)}(x, y) = \begin{cases} \boldsymbol{\eta}^{(j)+}(x, y) + \boldsymbol{\eta}^{(j)0}(x, y), & x > x^{(j)}, \\ \boldsymbol{\eta}^{(j)-}(x, y), & x < x^{(j)}, \end{cases} \tag{8.46}$$

with

$$\boldsymbol{\eta}^{(j)0}(x, y) =$$

$$= \mp \sum_{\alpha=1}^{8} [x - x^{(j)} + p_\alpha(y - y^{(i)})]\boldsymbol{\xi}_\alpha \left\{ \mathbf{V}_{F\alpha} \cdot \left[\sum_{\gamma=1}^{8} p_\gamma \mathbf{V}_{F\gamma} \otimes \mathbf{V}_{F\gamma} \right]^{-1} \boldsymbol{\Phi}_0^{(j)} \right\} / d, \tag{8.47}$$

$$\boldsymbol{\eta}^{(j)\pm}(x,y) =$$

$$= \pm \sum_{\alpha=1}^{8} \sum_{k_n^{\pm}} \boldsymbol{\xi}_\alpha A_{a\alpha}^{(j)}(k_n^{\pm}) \exp\left\{ik_n^{\pm}[x - x^{(j)} + p_\alpha(y - y^{(i)})]\right\}, \tag{8.48}$$

$$A_{a\alpha}^{(j)}(k) =$$

$$= \mathbf{V}_{F\alpha} \cdot \left\{\mathrm{Adj}\left[\sum_{\gamma=1}^{8} \mathbf{V}_{F\gamma} \otimes \mathbf{V}_{F\gamma} \exp\left(\mp ikp_\gamma d\right)\right]\right\} \boldsymbol{\Phi}_0^{(j)} / G_{VV}^{(aj)}(k), \tag{8.49}$$

$$G_{VV}^{(aj)}(k) = k\frac{\partial}{\partial k}\left|\sum_{\gamma=1}^{8} \mathbf{V}_{F\gamma} \otimes \mathbf{V}_{F\gamma} \exp\left(\mp ikp_\gamma d\right)\right|. \tag{8.50}$$

In eqs. (8.46)-(8.50) $j = 1, 2$, $i = 3 - j$; in (8.47), (8.49) and (8.50) the upper sign corresponds to $j = 1$. It is easily verified that the solution (8.45)-(8.48) satisfies both eq. (8.3) and the boundary conditions

$$[\mathbf{J}(x, y^{(j)}), \varphi(x, y^{(j)})]^t = [\pm\mathbf{F}^{(j)}, 0]^t \mathrm{H}(x - x^{(j)}), \tag{8.51}$$

where again the upper sign corresponds to $j = 1$.

(b) Prescribed electric charge distributions at both surfaces

In this subsection we are going to derive the Green's function describing the coupled fields in the strip shown in Fig. 8.1(*b*). Now the boundary conditions look:

$$\begin{aligned} [\mathbf{J}(x, y^{(1)}), D(x, y^{(1)})]^t &= [\mathbf{F}^{(1)}, -q^{(1)}]^t H(x - x^{(1)}), \\ [\mathbf{J}(x, y^{(2)}), D(x, y^{(2)})]^t &= -[\mathbf{F}^{(2)}, -q^{(2)}]^t H(x - x^{(2)}). \end{aligned} \tag{8.52}$$

The corresponding system for $(c)_\alpha$ replacing (8.32), (8.33) has the form:

$$\sum_{\alpha=1}^{8} c_\alpha \mathbf{V}_\alpha \exp[ikp_\alpha y^{(j)}] = \pm\boldsymbol{\Phi}_q^{(j)} \exp[-ikx^{(j)}]/2\pi ik, \quad j = 1, 2, \tag{8.53}$$

where the upper sign corresponds to $j = 1$, as before, and

$$\boldsymbol{\Phi}_q^{(j)} = [\mathbf{F}^{(j)}, -q^{(j)}]^t. \tag{8.54}$$

The procedure of solving this system as well as the further development is completely analogous to our previous considerations. Therefore we shall not repeat lengthy calculations and just present the final results. The Green's function has the form:

$$\begin{aligned} \boldsymbol{\eta}(x,y) = \frac{1}{2\pi i}\sum_{\alpha=1}^{8} \int_{-\infty}^{\infty} \frac{dk}{k} \boldsymbol{\xi}_\alpha \Big(& a_{b\alpha}^{(1)}(k) \exp\left\{ik[x - x^{(1)} + p_\alpha(y - y^{(1)})]\right\} \\ & - a_{b\alpha}^{(2)}(k) \exp\left\{ik[x - x^{(2)} + p_\alpha(y - y^{(2)})]\right\}\Big), \end{aligned} \tag{8.55}$$

where

$$a_{b\alpha}^{(1,2)}(k) = \mathbf{V}_\alpha \cdot \left\{ \sum_{\gamma=1}^{8} \mathbf{V}_\gamma \otimes \mathbf{V}_\gamma \exp[\mp ikd(p_\gamma - p_\alpha)] \right\}^{-1} \boldsymbol{\Phi}_q^{(1,2)} \quad (8.56)$$

and the subscript b relates our notation to the problem (b).

The Fourier integral in eq. (8.55) again can be analytically taken in the particular cases when, due to a symmetry, the eigenvalues p_α prove to be purely imaginary. The result of integration has the same form eqs. (8.45), (8.46) where now

$$\boldsymbol{\eta}^{(j)0}(x,y) =$$

$$= \mp \sum_{\alpha=1}^{8} [x - x^{(j)} + p_\alpha(y - y^{(i)})] \boldsymbol{\xi}_\alpha \left\{ \mathbf{V}_\alpha \cdot \left[\sum_{\gamma=1}^{8} p_\gamma \mathbf{V}_\gamma \otimes \mathbf{V}_\gamma \right]^{-1} \boldsymbol{\Phi}_q^{(j)} \right\} /d, \quad (8.57)$$

$$\boldsymbol{\eta}^{(j)\pm}(x,y) = \pm \sum_{\alpha=1}^{8} \sum_{k_n^\pm} \boldsymbol{\xi}_\alpha A_{b\alpha}^{(j)}(k_n^\pm) \exp\left\{ ik_n^\pm [x - x^{(j)} + p_\alpha(y - y^{(i)})] \right\} \quad (8.58)$$

$$A_{b\alpha}^{(j)}(k) = \mathbf{V}_\alpha \cdot \left\{ \mathrm{Adj} \left[\sum_{\gamma=1}^{8} \mathbf{V}_\gamma \otimes \mathbf{V}_\gamma \exp(\mp ikp_\gamma d) \right] \right\} \boldsymbol{\Phi}_q^{(j)} / G_{VV}^{(bj)}(k) \quad (8.59)$$

$$G_{VV}^{(bj)}(k) = k \frac{\partial}{\partial k} \left| \sum_{\gamma=1}^{8} \mathbf{V}_\gamma \otimes \mathbf{V}_\gamma \exp(\mp ikp_\gamma d) \right| . \quad (8.60)$$

and $k_n^\pm$ are the roots of the equation

$$\left| \sum_{\gamma=1}^{8} \mathbf{V}_\gamma \otimes \mathbf{V}_\gamma \exp(ikp_\gamma d) \right| = 0 \quad (8.61)$$

with positive (k_n^+) and negative (k_n^-) imaginary parts $(\mathrm{Im}k_n^- = -\mathrm{Im}k_n^+)$. As before, in eqs. (8.57) and (8.58) $j = 1,2$, $i = 3 - $j; and in (8.57), (8.59) and (8.60) the upper sign relates to $j = 1$.

(c) Prescribed electric potential distributions at both surfaces

The boundary problem is described by the conditions (Fig. 8.1(c))

$$\begin{aligned} [\mathbf{J}(x,y^{(1)}), \varphi(x,y^{(1)})]^t &= [\mathbf{F}^{(1)}, \Delta\varphi^{(1)}]^t \mathrm{H}(x - x^{(1)}), \\ [\mathbf{J}(x,y^{(2)}), \varphi(x,y^{(2)})]^t &= [-\mathbf{F}^{(2)}, \Delta\varphi^{(2)}]^t \mathrm{H}(x - x^{(2)}), \end{aligned} \quad (8.62)$$

which are transformed into a system coinciding with (8.53) after the replacements in the latter: $\mathbf{V}_\alpha \to \mathbf{V}_{F\alpha}$ and $\boldsymbol{\Phi}_q^{(j)} \to \boldsymbol{\Phi}_{\Delta\varphi}^{(j)}$, where

$$\boldsymbol{\Phi}_{\Delta\varphi}^{(1)} = [\mathbf{F}^{(1)}, \Delta\varphi^{(1)}]^t, \quad \boldsymbol{\Phi}_{\Delta\varphi}^{(2)} = [\mathbf{F}^{(2)}, -\Delta\varphi^{(2)}]^t. \tag{8.63}$$

Accordingly the above results eqs. (8.55)-(8.61) can be easily adapted for solving the problem eq. (8.62): one should just replace everywhere $\mathbf{V}_{\alpha,\gamma}$ by $\mathbf{V}_{F_\alpha,F_\gamma}$ and $\boldsymbol{\Phi}_q^{(j)}$ by $\boldsymbol{\Phi}_{\Delta\varphi}^{(j)}$. It is also supposed that here and in the forthcoming the previous index related to a number of a problem in Fig. 8.1 (a, b, c etc.) is replaced by a new one (in the considered case: $a_{b\alpha}^{(j)} \to a_{c\alpha}^{(j)}$, $A_{b\alpha}^{(j)} \to A_{c\alpha}^{(j)}$).

(d) Prescribed electric charge at one surface and prescribed electric potential at the other

In this case the boundary conditions represent a mixture of the problems (b) and (c), see Fig. 8.1:

$$\begin{aligned} [\mathbf{J}(x, y^{(1)}), \varphi(x, y^{(1)})]^t &= \boldsymbol{\Phi}_{\Delta\varphi}^{(1)} \mathrm{H}(x - x^{(1)}), \\ [\mathbf{J}(x, y^{(2)}), D(x, y^{(2)})]^t &= -\boldsymbol{\Phi}_q^{(2)} \mathrm{H}(x - x^{(2)}). \end{aligned} \tag{8.64}$$

The solution is again given by eq. (8.55), where one should replace $a_{b\alpha}^{(j)}(k)$ by $a_{d\alpha}^{(j)}(k)$ defined as

$$a_{d\alpha}^{(1)}(k) = \mathbf{V}_\alpha \cdot \left\{ \sum_{\gamma=1}^{8} \mathbf{V}_{F\gamma} \otimes V_\gamma \exp[-ikd(p_\gamma - p_\alpha)] \right\}^{-1} \boldsymbol{\Phi}_{\Delta\varphi}^{(1)}, \tag{8.65}$$

$$a_{d\alpha}^{(2)}(k) = \mathbf{V}_{F\alpha} \cdot \left\{ \sum_{\gamma=1}^{8} \mathbf{V}_\gamma \otimes \mathbf{V}_{F\gamma} \exp[ikd(p_\gamma - p_\alpha)] \right\}^{-1} \boldsymbol{\Phi}_q^{(2)}. \tag{8.66}$$

The results for the particular cases, when the eigenvalues p_α are purely imaginary, have the structure (8.45), (8.46) where this time

$$\begin{aligned} \boldsymbol{\eta}^{(1)0}(x, y) = &- \sum_{\alpha=1}^{8} \boldsymbol{\xi}_\alpha (\mathbf{l}_\alpha \cdot \mathbf{F}^{(1)})[x - x^{(1)} + p_\alpha(y - y^{(2)})]/d + \\ &+ [-\mathbf{l}_D, 1, \mathbf{0}, 0]^t (-\mathbf{l}_\varphi \cdot \mathbf{F}^{(1)} + \Delta\varphi^{(1)}), \end{aligned} \tag{8.67}$$

$$\begin{aligned} \boldsymbol{\eta}^{(2)0}(x, y) = &\sum_{\alpha=1}^{8} \boldsymbol{\xi}_\alpha (\mathbf{l}_\alpha \cdot \mathbf{F}^{(2)})[x - x^{(2)} + p_\alpha(y - y^{(1)})]/d + \\ &+ [\mathbf{l}_\varphi, 0, \mathbf{0}, -1]^t (\mathbf{l}_D \cdot \mathbf{F}^{(2)} + q^{(2)}), \end{aligned} \tag{8.68}$$

$$\mathbf{l}_D = \sum_{\beta=1}^{8} p_\beta D_\beta \mathbf{l}_\beta, \quad \mathbf{l}_\varphi = \sum_{\beta=1}^{8} p_\beta \varphi_\beta \mathbf{l}_\beta, \quad \mathbf{l}_\alpha = \mathbf{L}_\alpha \cdot \left[\sum_{\beta=1}^{8} p_\beta \mathbf{L}_\beta \otimes \mathbf{L}_\beta \right]^{-1}, \tag{8.69}$$

$$\boldsymbol{\eta}^{(j)\pm}(x, y) = \pm \sum_{\alpha=1}^{8} \sum_{k_n^\pm} \boldsymbol{\xi}_\alpha A_{d\alpha}^{(j)}(k_n^\pm) \exp\left\{ ik_n^\pm [x - x^{(j)} + p_\alpha(y - y^{(i)})] \right\}, \tag{8.70}$$

$$A_{d\alpha}^{(1)}(k) = \mathbf{V}_\alpha \cdot \left\{ \mathrm{Adj} \left[\sum_{\gamma=1}^{8} \mathbf{V}_{F\gamma} \otimes \mathbf{V}_\gamma \exp\left(-ikp_\gamma d\right) \right] \right\} \boldsymbol{\Phi}_{\Delta\varphi}^{(1)} / G_{VV}^{(d1)}(k), \quad (8.71)$$

$$A_{d\alpha}^{(2)}(k) = \mathbf{V}_{F\alpha} \cdot \left\{ \mathrm{Adj} \left[\sum_{\gamma=1}^{8} \mathbf{V}_\gamma \otimes \mathbf{V}_{F\gamma} \exp\left(ikp_\gamma d\right) \right] \right\} \boldsymbol{\Phi}_{q}^{(2)} / G_{VV}^{(d2)}(k), \quad (8.72)$$

$$G_{VV}^{(d1)}(k) = k \frac{\partial}{\partial k} \left| \sum_{\gamma=1}^{8} \mathbf{V}_{F\gamma} \otimes \mathbf{V}_\gamma \exp\left(-ikp_\gamma d\right) \right|, \quad (8.73)$$

$$G_{VV}^{(d2)}(k) = k \frac{\partial}{\partial k} \left| \sum_{\gamma=1}^{8} \mathbf{V}_\gamma \otimes \mathbf{V}_{F\gamma} \exp\left(ikp_\gamma d\right) \right|. \quad (8.74)$$

and $k_n^{\pm}$ $(\mathrm{Im}k_n^{+} = -\mathrm{Im}k_n^{-} \quad > 0)$ are the roots of the equation

$$\left| \sum_{\gamma=1}^{8} \mathbf{V}_\gamma \otimes \mathbf{V}_{F\gamma} \exp\left(ikp_\gamma d\right) \right| = 0. \quad (8.75)$$

(e) One surface charged and the other adjoined to a dielectric subspace

This case is a mixture of the problems (a) and (b) (Fig. 8.1). The boundary conditions are

$$\begin{gathered} [\mathbf{J}(k,y^{(1)}), D(k,y^{(1)}) + \mathrm{sgn}\,(k) i\varepsilon^{(1)}\varphi(k,y^{(1)})]^t = \boldsymbol{\Phi}_0^{(1)} \exp(-ikx^{(1)})/2\pi ik, \\ [\mathbf{J}(k,y^{(2)}), D(k,y^{(2)})]^t = -\boldsymbol{\Phi}_q^{(2)} \exp(-ikx^{(2)})/2\pi ik, \end{gathered} \quad (8.76)$$

The Green's function solving problem (8.76) is given by the above formulae (8.55), (8.65), (8.66) after the following replacements: in eq. (8.65) $\mathbf{V}_{F\gamma} \to \mathbf{V}_\gamma^{(1)\pm}$ and $\boldsymbol{\Phi}_{\Delta\varphi}^{(1)} \to \boldsymbol{\Phi}_0^{(1)}$; in eq. (8.66) $\mathbf{V}_{\mathbf{F}\alpha,\mathbf{F}\gamma} \to \mathbf{V}_{\alpha,\gamma}^{(1)\mp}$ (in both cases the upper sign corresponds to $k > 0$, and the lower sign to $k < 0$).

(f) One surface with a prescribed electric potential distribution and the other adjoined to a dielectric subspace

The last Green's function of the first series relates to the boundary problem (Fig. 8.1(f))

$$\begin{gathered} [\mathbf{J}(k,y^{(1)}), \varphi(k,y^{(1)})]^t = \boldsymbol{\Phi}_{\Delta\varphi}^{(1)} \exp(-ikx^{(1)})/2\pi ik, \\ [\mathbf{J}(k,y^{(2)}), D(k,y^{(2)}) - \mathrm{sgn}(k) i\varepsilon^{(2)}\varphi(k,y^{(2)})]^t = \\ = -\boldsymbol{\Phi}_0^{(2)} \exp(-ikx^{(2)})/2\pi ik, \end{gathered} \quad (8.77)$$

which is again solved by the same eqs. (8.55), (8.65), (8.66) if to replace $\boldsymbol{\Phi}_q^{(2)}$ by $\boldsymbol{\Phi}_0^{(2)}$, to change in eq. (8.65) $\mathbf{V}_{\alpha,\gamma}$ by $\mathbf{V}_{\alpha,\gamma}^{(2)\pm}$ and in eq. (8.66) $\mathbf{V}_\gamma$ by $\mathbf{V}_\gamma^{(2)\mp}$ (again with the upper sign for $k > 0$ and the lower sign for $k < 0$).

8.3 Strips with prescribed displacements at both surfaces

In this section we shall present the next series of six Green's functions united by the common mechanical conditions (8.9) at both surfaces. Physically these conditions correspond to a plate situated between two rigid substrates with dislocations at the interfaces. Mathematically a solution of such problem is a Green's function allowing a determination of electro-elastic fields in the strip with any distributions of elastic displacements at the surfaces. We again consider six non-equivalent combinations of various electrical conditions at the surfaces (Fig. 8.2).

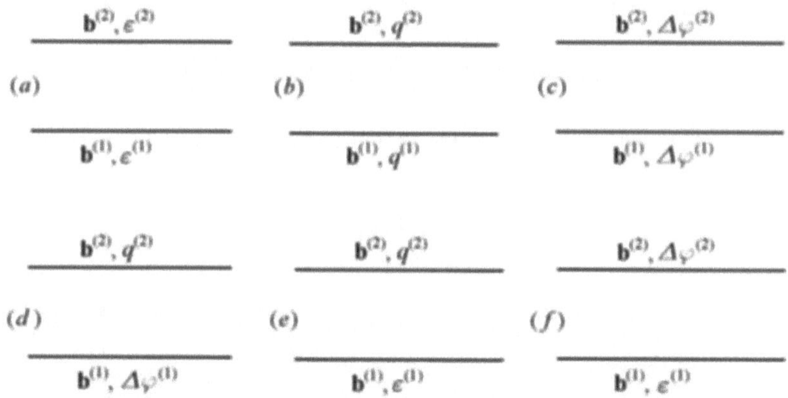

Fig. 8.2. Six non-equivalent boundary problems for a piezoelectric strip with prescribed displacements at both surfaces.

(a) Surfaces adjoined to dielectric media

In this case the boundary problem can be presented in the form of the condition (Fig. 8.2(*a*))

$$\begin{aligned} &[\mathbf{u}(k, y^{(j)}), \varphi(k, y^{(j)})] \pm \operatorname{sgn}(k) D(k, y^{(j)})/i\varepsilon^{(j)}]^t = \\ &= [\mathbf{b}^{(j)}, 0]^t \exp(-ikx^{(j)})/2\pi ik, \end{aligned} \tag{8.78}$$

taken at $j = 1$ (with the upper sign) and $j = 2$. Combining at $k > 0$ eqs. (8.78), $(8.25)_1$ and $(8.30)_1$ one obtains the system with respect to unknown coefficients c_α in eq. (8.6) which is very similar to eqs. (8.32) and (8.33):

$$\sum_{\alpha=1}^{8} c_\alpha \mathbf{U}_\alpha^{(1)+} \exp[ikp_\alpha y^{(1)}] = \mathbf{B}_0^{(1)} \exp[-ikx^{(1)}]/2\pi ik, \tag{8.79}$$

$$\sum_{\alpha=1}^{8} c_\alpha \mathbf{U}_\alpha^{(2)-} \exp[ikp_\alpha y^{(2)}] = \mathbf{B}_0^{(2)} \exp[-ikx^{(2)}]/2\pi ik, \tag{8.80}$$

with

$$\mathbf{B}_0^{(j)} = [\mathbf{b}^{(j)}, 0]^t, \quad j = 1, 2. \tag{8.81}$$

The further procedure is quite analogous to our considerations in Sec. 8.2 (a). The result is

$$\boldsymbol{\eta}(x,y) =$$

$$= \frac{1}{2\pi i} \sum_{\alpha=1}^{8} \int_{-\infty}^{\infty} \frac{dk}{k} \boldsymbol{\xi}_\alpha \left(c_\alpha^{(21)}(k) \exp\left\{ik[x - x^{(1)} + p_\alpha(y - y^{(1)})]\right\} + \right. \tag{8.82}$$
$$\left. + c_\alpha^{(12)}(k) \exp\left\{ik[x - x^{(2)} + p_\alpha(y - y^{(2)})]\right\} \right),$$

where $c_\alpha^{(12)}(k)$ and $c_\alpha^{(21)}(k)$ have the structure similar to (8.42) with $a_\alpha^{(ij)\pm}(k)$ replaced by

$$c_\alpha^{(ij)\pm}(k) =$$

$$= \mathbf{U}_\alpha^{(i)\pm} \cdot \left\{ \sum_{\gamma=1}^{8} \mathbf{U}_\gamma^{(j)\pm} \otimes \mathbf{U}_\gamma^{(i)\pm} \exp\left[ik(p_\gamma - p_\alpha)(y^{(j)} - y^{(i)})\right] \right\}^{-1} \mathbf{B}_0^{(j)}. \tag{8.83}$$

The Fourier integral in eq. (8.82) can be analytically evaluated at $\varepsilon^{(j)} \to \infty$ in the same particular cases, when purely imaginary eigenvalues p_α occur due to a symmetry of the plate. The structure of the solution is slightly different from eq. (8.45):

$$\boldsymbol{\eta}(x,y) = \boldsymbol{\eta}^{(1)}(x,y) + \boldsymbol{\eta}^{(2)}(x,y), \tag{8.84}$$

where again

$$\boldsymbol{\eta}^{(j)}(x,y) = \begin{cases} \boldsymbol{\eta}^{(j)+}(x,y) + \boldsymbol{\eta}^{(j)0}(x,y), & x > x^{(j)}, \\ \boldsymbol{\eta}^{(j)-}(x,y), & x < x^{(j)}, \end{cases} \tag{8.85}$$

with

$$\boldsymbol{\eta}^{(j)0}(x,y) =$$

$$= \mp \sum_{\alpha=1}^{8} [x - x^{(j)} + p_\alpha(y - y^{(i)})] \boldsymbol{\xi}_\alpha \left\{ \mathbf{U}_\alpha \cdot \left[\sum_{\gamma=1}^{8} p_\gamma \mathbf{U}_\gamma \otimes \mathbf{U}_\gamma \right]^{-1} \mathbf{B}_0^{(j)} \right\} /d, \tag{8.86}$$

$$\boldsymbol{\eta}^{(j)\pm}(x,y) = \pm \sum_{\alpha=1}^{8} \sum_{k_n^\pm} \boldsymbol{\xi}_\alpha C_{\alpha\alpha}^{(j)}(k_n^\pm) \exp\left\{ik_n^\pm[x - x^{(j)} + p_\alpha(y - y^{(i)})]\right\}, \tag{8.87}$$

$$C_{a\alpha}^{(j)}(k) = \mathbf{U}_\alpha \cdot \left\{ \mathrm{Adj}\left[\sum_{\gamma=1}^{8} \mathbf{U}_\gamma \otimes \mathbf{U}_\gamma \exp(\mp ikp_\gamma d) \right] \right\} \mathbf{B}_0^{(j)} / G_{UU}^{(aj)}(k), \quad (8.88)$$

$$G_{UU}^{(aj)}(k) = k\frac{\partial}{\partial k} \left| \sum_{\gamma=1}^{8} \mathbf{U}_\gamma \otimes \mathbf{U}_\gamma \exp\left(\mp ikp_\gamma d\right) \right| \quad (8.89)$$

and $k_n^{\pm}$ ($\mathrm{Im}k_n^+ = -\mathrm{Im}k_n^- > 0$) are the roots of the equation

$$\left| \sum_{\gamma=1}^{8} \mathbf{U}_\gamma \otimes \mathbf{U}_\gamma \exp\left(-ikp_\gamma d\right) \right| = 0. \quad (8.90)$$

In eqs. (8.85)-(8.89) $j = 1, 2,\ i = 3 - j$; in (8.86), (8.88) and (8.89) the upper sign corresponds to $j = 1$.

(b) Prescribed electric charge distributions at both surfaces

The boundary conditions are given by (Fig. 8.2(*b*))

$$[\mathbf{u}(x, y^{(j)}), D(x, y^{(j)})]^t = \mathbf{B}_q^{(j)} \mathrm{H}(x - x^{(j)}), \quad j = 1, 2, \quad (8.91)$$

where

$$\mathbf{B}_q^{(1)} = [\mathbf{b}^{(1)}, -q^{(1)}]^t, \quad \mathbf{B}_q^{(2)} = [\mathbf{b}^{(2)}, q^{(2)}]^t. \quad (8.92)$$

The solution is

$$\boldsymbol{\eta}(x,y) = \frac{1}{2\pi i} \sum_{\alpha=1}^{8} \int_{-\infty}^{\infty} \frac{dk}{k} \boldsymbol{\xi}_\alpha \left(c_{b\alpha}^{(1)}(k) \exp\left\{ ik[x - x^{(1)} + p_\alpha(y - y^{(1)})] \right\} + \right.$$
$$\left. + c_{b\alpha}^{(2)}(k) \exp\left\{ ik[x - x^{(2)} + p_\alpha(y - y^{(2)})] \right\} \right), \quad (8.93)$$

where

$$c_{b\alpha}^{(1,2)}(k) = \mathbf{U}_{F\alpha} \cdot \left\{ \sum_{\gamma=1}^{8} \mathbf{U}_{F\gamma} \otimes \mathbf{U}_{F\gamma} \exp[\mp ikd(p_\gamma - p_\alpha)] \right\}^{-1} \mathbf{B}_q^{(1,2)} \quad (8.94)$$

The particular symmetry cases, when the condition $\mathrm{Re}p_\alpha$=0 is fulfilled, are described by eqs.(8.84) and (8.85) where

$$\boldsymbol{\eta}^{(j)0}(x,y) =$$
$$= \mp \sum_{\alpha=1}^{8} [x - x^{(j)} + p_\alpha(y - y^{(i)})] \boldsymbol{\xi}_\alpha \left\{ \mathbf{U}_{F\alpha} \cdot \left[\sum_{\gamma=1}^{8} p_\gamma \mathbf{U}_{F\gamma} \otimes \mathbf{U}_{F\gamma} \right]^{-1} \mathbf{B}_q^{(j)} \right\} /d, \quad (8.95)$$

$$\boldsymbol{\eta}^{(j)\pm}(x,y) = \pm \sum_{\alpha=1}^{8} \sum_{k_n^{\pm}} \boldsymbol{\xi}_\alpha C_{b\alpha}^{(j)}(k_n^{\pm}) \exp\left\{ ik_n^{\pm}[x - x^{(j)} + p_\alpha(y - y^{(i)})] \right\}, \quad (8.96)$$

$$C_{b\alpha}^{(j)}(k) = \mathbf{U}_{F\alpha} \cdot \left\{ \mathrm{Adj} \left[\sum_{\gamma=1}^{8} \mathbf{U}_{F\gamma} \otimes \mathbf{U}_{F\gamma} \exp\left(\mp ikp_\gamma d\right) \right] \right\} \mathbf{B}_q^{(j)} / G_{UU}^{(bj)}(k), \tag{8.97}$$

$$G_{UU}^{(bj)}(k) = k \frac{\partial}{\partial k} \left| \sum_{\gamma=1}^{8} \mathbf{U}_{F\gamma} \otimes \mathbf{U}_{F\gamma} \exp\left(\mp ikp_\gamma d\right) \right| \tag{8.98}$$

and $k_n^{\pm}$ are the roots of the equation

$$\left| \sum_{\gamma=1}^{8} \mathbf{U}_{F\gamma} \otimes \mathbf{U}_{F\gamma} \exp\left(ikp_\gamma d\right) \right| = 0. \tag{8.99}$$

As before, in (8.95), (8.96) $j = 1, 2,\ i = 3 - j$; and in (8.95), (8.97) and (8.98) the upper sign relates to $j = 1$.

(c) Prescribed electric potential distributions at both surfaces

The boundary conditions are (Fig. 8.2(c))

$$[\mathbf{u}(x, y^{(j)}), \varphi(x, y^{(j)})]^t = \mathbf{B}_{\Delta\varphi}^{(j)} \mathrm{H}(x - x^{(j)}), \quad j = 1, 2, \tag{8.100}$$

where

$$\mathbf{B}_{\Delta\varphi}^{(j)} = [\mathbf{b}^{(j)}, \Delta\varphi^{(j)}]^t, \tag{8.101}$$

The solution is given by the formulae of the previous subsection, eqs. (8.93)-(8.99), if to replace there $\mathbf{U}_{F\alpha,F\gamma}$ by $\mathbf{U}_{\alpha,\gamma}$ and $\mathbf{B}_q^{(j)}$ by $\mathbf{B}_{\Delta\varphi}^{(j)}$, and to change notation in accordance with our system replacing subscripts b by c.

(d) Prescribed electric charge at one surface and prescribed electric potential at the other

The boundary problem is formulated as (Fig. 8.2(d))

$$\begin{aligned} [\mathbf{u}(x, y^{(1)}), \varphi(x, y^{(1)})]^t &= \mathbf{B}_{\Delta\varphi}^{(j)} \mathrm{H}(x - x^{(1)}), \\ [\mathbf{u}(x, y^{(2)}), D(x, y^{(2)})]^t &= \mathbf{B}_q^{(2)} \mathrm{H}(x - x^{(2)}). \end{aligned} \tag{8.102}$$

The solution is given by eq. (8.93), where one should replace $c_{b\alpha}^{(1,2)}(k)$ by $c_{d\alpha}^{(1,2)}(k)$ defined as

$$c_{d\alpha}^{(1)}(k) = \mathbf{U}_{F\alpha} \cdot \left\{ \sum_{\gamma=1}^{8} \mathbf{U}_\gamma \otimes \mathbf{U}_{F\gamma} \exp[-ikd(p_\gamma - p_\alpha)] \right\}^{-1} \mathbf{B}_{\Delta\varphi}^{(1)}, \tag{8.103}$$

$$c_{d\alpha}^{(2)}(k) = \mathbf{U}_\alpha \cdot \left\{ \sum_{\gamma=1}^{8} \mathbf{U}_{F\gamma} \otimes \mathbf{U}_\gamma \exp[ikd(p_\gamma - p_\alpha)] \right\}^{-1} \mathbf{B}_q^{(2)}. \tag{8.104}$$

The structure of this solution after the calculation of the Fourier integral for the particular case $\mathrm{Re}p_\alpha = 0$ has the form eqs. (8.84), (8.85) with

$$\boldsymbol{\eta}^{(1)0}(x,y) = -\sum_{\alpha=1}^{8} \boldsymbol{\xi}_\alpha(\mathbf{a}_\alpha \cdot \mathbf{b}^{(1)})[x - x^{(1)} + p_\alpha(y - y^{(2)})]/d + \\ +[\mathbf{0}, 1, -\mathbf{a}_D, 0]^t(-\mathbf{a}_\varphi \cdot \mathbf{b}^{(1)} + \Delta\varphi^{(1)}), \tag{8.105}$$

$$\boldsymbol{\eta}^{(2)0}(x,y) = \sum_{\alpha=1}^{8} \boldsymbol{\xi}_\alpha(\mathbf{a}_\alpha \cdot \mathbf{b}^{(2)})[x - x^{(2)} + p_\alpha(y - y^{(1)})]/d - \\ -[\mathbf{0}, 0, -\mathbf{a}_\varphi, 1]^t(\mathbf{a}_D \cdot \mathbf{b}^{(2)} + q^{(2)}), \tag{8.106}$$

$$\mathbf{a}_D = \sum_{\beta=1}^{8} p_\beta D_\beta \mathbf{a}_\beta, \quad \mathbf{a}_\varphi = \sum_{\beta=1}^{8} p_\beta \varphi_\beta \mathbf{a}_\beta, \quad \mathbf{a}_\alpha = \mathbf{A}_\alpha \cdot \left[\sum_{\beta=1}^{8} p_\beta \mathbf{A}_\beta \otimes \mathbf{A}_\beta\right]^{-1}, \tag{8.107}$$

$$\boldsymbol{\eta}^{(j)\pm}(x,y) = \pm \sum_{\alpha=1}^{8} \sum_{k_n^\pm} \boldsymbol{\xi}_\alpha C_{d\alpha}^{(j)}(k_n^\pm) \exp\left\{ik_n^\pm[x - x^{(j)} + p_\alpha(y - y^{(j)})]\right\}, \tag{8.108}$$

$$C_{d\alpha}^{(1)}(k) = \mathbf{U}_{F\alpha} \cdot \left\{\mathrm{Adj}\left[\sum_{\gamma=1}^{8} \mathbf{U}_\gamma \otimes \mathbf{U}_{F\gamma} \exp(-ikp_\gamma d)\right]\right\} \mathbf{B}_{\Delta\varphi}^{(1)} / G_{UU}^{(d1)}(k), \tag{8.109}$$

$$C_{d\alpha}^{(2)}(k) = \mathbf{U}_\alpha \cdot \left\{\mathrm{Adj}\left[\sum_{\gamma=1}^{8} \mathbf{U}_{F\gamma} \otimes \mathbf{U}_\gamma \exp(ikp_\gamma d)\right]\right\} \mathbf{B}_q^{(2)} / G_{UU}^{(d2)}(k), \tag{8.110}$$

$$G_{UU}^{(d1)}(k) = k\frac{\partial}{\partial k}\left|\sum_{\gamma=1}^{8} \mathbf{U}_\gamma \otimes \mathbf{U}_{F\gamma} \exp(-ikp_\gamma d)\right|, \tag{8.111}$$

$$G_{UU}^{(d2)}(k) = k\frac{\partial}{\partial k}\left|\sum_{\gamma=1}^{8} \mathbf{U}_{F\gamma} \otimes \mathbf{U}_\gamma \exp(ikp_\gamma d)\right|. \tag{8.112}$$

In eq. (8.108) $k_n^\pm$ ($\mathrm{Im}k_n^+ = -\mathrm{Im}k_n^- > 0$) are the roots of the equation

$$\left|\sum_{\gamma=1}^{8} \mathbf{U}_{F\gamma} \otimes \mathbf{U}_\gamma \exp(ikp_\gamma d)\right| = 0. \tag{8.113}$$

(e) One surface charged and the other adjoined to a dielectric subspace

The boundary conditions are (Fig. 8.2(*e*))

$$[\mathbf{u}(k, y^{(1)}), \varphi(k, y^{(1)}) + \mathrm{sgn}\,(k) D(k, y^{(1)})/i\varepsilon^{(1)}]^t = \mathbf{B}_0^{(1)} \exp(-ikx^{(1)})/2\pi ik, \\ [\mathbf{u}(k, y^{(2)}), D(k, y^{(2)})]^t = \mathbf{B}_q^{(2)} \exp(-ikx^{(2)})/2\pi ik. \tag{8.114}$$

The solutions can be obtained from eqs. (8.93), (8.103), (8.104), if to replace in eq. (8.103) $\mathbf{U}_\gamma$ by $\mathbf{U}_\gamma^{(1)\pm}$ and $\mathbf{B}_{\Delta\varphi}^{(1)}$ by $\mathbf{B}_0^{(1)}$; in eq. (8.104) $\mathbf{U}_{\alpha,\gamma}$ by $\mathbf{U}_{\alpha,\gamma}^{(1)\mp}$; and indices d by e (in both cases the upper sign corresponds to $k>0$, and the lower sign to $k<0$).

(f) One surface with a prescribed electric potential distribution and the other adjoined to a dielectric subspace

The last boundary problem of this series has the form (Fig. 8.2(f))

$$[\mathbf{u}(k,y^{(1)}),\varphi(k,y^{(1)})]^t=\mathbf{B}_{\Delta\varphi}^{(1)}\exp(-ikx^{(1)})/2\pi ik,$$

$$[\mathbf{u}(k,y^{(2)}),\varphi(k,y^{(2)})-\ \mathrm{sgn}(k)D(k,y^{(2)})/i\varepsilon^{(2)}]^t=\mathbf{B}_0^{(2)}\exp(-ikx^{(2)})/2\pi ik. \tag{8.115}$$

Its solution is again described by eqs. (8.93), (8.103), (8.104) after the appropriate replacements: in eq. (8.103) $\mathbf{U}_{F\alpha,F\gamma}$ by $\mathbf{U}_{\alpha,\gamma}^{(1,2)\pm}$; in eq. (8.104) $\mathbf{U}_{F\gamma}$ by $\mathbf{U}_\gamma^{(2)\mp}$ and $\mathbf{B}_q^{(2)}$ by $\mathbf{B}_0^{(2)}$ (again with the upper sign for $k>0$, and the lower sign for $k<0$).

8.4 Strips with prescribed displacements at one surface and loads at another

In this section we shall consider a series of problems with mixed mechanical boundary conditions, when at one surface a definite displacement distribution is prescribed and the other surface is supposed to be loaded:

$$\mathbf{u}(x,y^{(1)})=\mathbf{b}^{(1)}\mathrm{H}(x-x^{(1)}),\quad \mathbf{n}\boldsymbol{\sigma}(x,y^{(2)})=\mathbf{F}^{(2)}\delta(x-x^{(2)}). \tag{8.116}$$

With the considered above six variants of electrical boundary problems nine their non-equivalent combinations with mechanical conditions (8.116) are possible (see Fig. 8.3).

(a) Surfaces adjoined to dielectric media

In our standard terms the boundary conditions have the form (Fig. 8.3(a)) :

$$\begin{aligned}[\mathbf{u}(k,y^{(1)}),\varphi(k,y^{(1)})+\mathrm{sgn}\ (k)D(k,y^{(1)})/i\varepsilon^{(1)}]^t=\\ =\mathbf{B}_0^{(1)}\exp(-ikx^{(1)})/2\pi ik,\end{aligned} \tag{8.117}$$

$$\begin{aligned}[\mathbf{J}(k,y^{(2)}),D(k,y^{(2)})-\mathrm{sgn}\ (k)i\varepsilon^{(2)}\varphi(k,y^{(2)})]^t=\\ =-\boldsymbol{\Phi}_0^{(2)}\exp(-ikx^{(2)})/2\pi ik.\end{aligned} \tag{8.118}$$

The solution is

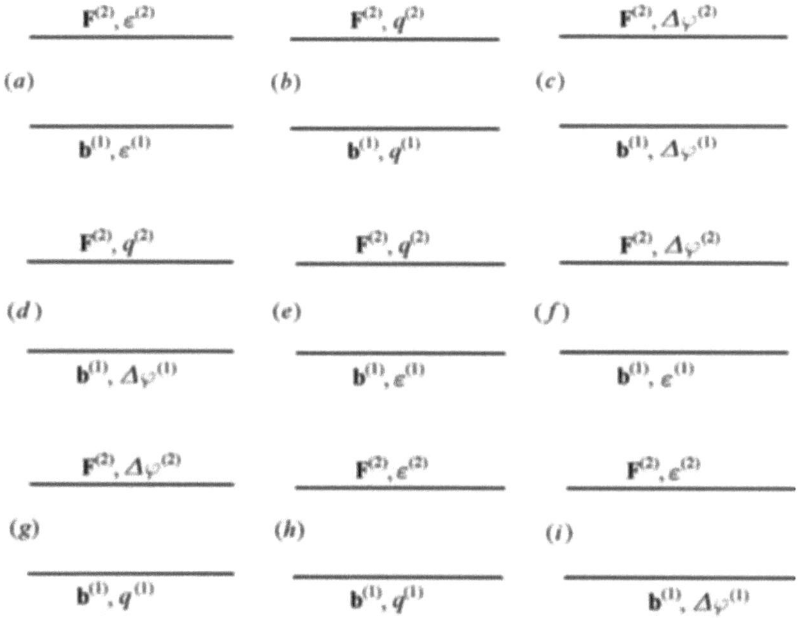

Fig. 8.3. Nine non-equivalent boundary problems for a piezoelectric strip with prescribed displacements at one surface and prescribed loads at another.

$$\begin{aligned}\boldsymbol{\eta}(x,y) = {} & \\ & \frac{1}{2\pi i}\sum_{\alpha=1}^{8}\int_{-\infty}^{\infty}\frac{dk}{k}\boldsymbol{\xi}_\alpha\left(b_{\alpha\alpha}^{(1)}(k)\exp\left\{ik[x-x^{(1)}+p_\alpha(y-y^{(1)})]\right\}-\right. \\ & \left.-\,b_{\alpha\alpha}^{(2)}(k)\exp\left\{ik[x-x^{(2)}+p_\alpha(y-y^{(2)})]\right\}\right),\end{aligned} \tag{8.119}$$

where

$$b_{\alpha\alpha}^{(1)}(k) = \mathbf{V}_\alpha^{(2)\pm}\cdot\left\{\sum_{\gamma=1}^{8}\mathbf{U}_\gamma^{(1)\pm}\otimes V_\gamma^{(2)\pm}\exp[-ikd(p_\gamma-p_\alpha)]\right\}^{-1}\mathbf{B}_0^{(1)} \tag{8.120}$$

$$b_{\alpha\alpha}^{(2)}(k) = \mathbf{U}_\alpha^{(1)\mp}\cdot\left\{\sum_{\gamma=1}^{8}\mathbf{V}_\gamma^{(2)\mp}\otimes\mathbf{U}_\gamma^{(1)\mp}\exp[ikd(p_\gamma-p_\alpha)]\right\}^{-1}\boldsymbol{\Phi}_0^{(2)} \tag{8.121}$$

with the upper sign for $k>0$, and the lower sign for $k<0$.

The integration in eq. (8.119) for $\varepsilon^{(j)}\to\infty$ and symmetry situations admitting the identically zero $\mathrm{Re}p_\alpha$ brings to the structure (8.45), (8.46) with

$$\boldsymbol{\eta}^{(1)0}(x,y) = [\mathbf{b}^{(1)},0,\mathbf{0},0]^t,\quad \boldsymbol{\eta}^{(2)0}(x,y) = [\mathbf{0},0,\mathbf{F}^{(2)},0]^t, \tag{8.122}$$

$$\boldsymbol{\eta}^{(j)\pm}(x,y) = \pm \sum_{\alpha=1}^{8} \sum_{k_n^{\pm}} \boldsymbol{\xi}_{\alpha} B_{a\alpha}^{(j)}(k_n^{\pm}) \exp\left\{ik_n^{\pm}[x - x^{(j)} + p_{\alpha}(y - y^{(i)})]\right\}, \tag{8.123}$$

$$B_{a\alpha}^{(1)}(k) = \mathbf{V}_{F\alpha} \cdot \left\{\mathrm{Adj}\left[\sum_{\gamma=1}^{8} \mathbf{U}_{\gamma} \otimes \mathbf{V}_{F\gamma} \exp\left(-ikp_{\gamma}d\right)\right]\right\} \mathbf{B}_0^{(1)} / G_{UV}^{(a1)}(k), \tag{8.124}$$

$$B_{a\alpha}^{(2)}(k) = \mathbf{U}_{\alpha} \cdot \left\{\mathrm{Adj}\left[\sum_{\gamma=1}^{8} \mathbf{V}_{F\gamma} \otimes \mathbf{U}_{\gamma} \exp\left(ikp_{\gamma}d\right)\right]\right\} \boldsymbol{\Phi}_0^{(2)} / G_{VU}^{(a2)}(k), \tag{8.125}$$

$$G_{UV}^{(a1)}(k) = k\frac{\partial}{\partial k}\left|\sum_{\gamma=1}^{8} \mathbf{U}_{\gamma} \otimes \mathbf{V}_{F\gamma} \exp\left(-ikp_{\gamma}d\right)\right|, \tag{8.126}$$

$$G_{VU}^{(a2)}(k) = k\frac{\partial}{\partial k}\left|\sum_{\gamma=1}^{8} \mathbf{V}_{F\gamma} \otimes \mathbf{U}_{\gamma} \exp\left(ikp_{\gamma}d\right)\right|. \tag{8.127}$$

In eq. (8.123) $k_n^{\pm}$ ($\mathrm{Im}k_n^{+} = -\mathrm{Im}k_n^{-} > 0$) are the roots of the equation

$$\left|\sum_{\gamma=1}^{8} \mathbf{V}_{F\gamma} \otimes \mathbf{U}_{\gamma} \exp\left(ikp_{\gamma}d\right)\right| = 0. \tag{8.128}$$

(b) Prescribed electric charge distributions at both surfaces

The boundary conditions are given by (Fig. 8.3(*b*))

$$\begin{aligned}
[\mathbf{u}(x,y^{(1)}), D(x,y^{(1)})]^t &= \mathbf{B}_q^{(1)} \mathrm{H}(x - x^{(1)}),\\
[\mathbf{J}(x,y^{(2)}), D(x,y^{(2)})]^t &= -\boldsymbol{\Phi}_q^{(2)} \mathrm{H}(x - x^{(2)}).
\end{aligned} \tag{8.129}$$

The solution is determined by eq. (8.119) with $b_{a\alpha}^{(j)}(k)$ replaced by

$$b_{b\alpha}^{(1)}(k) = \mathbf{V}_{\alpha} \cdot \left\{\sum_{\gamma=1}^{8} \mathbf{U}_{F\gamma} \otimes \mathbf{V}_{\gamma} \exp[-ikd(p_{\gamma} - p_{\alpha})]\right\}^{-1} \mathbf{B}_q^{(1)}, \tag{8.130}$$

$$b_{b\alpha}^{(2)}(k) = \mathbf{U}_{F\alpha} \cdot \left\{\sum_{\gamma=1}^{8} \mathbf{V}_{\gamma} \otimes \mathbf{U}_{F\gamma} \exp[ikd(p_{\gamma} - p_{\alpha})]\right\}^{-1} \boldsymbol{\Phi}_q^{(2)}. \tag{8.131}$$

After the integration for the mentioned above symmetry cases the result is again reduced to eqs.(8.45), (8.46) where

$$\begin{aligned}
&\boldsymbol{\eta}^{(1)0}(x,y) =\\
&= \sum_{\alpha=1}^{8} \boldsymbol{\xi}_{\alpha} \left\{\mathbf{L}_{\alpha} \cdot (\lambda_D \mathbf{b}^{(2)} + q^{(1)} \mathbf{a}_D) - D_{\alpha} \mathbf{l}_D \cdot \mathbf{b}^{(1)}\right.\\
&\left. - q^{(1)}[x - x^{(1)} + p_{\alpha}(y - y^{(2)})]/d\right\} / \lambda_D,
\end{aligned} \tag{8.132}$$

$$\boldsymbol{\eta}^{(2)0}(x,y) =$$

$$= \sum_{\alpha=1}^{8} \boldsymbol{\xi}_\alpha \left\{ \mathbf{A}_\alpha \cdot (\lambda_D \mathbf{F}^{(2)} + q^{(2)} \mathbf{l}_D) - D_\alpha \mathbf{a}_D \cdot \mathbf{F}^{(2)} \right. \tag{8.133}$$
$$\left. + q^{(2)} [x - x^{(2)} + p_\alpha (y - y^{(1)})]/d \right\} / \lambda_D,$$

$$\lambda_D = \sum_{\gamma=1}^{8} p_\gamma D_\gamma^2, \tag{8.134}$$

$$\boldsymbol{\eta}^{(j)\pm}(x,y) =$$

$$= \pm \sum_{\alpha=1}^{8} \sum_{k_n^\pm} \boldsymbol{\xi}_\alpha B_{b\alpha}^{(j)}(k_n^\pm) \exp\left\{ i k_n^\pm [x - x^{(j)} + p_\alpha (y - y^{(i)})] \right\}, \tag{8.135}$$

$$B_{b\alpha}^{(1)}(k) =$$

$$= \mathbf{V}_\alpha \cdot \left\{ \mathrm{Adj} \left[\sum_{\gamma=1}^{8} \mathbf{U}_{F\gamma} \otimes \mathbf{V}_\gamma \exp(-ikp_\gamma d) \right] \right\} \mathbf{B}_q^{(1)} / G_{UV}^{(b1)}(k), \tag{8.136}$$

$$B_{b\alpha}^{(2)}(k) =$$

$$= \mathbf{U}_{F\alpha} \cdot \left\{ \mathrm{Adj} \left[\sum_{\gamma=1}^{8} \mathbf{V}_\gamma \otimes \mathbf{U}_{F\gamma} \exp(ikp_\gamma d) \right] \right\} \boldsymbol{\Phi}_q^{(2)} / G_{VU}^{(b2)}(k), \tag{8.137}$$

$$G_{UV}^{(b1)}(k) = k \frac{\partial}{\partial k} \left| \sum_{\gamma=1}^{8} \mathbf{U}_{F\gamma} \otimes \mathbf{V}_\gamma \exp(-ikp_\gamma d) \right|, \tag{8.138}$$

$$G_{VU}^{(b2)}(k) = k \frac{\partial}{\partial k} \left| \sum_{\gamma=1}^{8} \mathbf{V}_\gamma \otimes \mathbf{U}_{F\gamma} \exp(ikp_\gamma d) \right|. \tag{8.139}$$

In eq. (8.135) $k_n^\pm$ ($\mathrm{Im}k_n^+ = -\mathrm{Im}k_n^- > 0$) are the roots of the equation

$$\left| \sum_{\gamma=1}^{8} \mathbf{V}_\gamma \otimes \mathbf{U}_{F\gamma} \exp(ikp_\gamma d) \right| = 0. \tag{8.140}$$

(c) Prescribed electric potential distributions at both surfaces

The boundary conditions in this case (Fig. 8.3(c))

$$[\mathbf{u}(x,y^{(1)}), \varphi(x,y^{(1)})]^t = \mathbf{B}_{\Delta\varphi}^{(1)} \mathrm{H}(x - x^{(1)}),$$
$$[\mathbf{J}(x,y^{(2)}), \varphi(x,y^{(2)})]^t = -\boldsymbol{\Phi}_{\Delta\varphi}^{(2)} \mathrm{H}(x - x^{(2)}), \tag{8.141}$$

are formally very similar to those, eq. (8.129), in the subsection 8.4 (b), which allows us to obtain the solution by some adaptation of the previous formulae (8.130)-(8.140). The necessary modifications are reduced to the replacements

$\mathbf{U}_{F\alpha,F\gamma}$ by $\mathbf{U}_{\alpha,\gamma}, \mathbf{V}_{\alpha,\gamma}$ by $\mathbf{V}_{F\alpha,F\gamma}$, $\mathbf{B}_q^{(1)}$ by $\mathbf{B}_{\Delta\varphi}^{(1)}, \boldsymbol{\Phi}_q^{(2)}$ by $\boldsymbol{\Phi}_{\Delta\varphi}^{(2)}$ and $q^{(1,2)}$ by $\mp\Delta\varphi^{(1,2)}$; and also $\mathbf{l}_D, \mathbf{a}_D$ and λ_D by $\mathbf{l}_\varphi, \mathbf{a}_\varphi$ and

$$\lambda_\varphi = \sum_{\gamma=1}^{8} p_\gamma \varphi_\gamma^2, \tag{8.142}$$

respectively.

(d) Prescribed electric charge at one surface and prescribed electric potential at the other

Now there are two non-equivalent combinations of these electrical conditions with the mechanical boundary problem (8.116) (see Figs. 8.3 (*d*) and (*g*)). The first one is given by

$$\begin{aligned} [\mathbf{u}(x,y^{(1)}), \varphi(x,y^{(1)})]^t &= \mathbf{B}_{\Delta\varphi}^{(1)} \mathrm{H}(x-x^{(1)}), \\ [\mathbf{J}(x,y^{(2)}), D(x,y^{(2)})]^t &= -\boldsymbol{\Phi}_q^{(2)} \mathrm{H}(x-x^{(2)}). \end{aligned} \tag{8.143}$$

The solution is obtained by the replacements $\mathbf{U}_{F\gamma} \to \mathbf{U}_\gamma$ and $\mathbf{B}_q^{(1)} \to \mathbf{B}_{\Delta\varphi}^{(1)}$ in eqs. (8.130), (8.131), (8.134)-(8.140) and by a radical simplification of eqs. (8.132), (8.133) for $\boldsymbol{\eta}^{(1)0}(x,y)$ and $\boldsymbol{\eta}^{(2)0}(x,y)$:

$$\boldsymbol{\eta}^{(1)0}(x,y) = [\mathbf{B}_{\Delta\varphi}^{(1)}, 0]^t, \quad \boldsymbol{\eta}^{(2)0}(x,y) = [\mathbf{0}, \boldsymbol{\Phi}_q^{(2)}]^t. \tag{8.144}$$

The second combination of boundary conditions (Fig. 8.3 (*g*)),

$$\begin{aligned} [\mathbf{u}(x,y^{(1)}), D(x,y^{(1)})]^t &= [\mathbf{B}_q^{(1)} \mathrm{H}(x-x^{(1)}), \\ [\mathbf{J}(x,y^{(2)}), \varphi(x,y^{(2)})]^t &= -\boldsymbol{\Phi}_{\Delta\varphi}^{(2)} \mathrm{H}(x-x^{(2)}), \end{aligned} \tag{8.145}$$

is solved by the same eqs. (8.130), (8.131), (8.134)-(8.140), (8.144) after replacing there $\mathbf{V}_\gamma$ by $\mathbf{V}_{F\gamma}$ and $\boldsymbol{\Phi}_q^{(2)}$ by $\boldsymbol{\Phi}_{\Delta\varphi}^{(2)}$ with the reverse change in eq. $(8.144)_1$ $\mathbf{B}_{\Delta\varphi}^{(1)} \to \mathbf{B}_q^{(1)}$.

(e) One surface charged and the other adjoined to a dielectric subspace

Again we must consider two different combinations of boundary conditions. Let us start from the boundary problem shown in Fig. 8.3 (*e*):

$$\begin{aligned} [\mathbf{u}(k,y^{(1)}), \varphi(k,y^{(1)}) + \ \mathrm{sgn}\ (k) D(k,y^{(1)})/i\varepsilon^{(1)}]^t &= \\ = \mathbf{B}_0^{(1)} \exp(-ikx^{(1)})/2\pi i k&, \\ [\mathbf{J}(k,y^{(2)}), D(k,y^{(2)})]^t = -\boldsymbol{\Phi}_q^{(2)} \exp(-ikx^{(1)})/2\pi i k&. \end{aligned} \tag{8.146}$$

The solution is again given by eqs. (8.130), (8.131) after replacing $\mathbf{U}_{F\gamma}$ by $\mathbf{U}_{\gamma}^{(1)\pm}$ and $\mathbf{B}_q^{(1)}$ by $\mathbf{B}_0^{(1)}$ in eq. (8.130), $\mathbf{U}_{F_\alpha,F_\gamma}$ by $\mathbf{U}_{\alpha\gamma}^{(1)\mp}$ in eq. (8.131) (as before, with the upper sign corresponds for $k > 0$ and the lower sign for $k < 0$).

The alternative combination of boundary conditions (Fig. 8.3 (h)),

$$
\begin{aligned}
&[\mathbf{u}(k,y^{(1)}), D(k,y^{(1)})]^t = \mathbf{B}_q^{(1)} \exp(-ikx^{(1)})/2\pi ik, \\
&[\mathbf{J}(k,y^{(2)}), \varphi(k,y^{(2)}) - \operatorname{sgn}(k) D(k,y^{(2)})/i\varepsilon^{(2)}]^t = \\
&\qquad = -\boldsymbol{\Phi}_0^{(2)} \exp(-ikx^{(2)})/2\pi ik.
\end{aligned}
\tag{8.147}
$$

brings to the solution determined by the same set of eqs. (8.130), (8.131) with the replacements: $\mathbf{V}_{\alpha,\gamma}$ by $\mathbf{V}_{F\alpha,F\gamma}^{(2)\pm}$ in eq. (8.130); $\mathbf{V}_\gamma$ by $\mathbf{V}_{F\gamma}^{(2)\mp}$ and $\boldsymbol{\Phi}_q^{(2)}$ by $\boldsymbol{\Phi}_0^{(2)}$ in eq. (8.131). The upper and the lower signs relate to the conditions $k > 0$ and to $k < 0$, respectively.

(f) One surface with a prescribed electric potential distribution and the other adjoined to a dielectric subspace

The boundary problem relating to Fig. 8.3 (f) is formulated as

$$
\begin{aligned}
&[\mathbf{u}(k,y^{(1)}), D(k,y^{(1)}) + \operatorname{sgn}(k) i\varepsilon^{(1)} \varphi(k,y^{(1)})]^t = \\
&\qquad \mathbf{B}_0^{(1)} \exp(-ikx^{(1)})/2\pi ik, \\
&[\mathbf{J}(k,y^{(2)}), \varphi(k,y^{(2)})]^t = -\boldsymbol{\Phi}_{\Delta\varphi}^{(2)} \exp(-ikx^{(2)})/2\pi ik.
\end{aligned}
\tag{8.148}
$$

The adaptation of eqs. (8.130), (8.131) to this case reduces to the replacements: $\mathbf{V}_{\alpha,\gamma}$ by $\mathbf{V}_{F\alpha,F\gamma}$ and $\mathbf{U}_{F\gamma}$ by $\mathbf{U}_{F\gamma}^{(1)\pm}$ in eq. (8.130); $\mathbf{V}_\gamma$ by $\mathbf{V}_{F\gamma}$ and $\mathbf{U}_{F\alpha,F\gamma}$ by $\mathbf{U}_{F\alpha,F\gamma}^{(1)\mp}$ in eq. (8.131); and in all relevant places $\mathbf{B}_q^{(1)}$ by $\mathbf{B}_0^{(1)}$ and $\boldsymbol{\Phi}_q^{(2)}$ by $\boldsymbol{\Phi}_{\Delta\varphi}^{(2)}$.

The last combination of boundary conditions (Fig. 8.3 (i))

$$
\begin{aligned}
&[\mathbf{u}(k,y^{(1)}), \varphi(k,y^{(1)})]^t = \mathbf{B}_{\Delta\varphi}^{(1)} \exp(-ikx^{(1)})/2\pi ik, \\
&[\mathbf{J}(k,y^{(2)}), D(k,y^{(2)}) - \operatorname{sgn}(k) i\varepsilon^{(2)} \varphi(k,y^{(2)})]^t = \\
&\qquad = -\boldsymbol{\Phi}_0^{(2)} \exp(-ikx^{(2)})/2\pi ik.
\end{aligned}
\tag{8.149}
$$

is solved by eqs. (8.130), (8.131) where $\mathbf{B}_q^{(1)}$ and $\boldsymbol{\Phi}_q^{(2)}$ changed by $\mathbf{B}_{\Delta\varphi}^{(1)}$ and $\boldsymbol{\Phi}_0^{(2)}$, respectively, and the replacements $\mathbf{V}_{\alpha,\gamma} \to \mathbf{V}_{\alpha,\gamma}^{(2)\pm}$, $\mathbf{U}_{F\gamma} \to \mathbf{U}_\gamma$ in eq. (8.130) and $\mathbf{V}_\gamma \to \mathbf{V}_\gamma^{(2)\mp}$, $\mathbf{U}_{F\alpha,F\gamma} \to \mathbf{U}_{\alpha,\gamma}$ in eq. (8.131) are made. The rule for a choice of signs in the above superscripts is the same as before (see the end of the previous subsection).

8.5 Conclusions

The Green's functions derived in this chapter form an analytical basis for a description of electro-elastic fields excited in an arbitrary infinite piezoelectric strip by any one-dimensional distributions of electromechanical sources at the surfaces. When instead of a line source at $x^{(j)}$ of the strength $\mathbf{b}^{(j)}, \Delta\varphi^{(j)}, \mathbf{F}^{(j)}$ or $q^{(j)}$ there is a distribution $\mathbf{b}^{(j)}(x)$, $\Delta\varphi^{(j)}(x), \mathbf{F}^{(j)}(x)$ or $q^{(j)}(x)$ one should just add an integration of the corresponding Green's function with respect to x.

In the general case of unrestricted anisotropy our results are presented in the form of Fourier integrals containing a set of eigenvectors and eigenvalues of the Stroh matrix. These are constant parameters dependent only on material constants, orientation of surfaces and direction of line sources. Of course their computation is a separate but purely technical problem, which is beyond the scope of this paper.

The Fourier integrals are taken for the particular cases of occurrence of such material symmetry elements, which provide purely imaginary values of the Stroh parameters p_α. For instance, this often realizes in materials of the plate with a reflection plane parallel to the surfaces or a 2-fold symmetry axis perpendicular to them. In fact, an analytical evaluation of the integrals is just reduced to solving some definite transcendental algebraic equations determining the poles k_n of the corresponding integrands. A computation of the roots k_n of those equations together with a summation of related residues might technically appear to be not more efficient for a computer than a direct calculation of the Fourier integrals. On the other hand, the presented analytical results have their own advantages related to an allowance for an analysis and better understanding of rather complex electro-elastic fields in anisotropic plates with different combinations of material constants, symmetry elements and boundary sources.

9

2D Electro-Elastic Fields in a Piezoelectric Layer-Substrate Structure

In the previous two chapters we have studied elasticity of piezoelectric plates considering their interactions with surrounding media through boundary conditions. In this chapter a more realistic situation will be studied, namely, we shall extend our theory for a description of 2D fields in piezoelectric films on substrates, which present in a large variety of modern electronic devices being their important element. Before our works [29, 30] such solutions were known only for purely elastic (non-piezoelectric) anisotropic media of the considered type in two cases: for a homogeneous film on a homogeneous substrate [52] and for an inhomogeneous (stratified or graded) coating on a homogeneous substrate [54]. Below we shall derive the corresponding Green's functions for arbitrary piezoelectric strips on substrates with line sources in the bulk or at the surface. The chapter is based on the publications [29, 30].

9.1 Basic equations

Consider the structure formed by a piezoelectric layer on a piezoelectric substrate, both of unrestricted anisotropy. The thickness of the layer is h. Elastic and piezoelectric moduli tensors and electric permittivities of the materials are denoted by $c^{(p)}_{ijkl}$, $e^{(p)}_{ikl}$, $\varepsilon^{(p)}_{ij}$ with $p = 1$ or 2 for the layer and the substrate, respectively. The coordinate system is chosen with the $y-$axis directed along the internal normal $\mathbf{n}$ to the surface (Fig. 9.1). The origin $y = 0$ is placed at the surface, so the interface is situated at $y = h$. The $z-$axis belonging to the surface is chosen collinear with the line source. In the above Cartesian system the general line source will be fixed by the coordinates (x', y'), which can belong both to the layer and to the substrate.

As we have seen, elasticity of piezoelectric media is conveniently described in terms of two tensor quantities, elastic stresses $\boldsymbol{\sigma}$ and distortions $\boldsymbol{\beta}$, and two vector characteristics, electric displacement $\mathbf{D}$ and electric field $\mathbf{E}$, which are related to each other by the constitutive equations

$$\sigma_{ij} = c_{ijlk}\beta_{kl} - e_{lij}E_l, \quad D_j = e_{jlk}\beta_{kl} + \varepsilon_{jl}E_l, \tag{9.1}$$

and obey the superposition of Kroener's [182] and Maxwell's equations.

$$\mathrm{div}(\boldsymbol{\sigma} + \boldsymbol{\sigma}^0) = 0, \qquad \mathrm{rot}(\boldsymbol{\beta} + \boldsymbol{\beta}^0) = 0, \tag{9.2}$$

$$\mathrm{div}(\mathbf{D} + \mathbf{D}^0) = 0, \qquad \mathrm{rot}(\mathbf{E} - \mathbf{E}^0) = 0. \tag{9.3}$$

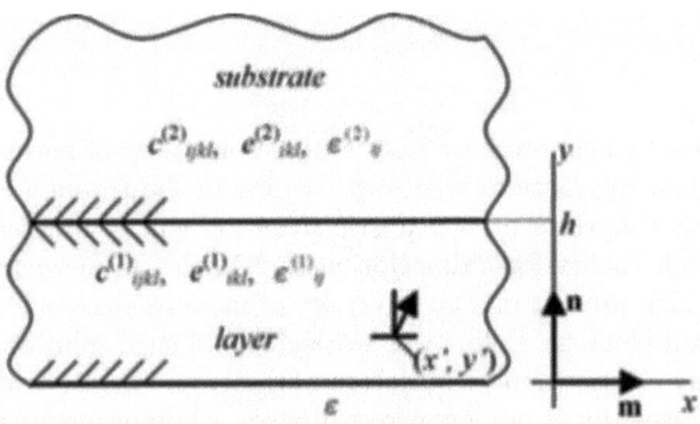

Fig. 9.1. Piezoelectric layer-substrate structure and the coordinate system.

The values $\boldsymbol{\sigma}^0, \boldsymbol{\beta}^0, \mathbf{D}^0$ and $\mathbf{E}^0$ in (9.2), (9.3) for a considered general line source are equal to [19, 63] (see also eqs. (7.5) and (7.14)

$$\begin{pmatrix} \boldsymbol{\sigma}^0 \\ \boldsymbol{\beta}^0 \\ \mathbf{D}^0 \\ \mathbf{E}^0 \end{pmatrix} = \begin{pmatrix} \mathbf{f} \otimes \mathbf{m} \\ -\mathbf{b} \otimes \mathbf{n} \\ -q\mathbf{m} \\ -\Delta\varphi\mathbf{n} \end{pmatrix} \mathrm{H}(x - x')\delta(y - y'), \tag{9.4}$$

where $\mathbf{m}$ and $\mathbf{n}$ are the unit vectors along the x and y axes (Fig. 9.1).

As usual, eqs. $(9.2)_2$ and $(9.3)_2$ can be excluded by introducing the displacement vector $\mathbf{u}$ and the electric potential φ:

$$\boldsymbol{\beta} + \boldsymbol{\beta}^0 = \mathrm{grad}\ \mathbf{u}, \quad \mathbf{E} - \mathbf{E}^0 = -\mathrm{grad}\ \varphi. \tag{9.5}$$

Standard further steps would be to substitute (9.1) with (9.5) into $(9.2)_1$, $(9.3)_1$ and then to deal with a system of four scalar differential equations of the second order with respect to the unknown 4-vector

$$\mathbf{U}(x, y) = [\mathbf{u}(x, y), \varphi(x, y)]^t. \tag{9.6}$$

However, as we now know, for the considering 2D problem it proves to be more practical to apply the alternative extended Stroh formalism [61] reducing the problem to a system of eight differential equations of the first order with respect to the unknown 8-component vector function

$$\boldsymbol{\eta}(x,y) = [\mathbf{U}(x,y), \mathbf{V}(x,y)]^t. \tag{9.7}$$

In (9.7) $\mathbf{U}(x,y)$ is the 4-vector (9.6) and $\mathbf{V}(x,y)$ is the stress function 4-vector

$$\mathbf{V}(x,y) = [\mathbf{J}(x,y), D(x,y)]^t, \tag{9.8}$$

defined so that

$$(\sigma + \sigma^0)\mathbf{m} = \partial_y \mathbf{J}, \quad (\mathbf{D} + \mathbf{D}^0) \cdot \mathbf{m} = \partial_y D; \quad \boldsymbol{\sigma}\mathbf{n} = -\partial_x \mathbf{J}, \quad \mathbf{D} \cdot \mathbf{n} = -\partial_x D. \tag{9.9}$$

One should not mix in (9.9) the electric displacement vector $D_i(x,y)$ and the scalar function $D(x,y)$. With such a choice of the stress function $\mathbf{V}(x,y)$ (9.8), (9.9) the equilibrium equations $(9.2)_1$, $(9.3)_1$ are satisfied automatically. And constitutive equations (9.1) are transformed into the Stroh-like equation

$$(\mathbf{I}\partial_y \quad - \mathbf{N}\partial_x)\boldsymbol{\eta}(x,y) = -\mathbf{g}\mathrm{H}(x - x')\delta(y - y'), \tag{9.10}$$

where $\mathbf{g}$ is the strength of the general line defect,

$$\mathbf{g} \ = [\mathbf{b}, \Delta\varphi, \ - \ \mathbf{f}, q]^t. \tag{9.11}$$

Thus, we have naturally come to the same equations as in Ch. 7 (see eqs. (7.8) and (7.9)). However, one should take into account that now we deal with two different media with distinct material moduli and accodingly different matrices $\mathbf{N}^{(1)}$ and $\mathbf{N}^{(2)}$. In further considerations, when necessary, we shall distinguish some quantities related to the layer and to the substrate by marking them, respectively, by the superscripts (1) and (2) in accordance with our notation of material constants (see Fig. 9.1).

After the Fourier transformation of the unknown function $\boldsymbol{\eta}(x,y)$ and the Heaviside function $\mathrm{H}(x - x')$ in (9.10) with respect to x,

$$\boldsymbol{\eta}(x,y) = \int_{-\infty}^{\infty} dk \exp(ikx)\boldsymbol{\eta}(k,y), \quad \mathrm{H}(x - x') = \frac{1}{2\pi i}\int_{-\infty}^{\infty} \frac{dk}{k} \exp\left[ik\left(x - x'\right)\right], \tag{9.12}$$

eq. (9.10) takes the form

$$(\mathbf{I}\partial_y - ik\mathbf{N})\boldsymbol{\eta}(k,y) = -\mathbf{g}(2\pi ik)^{-1}\exp(-ikx')\delta(y - y'). \tag{9.13}$$

For the sake of convergence of the integrals in (9.12), as before, it will be implied that

$$k = Rek - i\varepsilon, \qquad \varepsilon \to +0. \tag{9.14}$$

9.2 Boundary conditions at the surface and at the interface

Standard boundary conditions for electro-elastic fields at surfaces and interfaces are expressed in terms of four quantities: displacements $\mathbf{u}$, potential φ, and normal components of mechanical stresses $\mathbf{n}\boldsymbol{\sigma}$ and electric induction $\mathbf{n}\cdot\mathbf{D}$. As we know from Chs. 7 and 8, the Fourier amplitude $\boldsymbol{\eta}(k,y) = [\mathbf{U}(k,y),\ \mathbf{V}(k,y)]^t$ is directly related to the physical quantities involved in boundary conditions:

$$\mathbf{U}(k,y) = [\mathbf{u}(k,y), \varphi(k,y)]^t,$$
$$\mathbf{V}(k,y) = [\mathbf{J}(k,y), D(k,y)]^t = \frac{i}{k}[\mathbf{n}\boldsymbol{\sigma}(k,y), \mathbf{n}\cdot\mathbf{D}(k,y)]^t. \tag{9.15}$$

Thus, all four components of $\boldsymbol{\eta}(k,y)$ can be easily expressed in terms of given surface distributions of displacements, potential, forces and charges. On the other hand, at the interface $y = h$ all four quantities by definition must be continuous,

$$\boldsymbol{\eta}(k, h-0) = \boldsymbol{\eta}(k, h+0). \tag{9.16}$$

Now let us formulate the boundary conditions at the surface $y = 0$. Mechanically, we shall distinguish prescribed loads and prescribed displacements at the surface. In the first case

$$\mathbf{n}\boldsymbol{\sigma}(x,0) = -\mathbf{F}\delta(x - x_0) \tag{9.17}$$

(the minus at $\mathbf{F}$ relates to a choice an internal normal $\mathbf{n}$, Fig. 9.1), or in terms of (9.15) in the k-space,

$$\mathbf{n}\boldsymbol{\sigma}(k,0) = -\mathbf{F}\exp(-ikx_0)/2\pi, \quad \mathbf{J}(k,0) = \mathbf{F}\exp(-ikx_0)/2\pi ik. \tag{9.18}$$

At $\mathbf{F} = \mathbf{0}$ this is reduced to a free surface condition,

$$\mathbf{J}(k,0) = \mathbf{0}. \tag{9.19}$$

For the condition of prescribed displacements we choose the form, in the $x-$ and $k-$spaces,

$$\mathbf{u}(x,0) = \mathbf{u}_0\mathrm{H}(x - x_0), \mathbf{u}(k,0) = \mathbf{u}_0\exp(-ikx_0)/2\pi ik. \tag{9.20}$$

At $\mathbf{u}_0 = \mathbf{0}$ this leads to a clamped surface condition,

$$\mathbf{u}(k,0) = \mathbf{0}. \tag{9.21}$$

Electrically, the three types of boundary conditions will be considered: a prescribed charge or electric potential at the surface and an adjoined dielectric medium with the electric permittivity ε. In the first variant, similarly to (9.18), one obtains:

$$\mathbf{n} \cdot \mathbf{D}(k,0) = q_0 \exp(-ikx_0)/2\pi, \quad D(k,0) = -q_0 \exp(-ikx_0)/2\pi ik. \qquad (9.22)$$

At $q_0 = 0$ one obtains the particular case

$$D(k,0) = 0. \qquad (9.23)$$

The condition of a prescribed electric potential at the surface reads, in analogy with (9.20),

$$\varphi(x,0) = \varphi_0 H(x - x_0), \quad \varphi(k,0) = \varphi_0 \exp(-ikx_0)/2\pi ik. \qquad (9.24)$$

We shall be also interested in the particular case $\varphi_0 = 0$ related to a metallized surface,

$$\varphi(x,0) = 0, \quad \varphi(k,0) = 0. \qquad (9.25)$$

Finally, the condition of continuity of $\mathbf{n} \cdot \mathbf{D}(k,y)$ and $\varphi(k,y)$ at the boundary $y = 0$ with the adjoined isotropic dielectric subspace of the permittivity ε, is given by (see Ch. 8):

$$D(k,0) + i\varepsilon \text{ sgn } (k)\varphi(k,0) = 0. \qquad (9.26)$$

In further considerations, it will be convenient to present different combinations of the above mechanical and electrical boundary conditions (9.18)-(9.26) in a block 4-vector form directly related to the blocks of the unknown function. Following to our consideration in the two previous chapters, we shall use in addition to the blocks (9.15),

$$\mathbf{U}(k,0) = [\mathbf{u}(k,0), \varphi(k,0)]^t, \mathbf{V}(k,0) = [\mathbf{J}(k,0), D(k,0)]^t, \qquad (9.27)$$

the F-representation:

$$\mathbf{U}_F(k,0) = [\mathbf{u}(k,0), D(k,0)]^t, \mathbf{V}_F(k,0) = [\mathbf{J}(k,0), \varphi(k,0)]^t, \qquad (9.28)$$

and the boundary representation prepared for making use of condition (9.26):

$$\begin{aligned} \mathbf{U}^{(s)}(k,0) &= [\mathbf{u}(k,0), \varphi(k,0) + \text{ sgn } (k)D(k,0)/i\varepsilon]^t, \\ \mathbf{V}^{(s)}(k,0) &= [\mathbf{J}(k,0), D(k,0) + i\varepsilon \text{ sgn } (k)\varphi(k,0)]^t. \end{aligned} \qquad (9.29)$$

Below we shall consider separately the six boundary problems related to sources situated only in the interior of the piezoelectric structure (eqs. (9.10), (9.13) with $\mathbf{g} \neq \mathbf{0}$) :

$$\mathbf{U}(k,0) = \mathbf{0}, \quad \mathbf{U}_F(k,0) = \mathbf{0}, \quad \mathbf{U}^{(s)}(k,0) = \mathbf{0}, \qquad (9.30)$$

$$\mathbf{V}(k,0) = \mathbf{0}, \mathbf{V}_F(k,0) = \mathbf{0}, \quad \mathbf{V}^{(s)}(k,0) = \mathbf{0}, \qquad (9.31)$$

and the same number of problems when the bulk of the structure is free of sources ($\mathbf{g} = \mathbf{0}$) and the electro-elastic fields are excited from the surface:

$$\mathbf{U}(k,0) = \begin{pmatrix} \mathbf{u}_0 \\ \varphi_0 \end{pmatrix} \frac{e^{-ikx_0}}{2\pi ik}, \quad \mathbf{U}_F(k,0) = \begin{pmatrix} \mathbf{u}_0 \\ -q_0 \end{pmatrix} \frac{e^{-ikx_0}}{2\pi ik},$$
$$\mathbf{U}^{(s)}(k,0) = \begin{pmatrix} \mathbf{u}_0 \\ 0 \end{pmatrix} \frac{e^{-ikx_0}}{2\pi ik}, \tag{9.32}$$

$$\mathbf{V}(k,0) = \begin{pmatrix} \mathbf{F} \\ -q_0 \end{pmatrix} \frac{e^{-ikx_0}}{2\pi ik}, \quad \mathbf{V}_F(k,0) = \begin{pmatrix} \mathbf{F} \\ \varphi_0 \end{pmatrix} \frac{e^{-ikx_0}}{2\pi ik},$$
$$\mathbf{V}^{(s)}(k,0) = \begin{pmatrix} \mathbf{F} \\ 0 \end{pmatrix} \frac{e^{-ikx_0}}{2\pi ik}. \tag{9.33}$$

Finally, one should also remember the additional natural requirement to solutions of all formulated above boundary problems: the field $\boldsymbol{\eta}(k,y)$ must be finite at infinity $y \to \infty$.

9.3 General and partial solutions for $\boldsymbol{\eta}(k,y)$ and $\boldsymbol{\eta}(x,y)$

As usual, we shall look for solutions of eq. (9.13) in the form of a sum of a general solution of the corresponding homogeneous equation (with a vanishing right-hand side) and a partial solution of initial eq. (9.13). Taking into account that the considered piezoelectric structure is piece-wise homogeneous, the matrix $\mathbf{N}$ in (9.13) can be considered as a function

$$\mathbf{N}(y) = \begin{cases} \mathbf{N}^{(1)}, \ 0 < y < h; \\ \mathbf{N}^{(2)}, \ h < y < \infty. \end{cases} \tag{9.34}$$

Here $\mathbf{N}^{(p)}, p = 1,2$, are the matrices defined by eqs. (3.138),(3.15) and (3.51) where the material tensors must be supplied by appropriate superscripts. The corresponding homogeneous equation for an inhomogeneous medium,

$$(\mathbf{I}\partial_y \quad - ik\mathbf{N}(y))\boldsymbol{\eta}(k,y) = \mathbf{0}, \tag{9.35}$$

can be solved in terms of the propagator matrix $\mathbf{W}^{(k)}(y|y^0)$ [54]:

$$\boldsymbol{\eta}_{gen}(k,y) = \mathbf{W}^{(k)}(y|y^0)\mathbf{X}, \tag{9.36}$$

where $\mathbf{X}$ is an arbitrary 8-vector and y^0 is the arbitrary reference point. In our case it is convenient to choose the reference point at the interface, $y^0 = h$, which leads to the propagator matrix

$$\mathbf{W}^{(k)}(y|h) = \begin{cases} \sum\limits_{\alpha=1}^{8} \boldsymbol{\xi}_\alpha^{(1)} \otimes \mathbf{T}\boldsymbol{\xi}_\alpha^{(1)} \exp[ikp_\alpha^{(1)}(y-h)], & 0 \le y \le h; \\ \sum\limits_{\alpha=1}^{8} \boldsymbol{\xi}_\alpha^{(2)} \otimes \mathbf{T}\boldsymbol{\xi}_\alpha^{(2)} \exp[ikp_\alpha^{(2)}(y-h)], & h \le y < \infty. \end{cases} \tag{9.37}$$

In (9.37) the following notation is introduced: $\boldsymbol{\xi}_\alpha^{(j)}$ and $p_\alpha^{(j)}$ are, respectively, the eigenvectors and the eigenvalues of the matrix $\mathbf{N}^{(j)}$, $j=1,2$. We shall not consider particular cases when the matrices $\mathbf{N}^{(j)}$ are nonsemisimply degenerate [181].

The sets $\boldsymbol{\xi}_\alpha^{(j)}$ and $p_\alpha^{(j)}$ form the four pairs of complex conjugate

$$\boldsymbol{\xi}_{\alpha+4}^{(j)} = \boldsymbol{\xi}_\alpha^{(j)*}, \quad p_{\alpha+4}^{(j)} = p_\alpha^{(j)*}, \quad \alpha = 1,\ldots,4. \tag{9.38}$$

As before, we shall choose numbering so that

$$\mathrm{Im} p_\alpha^{(j)} > 0, \quad \alpha = 1,\ldots,4; \quad \mathrm{Im} p_\alpha^{(j)} < 0, \quad \alpha = 5,\ldots,8. \tag{9.39}$$

The Stroh eigenvectors $\boldsymbol{\xi}_\alpha^{(j)}$ are orthogonal and complete [61]:

$$\boldsymbol{\xi}_\alpha^{(j)} \cdot \mathbf{T}\boldsymbol{\xi}_\beta^{(j)} = \delta_{\alpha\beta}, \quad \sum_{\alpha=1}^{8} \boldsymbol{\xi}_\alpha^{(j)} \otimes \mathbf{T}\boldsymbol{\xi}_\alpha^{(j)} = \mathbf{I}. \tag{9.40}$$

Let us form the set of eight eigenfunctions

$$\boldsymbol{\xi}_{k\alpha}(y) = \mathbf{W}^{(k)}(y|h)\boldsymbol{\xi}_\alpha^{(2)}, \tag{9.41}$$

and replace general solution (9.36) by

$$\boldsymbol{\eta}_{gen}(k,y) = \sum_{\alpha=1}^{8} b_{k\alpha}\boldsymbol{\xi}_{k\alpha}(y), \tag{9.42}$$

where the eight indefinite coefficients $b_{k\alpha}$ are equivalent to the unknown 8-vector $\mathbf{X}$ in (9.36). In view of form (9.37) and properties (9.40), the eigenfunctions $\boldsymbol{\xi}_{k\alpha}$ can be written as

$$\boldsymbol{\xi}_{k\alpha}(y) = \begin{cases} \sum\limits_{\beta=1}^{8} \boldsymbol{\xi}_\beta^{(1)}[\boldsymbol{\xi}_\beta^{(1)} \cdot \mathbf{T}\boldsymbol{\xi}_\alpha^{(2)}] \exp[ikp_\beta^{(1)}(y-h)], \; 0 \le y \le h; \\ \boldsymbol{\xi}_\alpha^{(2)} \exp[ikp_\alpha^{(2)}(y-h)], \qquad h \le y < \infty. \end{cases} \tag{9.43}$$

It can be checked that due to properties (9.40) the set of eigenfunctions $\boldsymbol{\xi}_{k\alpha}(y)$ is also orthogonal and complete,

$$\boldsymbol{\xi}_{k\alpha}(y) \cdot \mathbf{T}\,\boldsymbol{\xi}_{-k\beta}(y) = \delta_{\alpha\beta}, \quad \sum_{\alpha=1}^{8} \boldsymbol{\xi}_{k\alpha}(y) \otimes \mathbf{T}\boldsymbol{\xi}_{-k\alpha}(y) = \mathbf{I}. \tag{9.44}$$

And, in view of (9.38),

$$\boldsymbol{\xi}_{k\alpha+4}(y) = [\boldsymbol{\xi}_{-k\alpha}(y)]^*. \tag{9.45}$$

Taking into account $(9.40)_2$, it is easily seen that the functions $\boldsymbol{\xi}_{k\alpha}(y)$ (9.43) are continuous at the interface $y=h$, i.e. they obey condition (9.16).

The partial solution of inhomogeneous eq. (9.13) can be chosen in the form

$$\boldsymbol{\eta}_{part}(k,y) = \frac{e^{-ikx'}}{2\pi ik}\sum_{\alpha=1}^{8}\boldsymbol{\xi}_{k\alpha}(y')[\boldsymbol{\xi}_{-k\alpha}(y')\cdot\mathbf{Tg}]\mathrm{H}(y'-y), \tag{9.46}$$

which clearly satisfies eq. (9.13) in view of $(9.44)_2$. This solution vanishing at $y > y'$ trivially guarantees a finiteness of the fields at $y \to \infty$. In order to provide the same property of general solution (9.42) and for some further conveniences let us renormalize the coefficients $b_{k\alpha}$ in (9.42) to

$$b_{k\alpha} = -c_{k\alpha}\frac{e^{-ikx'}}{2\pi ik}\mathrm{H}(k\mathrm{Im}p_\alpha^{(2)}). \tag{9.47}$$

Indeed, the behavior of $\boldsymbol{\eta}_{gen}(k,y)$ at $y \to \infty$ is determined by the function $\exp[ik\, p_\alpha^{(2)} y]$, which due to the form of (51) automatically becomes decaying.

The sum of (9.46) and (9.42) with (9.47) gives the basic solution for the Fourier amplitude $\boldsymbol{\eta}(k,y)$,

$$\boldsymbol{\eta}(k,y) = \frac{e^{-ikx'}}{2\pi ik}\sum_{\alpha=1}^{8}\boldsymbol{\xi}_{k\alpha}(y)\{\boldsymbol{\xi}_{-k\alpha}(y')\cdot\mathbf{Tg}\mathrm{H}(y'-y) - c_{k\alpha}\mathrm{H}(k\mathrm{Im}\, p_\alpha^{(2)})\}, \tag{9.48}$$

where the indefinite coefficients $c_{k\alpha}$ are supposed to be found from boundary conditions. Taking into account that boundary conditions (9.30)-(9.33) are prepared in a block form, let us also decompose 8-vectors $\boldsymbol{\xi}_{k\alpha}(0)$ similarly to (9.7) and (9.15):

$$\boldsymbol{\xi}_{k\alpha}(0) = [\mathbf{U}_{k\alpha}(0), \mathbf{V}_{k\alpha}(0)]^t, \tag{9.49}$$

where in view of (9.43):

$$\mathbf{U}_{k\alpha}(0) = \sum_{\beta=1}^{8}\mathbf{U}_\beta^{(1)}[\boldsymbol{\xi}_\beta^{(1)}\cdot\mathbf{T}\boldsymbol{\xi}_\alpha^{(2)}]\exp\left(-ikp_\beta^{(1)}h\right), \tag{9.50}$$

$$\mathbf{V}_{k\alpha}(0) = \sum_{\beta=1}^{8}\mathbf{V}_\beta^{(1)}[\boldsymbol{\xi}_\beta^{(1)}\cdot\mathbf{T}\boldsymbol{\xi}_\alpha^{(2)}]\exp\left(-ikp_\beta^{(1)}h\right). \tag{9.51}$$

Here $\mathbf{U}_\beta^{(1)}$ and $\mathbf{V}_\beta^{(1)}$ are the block 4-vectors of the decompositions

$$\boldsymbol{\xi}_\alpha^{(1)} = \begin{pmatrix}\mathbf{U}_\alpha^{(1)}\\ \mathbf{V}_\alpha^{(1)}\end{pmatrix}, \quad \mathbf{U}_\alpha^{(1)} = \begin{pmatrix}\mathbf{A}_\alpha^{(1)}\\ \varphi_\alpha^{(1)}\end{pmatrix}, \quad \mathbf{V}_\alpha^{(1)} = \begin{pmatrix}\mathbf{L}_\alpha^{(1)}\\ D_\alpha^{(1)}\end{pmatrix}. \tag{9.52}$$

Quite similarly let us introduce the block 4-vectors adapted to the other boundary conditions

$$\mathbf{U}_{Fk\alpha}(0) = \sum_{\beta=1}^{8}\mathbf{U}_{F\beta}^{(1)}[\boldsymbol{\xi}_\beta^{(1)}\cdot\mathbf{T}\boldsymbol{\xi}_\alpha^{(2)}]\exp\left(-ikp_\beta^{(1)}h\right), \tag{9.53}$$

$$\mathbf{V}_{Fk\alpha}(0)=\sum_{\beta=1}^{8}\mathbf{V}_{F\beta}^{(1)}[\boldsymbol{\xi}_{\beta}^{(1)}\cdot\mathbf{T}\boldsymbol{\xi}_{\alpha}^{(2)}]\exp\left(-ikp_{\beta}^{(1)}h\right), \tag{9.54}$$

$$\mathbf{U}_{k\alpha}^{s}(0)=\sum_{\beta=1}^{8}\mathbf{U}_{\beta}^{s}[\boldsymbol{\xi}_{\beta}^{(1)}\cdot\mathbf{T}\boldsymbol{\xi}_{\alpha}^{(2)}]\exp\left(-ikp_{\beta}^{(1)}h\right), \tag{9.55}$$

$$\mathbf{V}_{k\alpha}^{s}(0)=\sum_{\beta=1}^{8}\mathbf{V}_{\beta}^{s}[\boldsymbol{\xi}_{\beta}^{(1)}\cdot T\boldsymbol{\xi}_{\alpha}^{(2)}]\exp\left(-ikp_{\beta}^{(1)}h\right), \tag{9.56}$$

where $\mathbf{U}_{F\beta}^{(1)}$, $\mathbf{V}_{F\beta}^{(1)}$, $\mathbf{U}_{\beta}^{s}$ and $\mathbf{V}_{\beta}^{s}$ are obtained by some rearrangements in (9.52):

$$\mathbf{U}_{F\beta}^{(1)}=\begin{pmatrix}\mathbf{A}_{\beta}^{(1)}\\ D_{\beta}^{(1)}\end{pmatrix},\quad \mathbf{V}_{F\beta}^{(1)}=\begin{pmatrix}\mathbf{L}_{\beta}^{(1)}\\ \varphi_{\beta}^{(1)}\end{pmatrix}; \tag{9.57}$$

$$\mathbf{U}_{\beta}^{s}=\begin{pmatrix}\mathbf{A}_{\beta}^{(1)}\\ \varphi_{\beta}^{(1)}+\ \mathrm{sgn}\ (k)D_{\beta}^{(1)}/i\varepsilon\end{pmatrix},\quad \mathbf{V}_{\beta}^{s}=\begin{pmatrix}\mathbf{L}_{\beta}^{(1)}\\ D_{\beta}^{(1)}+\ \mathrm{sgn}(k)i\varepsilon\varphi_{\beta}^{(1)}\end{pmatrix}. \tag{9.58}$$

9.4 Electro-elastic fields excited by the line source in the interior of the structure

(a) Clamped metallized surface, eq. (9.30)$_1$

i) Derivation of the field $\eta(x,y)$

Simultaneous vanishing of the displacements $\mathbf{u}$ and the electric potential φ at the surface, described by the equation $\mathbf{U}(k,0)=0$, in terms of eq. (9.48), means that its upper four-dimensional block taken at $y=0$ must be equal zero. This leads to the following equation with respect to the unknown coefficients $c_{k\alpha}$:

$$\sum_{\alpha=1}^{8}\mathbf{U}_{k\alpha}(0)\{\boldsymbol{\xi}_{-k\alpha}(y')\cdot\mathbf{Tg}-c_{k\alpha}H(k\mathrm{Im}\ p_{\alpha}^{(2)})\}=0. \tag{9.59}$$

In view of the numbering rule (9.39), at $k>0$ only the coefficients $c_{k\alpha}$ for $\alpha=1,\ldots,4$ act in (9.48), (9.59), and similarly at $k<0$ only $c_{k\alpha+4}$ $(\alpha=1,\ldots,4)$ are involved into them. It will be also essential that identity (9.45) is valid. With these remarks, the equations for $c_{k\alpha}$ and $c_{k\alpha+4}$ $(\alpha=1,\ldots,4)$ are given by

$$\sum_{\alpha=1}^{4}\mathbf{U}_{k\alpha}(0)c_{k\alpha}=\sum_{\alpha=1}^{4}\{\mathbf{U}_{k\alpha}(0)[\boldsymbol{\xi}_{-k\alpha}(y')\cdot\mathbf{Tg}]+\mathbf{U}_{-k\alpha}^{*}(0)[\boldsymbol{\xi}_{-k\alpha+4}(y')\cdot\mathbf{Tg}]\}, \tag{9.60}$$

$$\sum_{\alpha=1}^{4} \mathbf{U}^*_{-k\alpha}(0)c_{k\alpha+4} = \sum_{\alpha=1}^{4} \{\mathbf{U}^*_{-k\alpha}(0)[\boldsymbol{\xi}_{-k\alpha}(y')\cdot\mathbf{Tg}] + \mathbf{U}_{k\alpha}(0)[\boldsymbol{\xi}_{-k\alpha}(y')\cdot\mathbf{Tg}]\}. \tag{9.61}$$

Let us introduce the 4×4 matrix $\mathbf{U}^{(k)}$ built of the four columns $\mathbf{U}_{k\alpha}(0)$ ($\alpha = 1,\ldots,4$) given by (9.50):

$$\mathbf{U}^{(k)} = [\mathbf{U}_{k1}(0), \mathbf{U}_{k2}(0), \mathbf{U}_{k3}(0), \mathbf{U}_{k4}(0)], \tag{9.62}$$

and the 4-vectors $\bar{\mathbf{U}}_{k\alpha}$ defined as the lines of the inverse matrix $[\mathbf{U}^{(k)}]^{-1}$:

$$[\mathbf{U}^{(k)}]^{-1} = [\bar{U}_{k1}, \bar{U}_{k2}, \bar{U}_{k3}, \bar{U}_{k4}]^t. \tag{9.63}$$

Note that by definition,

$$\bar{\mathbf{U}}_{k\alpha}\cdot\ \mathbf{U}_{k\beta}(0) = \delta_{\alpha\beta}. \tag{9.64}$$

In these terms the solutions of eqs. (9.60) and (9.61) have the following form:

$$c_{k\alpha} = \boldsymbol{\xi}_{-k\alpha}(y')\cdot\mathbf{Tg} + \sum_{\beta=1}^{4} \{[\mathbf{U}^{(k)}]^{-1}\mathbf{U}^{(-k)*}\}_{\alpha\beta}[\boldsymbol{\xi}_{-k\beta+4}(y')\cdot\mathbf{Tg}], \quad \alpha = 1,\ldots,4; \tag{9.65}$$

$$c_{k\alpha+4} = \boldsymbol{\xi}_{-k\alpha+4}(y')\cdot\mathbf{Tg} + \sum_{\beta=1}^{4} \{[\mathbf{U}^{(-k)}*]^{-1}\mathbf{U}^{(k)}\}_{\alpha\beta}[\boldsymbol{\xi}_{-k\beta}(y')\cdot\mathbf{Tg}], \quad \alpha = 1,\ldots,4. \tag{9.66}$$

In accordance with definitions (9.62), (9.63), the $\alpha\beta$-components of the matrices in (9.65), (9.66) are equal

$$\begin{aligned} \{[\mathbf{U}^{(k)}]^{-1}\mathbf{U}^{(-k)*}\}_{\alpha\beta} &= \bar{\mathbf{U}}_{k\alpha}\cdot[\mathbf{U}_{-k\beta}(0)]*, \\ \{[\mathbf{U}^{(-k)*}]^{-1}\ \mathbf{U}^{(k)}\}_{\alpha\beta} &= [\bar{\mathbf{U}}_{-k\alpha}]^*\cdot\ \mathbf{U}_{k\beta}(0). \end{aligned} \tag{9.67}$$

The combined form for $c_{k\alpha}$ ($\alpha = 1,\ldots,8$) following from (9.65), (9.66) reads:

$$c_{k\alpha} = \boldsymbol{\xi}_{-k\alpha}(y')\cdot\mathbf{Tg} + \sum_{\beta=1}^{8}\{\mathbf{Y}(\mathbf{U}^{(k)})\}_{\alpha\beta}[\boldsymbol{\xi}_{-k\beta}(y')\cdot\mathbf{Tg}], \alpha = 1,\ldots,8, \tag{9.68}$$

where the new 8×8 matrix $\mathbf{Y}(\mathbf{U}^{(k)})$ is introduced:

$$\mathbf{Y}\ (\mathbf{U}^{(k)}) = \begin{pmatrix} \mathbf{0} & [\mathbf{U}^{(k)}]^{-1}\mathbf{U}^{(-k)*} \\ [\mathbf{U}^{(-k)*}]^{-1}\mathbf{U}^{(k)} & \mathbf{0} \end{pmatrix}. \tag{9.69}$$

Finally, combining eqs. (9.48), (9.68) and substituting the result into Fourier integral $(15)_1$, we obtain the desired solution for the Green function describing electro-elastic fields excited by the general line defect in the considered piezoelectric structure with a clamped metallized surface:

$$\boldsymbol{\eta}(x,y) = -\frac{1}{2\pi i}\sum_{\alpha,\beta=1}^{8}\int_{-\infty}^{\infty}\frac{dk}{k}\exp[ik(x-x')]\boldsymbol{\xi}_{k\alpha}(y)[\boldsymbol{\xi}_{-k\alpha}(y')\cdot\mathbf{Tg}]$$
$$\times\{\mathrm{sgn}(k\mathrm{Im}p_\alpha^{(2)})\mathrm{H}[k\mathrm{Im}p_\alpha^{(2)}(y-y')]\delta_{\alpha\beta} + \{\,\mathbf{Y}(\mathbf{U}^{(k)})\}_{\alpha\beta}\mathrm{H}(k\mathrm{Im}p_\alpha^{(2)})\}, \tag{9.70}$$

where we have used the identity (3.157)

$$\mathrm{H}(a) - \mathrm{H}(-b) = \mathrm{sgn}(a)\mathrm{H}(ab) = \ \mathrm{sgn}(b)\mathrm{H}(ab). \tag{9.71}$$

ii) Analysis of convergence of the Fourier integral

Let us explicitly demonstrate convergence of the integral in (9.70). Just for a definiteness, consider the field in the substrate for the situation when $y > y' > h$. In this particular case general eq. (9.70) is easily transformed to the form

$$\boldsymbol{\eta}(x,y) = -\frac{1}{2\pi i}\int_{-\infty}^{\infty}\frac{dk}{k}\exp[ik(x-x')]$$
$$\left\{\sum_{\alpha=1}^{8}\boldsymbol{\xi}_\alpha^{(2)}[\boldsymbol{\xi}_\alpha^{(2)}\cdot\mathbf{Tg}]\exp[ikp_\alpha^{(2)}(y-y')]\mathrm{H}(k\mathrm{Im}\ p_\alpha^{(2)})\right.$$
$$+\sum_{\alpha=1}^{4}\mathrm{H}\left(k\mathrm{Im}\,p_\alpha^{(2)}\right)\exp[ikp_\alpha^{(2)}(y-h)]\times$$
$$\sum_{\beta=5}^{8}\exp[-ikp_\beta^{(2)}(y'-h)]\left\{\mathbf{Y}(\mathbf{U}^{(k)})\right\}_{\alpha\beta}\boldsymbol{\xi}_\alpha^{(2)}\left[\boldsymbol{\xi}_\alpha^{(2)}\cdot\mathbf{Tg}\right] \tag{9.72}$$
$$+\sum_{\alpha=5}^{8}\mathrm{H}\left(k\mathrm{Im}\,p_\alpha^{(2)}\right)\exp[ikp_\alpha^{(2)}(y-h)]\times$$
$$\left.\sum_{\beta=1}^{4}\exp[-ikp_\beta^{(2)}(y'-h)]\left\{\mathbf{Y}(\mathbf{U}^{(k)})\right\}_{\alpha\beta}\boldsymbol{\xi}_\alpha^{(2)}\left[\boldsymbol{\xi}_\alpha^{(2)}\cdot\mathbf{Tg}\right]\right\},$$

where we have taken into account "anti-diagonal" form (9.69) of the matrix $\mathbf{Y}(\mathbf{U}^{(k)})$ and definition (9.43) of the vector-function $\boldsymbol{\xi}_{k\alpha}(y)$. We are interested in an asymptotic behavior of the integrand in (9.72) at $k \to \pm\infty$. The first term in (9.72), due to the presence of the Heaviside function and the condition $y - y' > 0$, is automatically exponentially small at both infinities $k \to \pm\infty$. The second term, in view of (9.39), at $k < 0$ is an identical zero, and at $k \to \infty$ is exponentially small, because, as it follows from eqs. (9.69), (9.62), (9.50), the matrix $\mathbf{Y}(\mathbf{U}^{(k)})$ remains finite at $k \to \pm\infty$. Similarly, the third term vanishes at $k > 0$ being exponentially small at $k \to -\infty$. If to add that with our choice (9.14) of the integration path the point $k = 0$ is not singular, there is no reasons for a divergence of the integral (9.72).

We shall not proceed with a consideration of the other cases for different boundary problems and various positions of y and y'. They can be analyzed in quite a similar manner with the same conclusion: all the integrals determining our Green's functions are convergent.

iii) Solutions for physical quantities

In complete analogy with Ch. 7, the deduced solution (9.70) presenting the result in terms of "potential" functions, $\boldsymbol{\eta}(x,y) = [\mathbf{u}(x,y),\ \varphi(x,y),\ \mathbf{J}(x,y),\ D(x,y)]$, can be easily transformed for finding the corresponding physical quantities: the elastic distortion tensor $\boldsymbol{\beta}(x,y)$, eq. $(9.5)_1$, the electric field $\mathbf{E}(x,y)$, eq.$(9.5)_2$, the stress tensor $\sigma(x,y)$ and the electric displacement $\mathbf{D}(x,y)$, eqs. $(9.9)_{1-4}$. These fields are obtained by some differentiations, eq. (7.70),

$$\begin{pmatrix}\boldsymbol{\beta}\\ \mathbf{E}\end{pmatrix} = \begin{pmatrix}\dfrac{\partial \mathbf{u}}{\partial x}\otimes\mathbf{m} + \dfrac{\partial \mathbf{u}}{\partial y}\otimes\mathbf{n}\\ -\dfrac{\partial \varphi}{\partial x}\mathbf{m} - \dfrac{\partial \varphi}{\partial y}\mathbf{n}\end{pmatrix} - \begin{pmatrix}\boldsymbol{\beta}^0\\ \mathbf{E}^0\end{pmatrix},$$
$$\begin{pmatrix}\boldsymbol{\sigma}\\ \mathbf{D}\end{pmatrix} = \begin{pmatrix}\dfrac{\partial \mathbf{J}}{\partial y}\otimes\mathbf{m} - \dfrac{\partial \mathbf{J}}{\partial x}\otimes\mathbf{n}\\ \dfrac{\partial D}{\partial y}\mathbf{m} - \dfrac{\partial D}{\partial x}\mathbf{n}\end{pmatrix} - \begin{pmatrix}\boldsymbol{\sigma}^0\\ \mathbf{D}^0\end{pmatrix}, \tag{9.73}$$

with $\boldsymbol{\beta}^0, \mathbf{E}^0, \boldsymbol{\sigma}^0$ and $\mathbf{D}^0$ defined by eq. (9.4). Note that the stress tensor $\boldsymbol{\sigma}(x,y)$ is determined by $(9.73)_2$ apart from the component σ_{zz}. All components of $\boldsymbol{\sigma}$ (including σ_{zz}) can be alternatively found by combining of $(9.73)_1$ with constitutive equation $(9.1)_1$.

Substituting (9.70) into (9.73) with making use of (9.43) one can easily obtain the desired analytical presentations for the above physical fields. In view of the form of eq. (9.43), these fields are described by different forms, depending on relations between y, y' and h. We shall limit ourselves by presenting the result only for the case, when both the source and the observation point belong to the substrate ($y, y' > h$):

$$\begin{pmatrix}\boldsymbol{\beta}\\ \mathbf{E}\end{pmatrix} = \sum_{\alpha,\beta=1}^{8}\begin{pmatrix}\mathbf{A}_\alpha^{(2)}\\ -\boldsymbol{\varphi}_\alpha^{(2)}\end{pmatrix}\otimes(\mathbf{m}+p_\alpha^{(2)}\mathbf{n})\left(\boldsymbol{\xi}_\beta^{(2)}\cdot\mathbf{Tg}\right)I_{\alpha\beta}(x,y|x',y'), \tag{9.74}$$

$$\begin{pmatrix}\boldsymbol{\sigma}\\ \mathbf{D}\end{pmatrix} = \sum_{\alpha,\beta=1}^{8}\begin{pmatrix}\mathbf{L}_\alpha^{(2)}\\ D_\alpha^{(2)}\end{pmatrix}\otimes(p_\alpha^{(2)}\mathbf{m}-\mathbf{n})\left(\boldsymbol{\xi}_\beta^{(2)}\cdot\mathbf{Tg}\right)I_{\alpha\beta}(x,y|x',y'), \tag{9.75}$$

with

$$I_{\alpha\beta}(x,y|x',y') =$$
$$= -\frac{1}{2\pi}\int_{-\infty}^{\infty}dk\ \exp\left\{ik[x-x'+p_\alpha^{(2)}(y-h)-p_\beta^{(2)}(y'-h)]\right\}$$
$$\times\{\mathrm{sgn}(k\ \mathrm{Im}p_\alpha^{(2)})\mathrm{H}[k\ \mathrm{Im}p_\alpha^{(2)}(y-y')]\delta_{\alpha\beta}+$$
$$\{\mathbf{Y}(\mathbf{U}^{(k)})\}_{\alpha\beta}\mathrm{H}(k\ \mathrm{Im}p_\alpha^{(2)})\}. \tag{9.76}$$

The integrand in (9.76) differs from that in (9.70) (for the same case $y, y' > h$) only by the absence of k in the denominator. Accordingly the convergence of $I_{\alpha\beta}$ is guaranteed by the consideration in the above subsection. The first term in integral (9.76), diagonal with respect to α, β, can be easily evaluated:

$$I^{\infty}_{\alpha\beta}(x, y|x', y') = \delta_{\alpha\beta}\frac{\mathrm{sgn}(\mathrm{Im}\; p^{(2)}_{\alpha})}{2\pi i[x - x' + p^{(2)}_{\alpha}(y - y')]}. \tag{9.77}$$

It can be verified that (9.77) with (9.74), (9.75) determines the fields of the general line defect in the infinite medium with the material moduli of the substrate. For instance, the corresponding elastic distortion and electric potential fields created by the dislocation alone (in this case $\boldsymbol{\xi}^{(2)}_{\alpha} \cdot \mathbf{Tg} = \mathbf{L}^{(2)}_{\alpha} \cdot \mathbf{b}$) are given by

$$\begin{pmatrix} \boldsymbol{\beta}^{\infty}(x, y) \\ \mathbf{E}^{\infty}(x, y) \end{pmatrix} =$$
$$\frac{1}{2\pi i}\sum_{\alpha=1}^{8}\begin{pmatrix} \mathbf{A}^{(2)}_{\alpha} \\ -\varphi^{(2)}_{\alpha} \end{pmatrix} \otimes (\mathbf{m} + p^{(2)}_{\alpha}\mathbf{n})\frac{\mathrm{sgn}(\mathrm{Im}\, p^{(2)}_{\alpha})(\mathbf{L}^{(2)}_{\alpha} \cdot \mathbf{b})}{x - x' + p^{(2)}_{\alpha}(y - y')}, \tag{9.78}$$

which coincides with the known expression for a straight dislocation in an infinite piezoelectric medium [61]. The second term in (9.76) relates evidently to the image field,

$$I^{im}_{\alpha\beta}(x, y|x', y') =$$
$$= -\frac{1}{2\pi}\int_{-\infty}^{\infty} dk \exp\left\{ik[x - x' + p^{(2)}_{\alpha}(y - h) - p^{(2)}_{\beta}(y' - h)]\right\}\times \tag{9.79}$$
$$\times\left\{\mathbf{Y}(\mathbf{U}^{(k)})\right\}_{\alpha\beta}\mathrm{H}(k\; \mathrm{Im}\; p^{(2)}_{\alpha}),$$

which is not singular at $\mathbf{r} \to \mathbf{r}'$ and describes a contribution of inhomogeneity, surface, interface *etc.* Taking into account that the self-action of a straight dislocation in an infinite homogeneous medium vanishes, the interaction of the considered dislocation with its own stress field $\boldsymbol{\sigma}$ is completely determined by the image component $\boldsymbol{\sigma}^{im}$ of the latter. The corresponding Pitch-Koeller force is equal to $\mathbf{f}^{PK} = [\mathbf{b}\boldsymbol{\sigma}^{im}(x', y')] \times \boldsymbol{\tau}$, where $\boldsymbol{\tau} = \mathbf{m} \times \mathbf{n}$ is the unit vector directed along the dislocation, i.e. parallel to the z-axis. It is clear that the component σ_{zz} does not contribute to $\mathbf{f}^{PK}$, therefore eq. (9.75) can be used for its calculation. From general considerations the vector $\mathbf{f}^{PK}$ cannot have a non-vanishing component on $\mathbf{m}$, because a translation of the dislocation along $\mathbf{m}$ retains its energy unchanged. Therefore we are interested only in the projection

$$\mathbf{f}^{PK}\cdot\mathbf{n}=\mathbf{b}\cdot\boldsymbol{\sigma}^{im}(x',y')\mathbf{m}=-\mathbf{b}\cdot\left[\frac{1}{2\pi}\sum_{\alpha,\beta=1}^{8}p_\alpha^{(2)}\left(\mathbf{L}_\alpha^{(2)}\otimes\mathbf{L}_\beta^{(2)}\right)\times\right.$$
$$\left.\times\int\limits_{-\infty}^{\infty}dk e^{ik(p_\alpha^{(2)}+p_\beta^{(2)})(y'-h)}\{\mathbf{Y}(\mathbf{U}^{(k)})\}_{\alpha\beta}\mathrm{H}(k\mathrm{Im}\,p_\alpha^{(2)})\right]\mathbf{b}. \tag{9.80}$$

Of course, a division of the found fields into singular and image parts can be done also for a source position in the layer ($y'<h$). Moreover, the potential field (9.70) quite similarly can be divided into two analogous parts:

$$\boldsymbol{\eta}(x,y)=\boldsymbol{\eta}^{\infty(j)}(x,y)+\boldsymbol{\eta}^{im}(x,y), \tag{9.81}$$

where the first term relates to the field of the general line defect in an infinite medium coinciding in its material parameters with the layer ($j=1,\ y'<h$) or the substrate ($j=2,\ y'>h$):

$$\boldsymbol{\eta}^{\infty(j)}(x,y)=$$
$$=\frac{1}{2\pi i}\sum_{\alpha=1}^{8}\mathrm{sgn}\left(\mathrm{Im}\,p_\alpha^{(j)}\right)\boldsymbol{\xi}_\alpha^{(j)}\left[\boldsymbol{\xi}_\alpha^{(j)}\cdot\mathbf{Tg}\right]\ln[x-x'+p_\alpha^{(j)}(y-y')]. \tag{9.82}$$

In fact, obtained solutions (9.70),(9.73)-(9.75) allow finding the electro-elastic fields excited by any 2D distributions of the considered sources. For example, the electric field $\mathbf{E}(x,y)$ produced by the distribution of the body forces $\mathbf{f}(x,y)$ in the substrate, is given at $y>h$ by

$$\mathbf{E}(x,y)=$$
$$=\sum_{\alpha,\beta=1}^{8}\varphi_\alpha^{(2)}(\mathbf{m}+p_\alpha^{(2)}\mathbf{n})\int\limits_{-\infty}^{\infty}dx'\int\limits_{h}^{\infty}dy'\left(\mathbf{A}_\beta^{(2)}\cdot\mathbf{f}(x',y')\right)I_{\alpha\beta}(x,y|x',y'). \tag{9.83}$$

That is why the above solutions can be considered as Green's functions.

(b) Free metallized surface, eq. (9.31)$_2$

This time the boundary condition $\mathbf{V}_F(k,0)=0$ requires to start from a transformation of the general solution $\boldsymbol{\eta}(k,y)$ (9.48) to its $F-$representation, eqs. (9.28), (9.53), (9.54), which, in fact, differs from the initial representation just by the order in the list of the same fields. Then one should equate the lower 4-vector block $\mathbf{V}_F(k,0)$, eq.(9.28)$_2$, to zero. The further development is completely analogous to the above derivation of eq. (9.70).The result can be expressed in the same form, eq. (9.70), where the matrix $\mathbf{Y}(\mathbf{U}^{(k)})$ is replaced by $\mathbf{Y}(\mathbf{V}_F^{(k)})$. In its turn, the 8×8 matrix $\mathbf{Y}(\mathbf{V}_F^{(k)})$ has the same structure as $\mathbf{Y}(\mathbf{U}^{(k)})$, eq. (9.69), where one should replace all sub-matrices $\mathbf{U}^{(k)}$ and

$\mathbf{U}^{(-k)*}$ by $\mathbf{V}_F^{(k)}$ and $\mathbf{V}_F^{(-k)*}$, respectively, with the 4×4 matrix $\mathbf{V}_F^{(k)}$ defined by the four columns $\mathbf{V}_{Fk\alpha}(0)$, (9.54):

$$\mathbf{V}_F^{(k)} = [\mathbf{V}_{Fk1}(0), \mathbf{V}_{Fk2}(0), \mathbf{V}_{Fk3}(0), \mathbf{V}_{Fk4}(0)]. \tag{9.84}$$

(c) Clamped surface with vanishing charge density, eq. (9.30)$_2$

The boundary problem $\mathbf{U}_F(k, 0) = 0$ is again solved by eq. (9.70), where the matrix $\mathbf{Y}(\mathbf{U}^{(k)})$ is replaced by $\mathbf{Y}(\mathbf{U}_F^{(k)})$ of the same structure (9.69) with the matrix $\mathbf{U}_F^{(k)}$ built of the four columns $\mathbf{U}_{Fk\alpha}(0)$, eq. (9.53):

$$\mathbf{U}_F^{(k)} = [\mathbf{U}_{Fk1}(0), \mathbf{U}_{Fk2}(0), \mathbf{U}_{Fk3}(0), \mathbf{U}_{Fk4}(0)]. \tag{9.85}$$

(d) Free surface with vanishing charge density, eq. (9.31)$_1$

The boundary conditions are given by the block equation $\mathbf{V}(k, 0) = 0$. The solution has again form (9.70) with $\mathbf{Y}(\mathbf{U}^{(k)}) \to \mathbf{Y}(\mathbf{V}^{(k)})$,

$$\mathbf{V}^{(k)} = \{\ \mathbf{V}_{k1}(0), \mathbf{V}_{k2}(0), \mathbf{V}_{k3}(0), \mathbf{V}_{k4}(0)\}, \tag{9.86}$$

where the 4-vectors $\mathbf{V}_{k\alpha}(0)$ are defined by eq. (9.51).

(e) Clamped surface adjoined to an isotropic dielectric medium, eq. (9.30)$_3$

The boundary problem $\mathbf{U}^s(k, 0) = 0$ is solved as before by eq.(9.70) after the replacement $\mathbf{Y}(\mathbf{U}^{(k)}) \to \mathbf{Y}(\mathbf{U}^{s(k)})$, eq. (9.69), where

$$\mathbf{U}^{s(k)} = \{\mathbf{U}_{k1}^s(0),\ \mathbf{U}_{k2}^s(0),\ \mathbf{U}_{k3}^s(0),\ \mathbf{U}_{k4}^s(0)\}, \tag{9.87}$$

with the 4-vectors $\mathbf{U}_{k\alpha}^s(0)$ given by (9.55).

(f) Free surface adjoined to an isotropic dielectric medium, eq. (9.31)$_3$

In this case again the Green's function solving the boundary problem $\mathbf{V}^s(k, 0) = 0$ has the structure given by eqs. (9.70), (9.69) where one should make the following transformations: $\mathbf{Y}(\mathbf{U}^{(k)}) \to \mathbf{Y}(\mathbf{V}^{s(k)})$, $\mathbf{U}^{(k)} \to \mathbf{V}^{s(k)}$, with the matrix $\mathbf{V}^{s(k)}$ defined by the columns $\mathbf{V}_{k\alpha}^s(0)$, eq. (9.56):

$$V^{s(k)} = \{\mathbf{V}_{k1}^s(0),\ \mathbf{V}_{k2}^s(0),\ \mathbf{V}_{k3}^s(0),\ \mathbf{V}_{k4}^s(0)\}. \tag{9.88}$$

9.5 Electro-elastic fields excited at the surface of the structure

The boundary problems of the series (9.32), (9.33) could be solved, basing on the above solutions for the fields excited by body sources, by means of a limiting procedure with tending positions of appropriate combinations of sources to the surface, $x' \to x_0$, $y' \to 0$, with subsequent replacements: $\mathbf{b} \to -\mathbf{u}_0$, $\Delta\varphi \to -\varphi_0$, $\mathbf{f} \to \mathbf{F}$, $q \to q_0$. However, technically it will be much easier for us to find these solutions directly than to transform six variants of eq. (9.70). Anyway, of course, the final results in both cases are identical.

Below we shall look for a solution in the form replacing the combination of (9.48) with (9.12):

$$\boldsymbol{\eta}(x,y) = \frac{1}{2\pi i} \sum_{\alpha,\beta=1}^{8} \int_{-\infty}^{\infty} \frac{dk}{k} \exp[ik(x-x')] \boldsymbol{\xi}_{k\alpha}(y) c_{k\alpha} \mathrm{H}(k \mathrm{Im}\, p_\alpha^{(2)}), \quad (9.89)$$

which is organized so that, due to (9.39) and (9.43), again provides finiteness of the solution at infinity $y \to \infty$ by excluding the "bad" large terms related to $\alpha = 1, \ldots, 4$ at $k < 0$ and $\alpha = 5, \ldots, 8$ at $k > 0$.

The further formulations of boundary problems will be based on the three block representations obtained by means of the explained above rearrangements of the function $\boldsymbol{\eta}(k,y)$:

$$\begin{pmatrix} \mathbf{U}(k,0) \\ \mathbf{V}(k,0) \end{pmatrix} = \frac{e^{-ikx_0}}{2\pi i k} \sum_{\alpha=1}^{8} \begin{pmatrix} \mathbf{U}_{k\alpha}(0) \\ \mathbf{V}_{k\alpha}(0) \end{pmatrix} c_{k\alpha} \mathrm{H}(k \ \mathrm{Im}\ p_\alpha^{(2)}), \quad (9.90)$$

$$\begin{pmatrix} \mathbf{U}_F(k,0) \\ \mathbf{V}_F(k,0) \end{pmatrix} = \frac{e^{-ikx_0}}{2\pi i k} \sum_{\alpha=1}^{8} \begin{pmatrix} \mathbf{U}_{Fk\alpha}(0) \\ \mathbf{V}_{Fk\alpha}(0) \end{pmatrix} c_{k\alpha} \mathrm{H}(k \ \mathrm{Im}\ p_\alpha^{(2)}), \quad (9.91)$$

$$\begin{pmatrix} \mathbf{U}^{(s)}(k,0) \\ \mathbf{V}^{(s)}(k,0) \end{pmatrix} = \frac{e^{-ikx_0}}{2\pi i k} \sum_{\alpha=1}^{8} \begin{pmatrix} \mathbf{U}^s_{k\alpha}(0) \\ \mathbf{V}^s_{k\alpha}(0) \end{pmatrix} c_{k\alpha} \mathrm{H}(k \ \mathrm{Im}\ p_\alpha^{(2)}). \quad (9.92)$$

(a) Displacements and electric potential prescribed at the surface, eq. (9.32)$_1$

The boundary conditions (9.20),(9.24) are formulated in the form

$$\mathbf{U}(k,0) = \mathbf{U}_\varphi \exp(-ikx_0)/2\pi i k, \mathbf{U}_\varphi = [\mathbf{u}_0, \varphi_0]^t. \quad (9.93)$$

Combining (9.93) with the upper block of (9.90) one obtains the equations for $c_{k\alpha}$

$$\sum_{\alpha=1}^{4} \mathbf{U}_{k\alpha}(0) c_{k\alpha} = \mathbf{U}_\varphi, \quad \sum_{\alpha=1}^{4} \mathbf{U}^*_{-k\alpha}(0) c_{k\alpha+4} = \mathbf{U}_\varphi, \quad (9.94)$$

with solutions

$$c_{k\alpha} = \bar{\mathbf{U}}_{k\alpha} \cdot \mathbf{U}_\varphi, \quad c_{k\alpha+4} = [\bar{\mathbf{U}}_{-k\alpha}]^* \cdot \mathbf{U}_\varphi, \quad \alpha = 1, \ldots, 4, \quad (9.95)$$

where the 4-vectors $\bar{\mathbf{U}}_{k\alpha}$ are defined by (9.63) as lines of the inverse matrix $[\mathbf{U}^{(k)}]^{-1}$ (the matrix $\mathbf{U}^{(k)}$ is built of the columns $\mathbf{U}_{k\alpha}(0)$, $\alpha = 1, \ldots, 4$, see (9.62)).

Thus, the desired Green's function solving the boundary problem (9.93) is given by eq.(9.89) with coefficients $c_{k\alpha}$ defined by (9.95). The convergence of the Fourier integral is easily checked quite similarly to our consideration in Sec. 9.4 (a – ii).

(b) Loads and electric potential prescribed at the surface, eq. (9.33)$_2$

The boundary problem defined by eqs. (9.18), (9.24) is stated as

$$\mathbf{V}_F(k,0) = \mathbf{V}_\varphi \exp(-ikx_0)/2\pi ik, \quad \mathbf{V}_\varphi = [\mathbf{F}, \varphi_0]^t. \tag{9.96}$$

The procedure is practically identical to that in the previous section. The coefficients $c_{k\alpha}$ in eq. (9.89) this time equal to

$$c_{k\alpha} = \bar{\mathbf{V}}^F_{k\alpha} \cdot \mathbf{V}_\varphi, \quad c_{k\alpha+4} = [\bar{\mathbf{V}}^F_{-k\alpha}]^* \cdot \mathbf{V}_\varphi, \quad \alpha = 1, \ldots, 4, \tag{9.97}$$

where $\bar{\mathbf{V}}^F_{k\alpha}$ are the 4-vectors defined by the lines of the 4×4 matrix inverse to $\mathbf{V}^{(k)}_F$, eq. (9.84):

$$\left[\mathbf{V}^{(k)}_F\right]^{-1} = \left[\bar{\mathbf{V}}^F_{k1},\ \bar{\mathbf{V}}^F_{k2},\ \bar{\mathbf{V}}^F_{k3},\ \bar{\mathbf{V}}^F_{k4}\right]^t. \tag{9.98}$$

(c) Displacements and charge density prescribed at the surface, eq. (9.32)$_2$
The boundary conditions (9.20),(9.22) are combined to

$$\mathbf{U}_F(k,0) = \mathbf{U}_q \exp(-ikx_0)/2\pi ik, \quad \mathbf{U}_q = [\mathbf{u}_0, -q_0]^t. \tag{9.99}$$

The problem is solved by the substitution into (9.89) the coefficients

$$c_{k\alpha} = \bar{\mathbf{U}}^F_{k\alpha} \cdot \mathbf{U}_q, \quad c_{k\alpha+4} = [\bar{\mathbf{U}}^F_{-k\alpha}]^* \cdot \mathbf{U}_q, \quad \alpha = 1, \ldots, 4, \tag{9.100}$$

with the 4-vectors $\bar{\mathbf{U}}^F_{k\alpha}$ defined, as before, as lines of the matrix inverse to $\mathbf{U}^{(k)}_F$ (9.85):

$$\left[\mathbf{U}^{(k)}_F\right]^{-1} = \left[\bar{\mathbf{U}}^F_{k1},\ \bar{\mathbf{U}}^F_{k2},\ \bar{\mathbf{U}}^F_{k3},\ \bar{\mathbf{U}}^F_{k4}\right]^t. \tag{9.101}$$

(d) Loads and charge density prescribed at the surface, eq. (9.33)$_1$

The conditions at the surface (9.18), (9.22) are expressed by

$$\mathbf{V}(k,0) = \mathbf{V}_q \exp(-ikx_0)/2\pi ik, \quad \mathbf{V}_q = [\mathbf{F}, -q_0]^t. \tag{9.102}$$

The solution (9.89) is determined by the coefficients

$$c_{k\alpha} = \bar{\mathbf{V}}_{k\alpha} \cdot \mathbf{V}_q, \quad c_{k\alpha+4} = [\bar{\mathbf{V}}_{-k\alpha}]^* \cdot \mathbf{V}_q, \quad \alpha = 1, \ldots, 4, \tag{9.103}$$

where $\bar{\mathbf{V}}_{k\alpha}$ are the 4-vectors equal to the lines of the matrix inverse to $\mathbf{V}^{(k)}$, eq. (9.86):

$$\left[\mathbf{V}^{(k)}\right]^{-1} = \left[\bar{\mathbf{V}}_{k1},\ \bar{\mathbf{V}}_{k2},\ \bar{\mathbf{V}}_{k3},\ \bar{\mathbf{V}}_{k4}\right]^t. \tag{9.104}$$

(e) Displacements prescribed at the surface adjoined to an isotropic dielectric medium, eq. (9.32)$_3$

The boundary problem (9.20), (9.26) is reduced to

$$\mathbf{U}^{(s)}(k,0) = \mathbf{U}_0 \exp(-ikx_0)/2\pi ik, \quad \mathbf{U}_0 = [\mathbf{u}_0, 0]^t, \tag{9.105}$$

and is solved by eq. (9.89) with coefficients

$$c_{k\alpha} = \bar{\mathbf{U}}^s_{k\alpha} \cdot \mathbf{U}_0, \quad c_{k\alpha+4} = [\bar{\mathbf{U}}^s_{-k\alpha}]^* \cdot \mathbf{U}_0, \quad \alpha = 1, \ldots, 4. \tag{9.106}$$

The 4-vectors $\bar{\mathbf{U}}^s_{k\alpha}$ are defined by the lines of the inverse matrix

$$\left[\mathbf{U}^{s(k)}\right]^{-1} = \left[\bar{\mathbf{U}}^s_{k1}, \ \bar{\mathbf{U}}^s_{k2}, \ \bar{\mathbf{U}}^s_{k3}, \ \bar{\mathbf{U}}^s_{k4}\right]^t. \tag{9.107}$$

where $\mathbf{U}^{s(k)}$ is matrix (9.87).

(f) Loads prescribed at the surface adjoined to an isotropic dielectric medium, eq. (9.33)$_3$

The boundary conditions (9.18), (9.26), or

$$\mathbf{V}^{(s)}(k,0) = \mathbf{V}_0 \exp(-ikx_0)/2\pi ik, \quad \mathbf{V}_0 = [\mathbf{F}, 0]^t \tag{9.108}$$

are satisfied by the solution (9.89) with the coefficients

$$c_{k\alpha} = \bar{\mathbf{V}}^s_{k\alpha} \cdot \mathbf{V}_0, \quad c_{k\alpha+4} = [\bar{\mathbf{V}}^s_{-k\alpha}]^* \cdot \mathbf{V}_0, \quad \alpha = 1, \ldots, 4. \tag{9.109}$$

The 4-vectors $\bar{\mathbf{V}}^s_{k\alpha}$ are given by the lines of the inverse matrix

$$\left[\mathbf{V}^{s(k)}\right]^{-1} = \left[\bar{\mathbf{V}}^s_{k1}, \ \bar{\mathbf{V}}^s_{k2}, \ \bar{\mathbf{V}}^s_{k3}, \ \bar{V}^s_{k4}\right]^t. \tag{9.110}$$

where $\mathbf{V}^{s(k)}$ is defined by eq. (9.88).

9.6 Conclusions

Thus, we have constructed the series of Green's functions describing 2D electro-elastic fields excited by line sources in the interior or at the surface of the piezoelectric layer-substrate structure. The solutions of twelve boundary problems (9.30)-(9.33) are expressed in the form of convergent Fourier integrals in terms of eigenvectors and eigenvalues of the 8 × 8 Stroh matrices for the layer and the substrate. Finding these two sets of parameters, $\boldsymbol{\xi}^{(1)}_\alpha$, $p^{(1)}_\alpha$ and $\boldsymbol{\xi}^{(2)}_\alpha$, $p^{(2)}_\alpha$ ($\alpha = 1, \ldots, 8$), needs additional (fortunately, rather trivial) computations which cannot be done explicitly. On the other hand, that appears to be quite a modest price for the obtained general analytical results for arbitrary piezoelectric media of unrestricted anisotropy.

The solutions found in terms of potentials $\boldsymbol{\eta}(x, y) = [\mathbf{u}(x, y),\ \varphi(x, y),\ \mathbf{J}(x, y),\ D(x, y)]$ by simple differentiations (9.73) can be transformed to the elastic distortions $\beta(x, y)$ and stresses $\sigma(x, y)$, together with the electric field $\mathbf{E}(x, y)$ and the electric displacement $\mathbf{D}(x, y)$. The corresponding illustrations presented in Sec.9.4 (a – iii) can be straightforwardly adapted for the other boundary problems. And, as was mentioned above, with found solutions for line sources one can easily obtain electro-elastic fields produced by any 2D distributed sources of the considered types (see, for instance, eq. (9.83)). In this sense the constructed general solutions for $\boldsymbol{\eta}(x, y|x', y')$ are Green's functions.

10

Acoustic Waves in Piezomagnetic and Piezoelectric Structures

In previous chapters of this part of the book we considered only static problems related to a description of electro-elastic coupling fields induced by various sources in a series of bounded structures. Below we shall discuss a principally different type of coupled fields, which arise not due to sources but as eigenwaves in structures of the same types. We shall consider two examples of such eigenwaves: magneto-elastic surface waves in a bicrystalline gap structure and electro-elastic waveguide SH waves in a piezoelectric plate. Our attention will be concentrated on the very peculiar resonance phenomenon in the vicinity of the corresponding leaky wave solutions which provide an efficient method of excitation of both types of eigenwaves.

The content of this chapter is based on the papers [68, 69].

10.1 Bleustein-Gulyaev type surface waves in piezomagnetics

(a) Statement of the problem

As was explained in Ch. 1, the only practically important type of magneto-mechanical coupling is provided by magnetostriction which is present even in centrosymmetric media being described by the moduli tensor of the 4th rank, q_{ijkl}. Under high enough bias magnetic field $\mathbf{H}_0$ the corresponding linearized effect can be described as effective piezomagnetic coupling with piezomoduli tensor $m'_{kij} = H_l^0 q_{lkij}$, eqn. (1.101). In this chapter speaking about magneto-mechanical interaction we shall mean just this linearized piezomagnetism.

Let us consider a free semi-infinite medium belonging to one of the magnetic symmetry classes:

$$\begin{gathered} \infty/mm' \quad \infty 2', \quad \infty m', \\ 6/mm'm' \quad 62'2', \quad 6m'm', \quad \bar{6}m2', \\ 4/mm'm', \quad 42'2', \quad 4m'm', \quad \bar{4}m2'. \end{gathered} \tag{10.1}$$

The principal axis (z-axis of the corresponding crystallophysical coordinate system) is supposed to be parallel to the surface, and the sagittal plane of wave propagation is chosen in the basal plane xy of the medium orthogonal to its principal axis, the y-axis being along the internal unit normal $\mathbf{n}$ to the surface. The bias magnetic field $\mathbf{H}_0$ is applied along z-axis. We shall be interested by only the transverse wave family characterized by mechanical displacements $\mathbf{u}$ perpendicular to the plane of propagation (i.e. along the principal axis: $\mathbf{u} \parallel z$). With such a choice of symmetry of the half-space, the direction of the field $\mathbf{H}_0$ and the geometry of wave propagation, the medium will behave in our problem as transversely isotropic. In particular, the only nonvanishing piezomagnetic moduli m'_{ikl} (1.101)

$$\begin{aligned} m_{15} &\equiv m'_{131} = q_{3131} H_{03} \equiv q_{55} H_0, \\ m_{24} &\equiv m'_{223} = q_{3232} H_{03} \equiv q_{44} H_0 \end{aligned} \tag{10.2}$$

must coincide, $m_{15} = m_{24}$, because $q_{44} = q_{55}$. In the same reduced suffix notation the elastic moduli $c_{\alpha\beta}$ (α, $\beta = 1,\ldots, 6$) and the magnetic permeability μ_α will contribute to our equations only by components

$$c_{44} = c_{55}, \quad \mu_1 = \mu_2 \equiv \mu. \tag{10.3}$$

(b) Basic equations and boundary conditions

The magnetoelastic dynamics of the of the considered medium is described by the system of the equation of the dynamic theory of elasticity (1.40)

$$\rho \ddot{\mathrm{u}} = \mathrm{div}\boldsymbol{\sigma} \quad (y > 0) \tag{10.4}$$

and Maxwell's equations (in the quasistatic approximation, see Ch. 1)

$$\mathrm{div}\mathbf{B} = 0 \ (y > 0), \quad \mathrm{div}\mathbf{B}^{(v)} = 0 \ (y < 0), \tag{10.5}$$

where $\mathbf{B}$ and $\mathbf{B}^{(v)}$ are the magnetic inductions in the medium and in the adjoined vacuum, respectively. By (1.92), (1.93) the mechanical stress tensor $\boldsymbol{\sigma}$ and the magnetic inductions $\mathbf{B}$ and $\mathbf{B}^{(v)}$ are determined by the constitutive equations

$$\sigma_{ij} = c_{ijkl} u_{l,k} - m'_{kij} H_k, \tag{10.6}$$

$$B_i = m'_{ikl} u_{l,k} + \mu_{ik} H_k, \tag{10.7}$$

$$B_i^{(\mathrm{V})} = \mu_0 H_i^{(\mathrm{V})}, \tag{10.8}$$

where μ_0 is the magnetic permeability of vacuum.

The above equations (10.4)-(10.8) must be added by the boundary conditions at the surface $y = 0$: the free surface condition

$$\boldsymbol{\sigma}\mathbf{n}|_{y=0} = 0, \tag{10.9}$$

and the condition of a continuity of the normal components of magnetic inductions,

$$(\mathbf{B} - \mathbf{B}^{(v)}) \cdot \mathbf{n}|_{y=0} = 0, \tag{10.10}$$

and of the tangential components of magnetic fields,

$$(\mathbf{H} - \mathbf{H}^{(v)}) \times \mathbf{n}|_{y=0} = 0. \tag{10.11}$$

In fact, the presented cumbersome system (10.4)-(10.11) becomes much more simple and solvable for our particular choice of symmetry and geometry of the problem. However, we shall not go to details of calculations and limit ourselves below by presenting the final results without derivation.

(c) The surface wave solution of the Bleustein-Gulyaev type

When switching off the magnetoelastic coupling in our medium by putting $H_0 = 0$ the eigenwave field polarized along z is formed by the bulk waves, the so-called SH (*shear horizontal*) modes, propagating with the phase speed

$$v_t^{(0)} = (c_{44}/\rho)^{\frac{1}{2}}. \tag{10.12}$$

As is well known, this wave family automatically satisfies the free surface condition (10.9).

Switching on the magnetoelastic interaction changes the bulk wave speed (10.12) to

$$v_t^2 = c_{44}(1 + m^2)/\rho, \qquad (10.12')$$

where m^2 is the magnetomechanical coupling coefficient defined by

$$m^2 = m_{15}^2/\mu c_{44}. \tag{10.13}$$

These modified bulk waves form eigenwave spectrum for an unbounded medium or determine pairs of incident and reflected waves of the reflection problem at the surface of a semi-infinite medium. However they cannot propagate along the surface anymore – they do not satisfy the boundary conditions (10.9)-(10.11). This reflects the well known fact that the magnetoelastic coupling brings to a localization of the above bulk eigenmode transforming it into a surface acoustic wave of the Bleustein-Gulyaev type [183-186] (Fig. 10.1). This is again a one-partial wave with the displacement vector $\mathbf{u} \,||z$, however with the different speed $v_{BG} < v_t$ and with the amplitude decreasing exponentially in the depth of the medium:

$$u_{BG}(\mathbf{r}, t) = Ae^{-q_0 m^2 ky} \exp[ik(x - v_{BG}t)], \tag{10.14}$$

where A is the amplitude of the wave on the surface, k is a wavenumber, and

$$q_0 = (1 + \mu/\mu_0)^{-1}, \quad v_{BG} = v_t(1 - q_0^2 m^4/2). \tag{10.15}$$

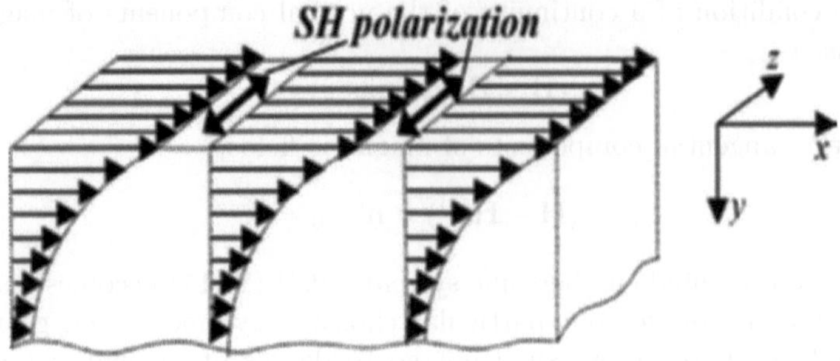

Fig. 10.1. Surface acoustic wave of the Bleustein-Gulyaev type.

The surface wave of displacements (10.14) is accompanied by the wave of quasistatic magnetic field localized at the surface from both sides of the latter. In particular, the wave of the magnetic potential φ^H (1.4) in the adjoined vacuum ($y < 0$) is given by

$$\varphi^H(r,t) = \Phi_0 e^{ky} \exp[ik(x - v_{BG}t)], \tag{10.16}$$

where Φ_0 is the amplitude of this wave.

The presented solution is a complete analog of the known Bleustein-Gulyaev surface eigenwave in piezoelectrics [66, 67].

10.2 Magnetoelastic SH waves in a bicrystalline gap structure

(a) The geometry of wave propagation and some preliminary remarks

Consider the bicrystalline structure consisting of two different semi-infinite media of symmetry (10.1) devided by a finite vacuum gap of the thickness h (Fig. 10.2). The principal axes of media are parallel both to each other and to the surface. They determine the direction of the z-axis of our coordinate system. The sagittal plane of wave propagation is again chosen in the coinciding basal planes (xy) of the halves of the structure orthogonal to their principal axes. And the bias magnetic field $\mathbf{H}_0$, as before, is applied along the z-axis.

In the considered bicrystalline gap structure the SH polarized acoustic waves will again "perceive" the both media as transversely isotropic. We shall distinguish material constants in these media by adding a wavy "hat" to each characteristic ($\tilde{c}_{44}$, $\tilde{m}_{15}$, $\tilde{\mu}$ *etc.*) for one of the media (see Fig. 10.2).

It is rather evident that for a large enough gap h as compared with the wavelength, when $kh \gg 1$ and the parameter exp(-kh) is extremely small, the SH polarized eigenwaves in both media of our structure reduce in zero approximation to the above presented Bleustein-Gulyaev localized waves of the type (10.14)-(10.16). However at finite kh the acoustic fields in the upper and lower half-spaces become more sensitive to each other due to the magnetomechanical coupling. There are well known analogous effects of elastic wave penetration through the gap of similar piezoelectric structures [187-190]. Usually they make use of a relative weakness of considered coupling effects which allows them to apply the perturbation theory. We shall use the same idea supposing m^2 and $\tilde{m}^2$ to be small parameters. On this basis one can easily obtain the coupled solutions for SH acoustic fields transforming the above zero approximation in the vicinity of the two tracing speeds: v_{BG} (10.15) in the upper half-space in Fig. 10.2, and

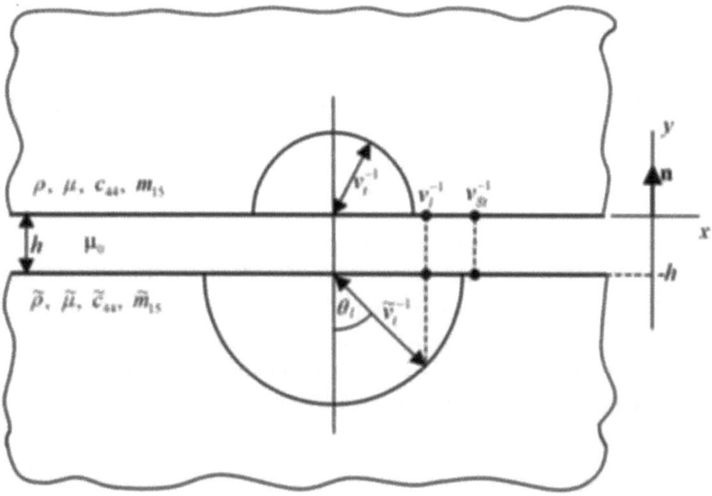

Fig. 10.2. The bicrystalline gap structure and its basic material parameters and acoustic characteristics.

$$\tilde{v}_{BG} = \tilde{v}_t(1 - \tilde{q}_0^2\tilde{m}^4/2), \quad \tilde{q}_0 = (1 + \tilde{\mu}/\mu_0)^{-1} \tag{10.17}$$

in the lower half-space. The results of this transformation turn out to be completely different for the above two waves. We start from the perturbation of the surface wave in the lower medium.

(b) The Stoneley type localized SH acoustic wave

Let us suppose for definiteness that $\tilde{v}_t < v_t$ as in Fig. 10.2. In this case the surface wave in the lower simi-infinite medium belongs to a subsonic speed region $v < v_t$ with respect to the upper medium. Accordingly the accompanying dynamic field in this medium must be also localized at the surface. The complete solution is contained by two surface waves propagating in both media with the same tracing speed, i.e. the wave field is of the Stoneley type (Fig. 10.3):

$$u_{St}(\mathbf{r},t) = u_{St}(y)\exp[ik(x - v_{St}t)], \tag{10.18}$$

$$u_{St}(y) = \begin{cases} A_{St}\exp(-pky), & y \geq 0, \\ \tilde{A}_{St}\exp[\tilde{m}^2\tilde{q}k(y+h)], & y \leq -h, \end{cases} \tag{10.19}$$

where the localization is determined by the parameters

$$p = \left(1 - \frac{\tilde{v}_t^2}{v_t^2}\right)^{\frac{1}{2}}, \quad \tilde{q} = \left(1 + \frac{\tilde{\mu}}{\mu_0}\frac{\mu_0 + \mu\tanh kh}{\mu + \mu_0\tanh kh}\right)^{-1}, \tag{10.20}$$

and the relation between the amplitudes on the surfaces $y = 0$ and $y = -h$ is given by

$$\frac{A_{St}}{\tilde{A}_{St}} = -e^{-kh}\frac{m_{15}\tilde{m}_{15}}{c_{44}\tilde{\mu}}\frac{\mu_0 + (\tilde{\mu} - \mu_0)\tilde{q}}{(\mu + \mu_0)p}. \tag{10.21}$$

The tracing speed of propagation is again very close to the speed $\tilde{v}_t$ of SH bulk waves in the lower sub-space:

$$v_{St} = \tilde{v}_t(1 - \tilde{m}^4\tilde{q}^2/2), \tag{10.22}$$

though the quasi-bulk character of the wave field occurs only at the lower half of the structure. At $y > 0$ the wave is localized in the much narrower layer of the depth of order of several wavelengths. However, as was expected, this accompanied wave has rather small amplitude proportional to $m\tilde{m}\exp(-kh)$. At $kh \to \infty$ the solution (10.18)-(10.22) is naturally transformed back to an ordinary Bleustein-Gulyaev wave in the lower medium.

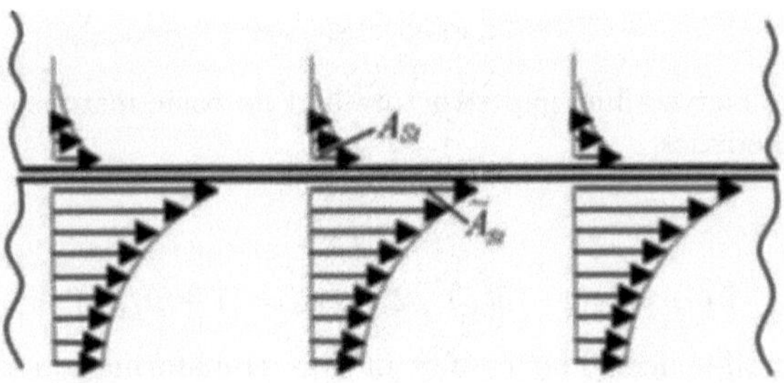

Fig. 10.3. A Stoneley type wave in the bicrystalline gap structure.

(c) The leaky wave solution in the vicinity of v_{BG}

The results of the perturbation of the other initial Bleustein-Gulyaev wave in the upper medium must be qualitatively different because its speed v_{BG} belongs to a supersonic region for the other medium ($v_{BG} > \tilde{v}_t$, see Fig. 10.2). In other words, the accompanied magnetoinduced acoustic penetration into this medium must have a form of the weak bulk wave outgoing from the surface, i.e. providing a leakage of energy from the upper medium (Fig. 10.4).

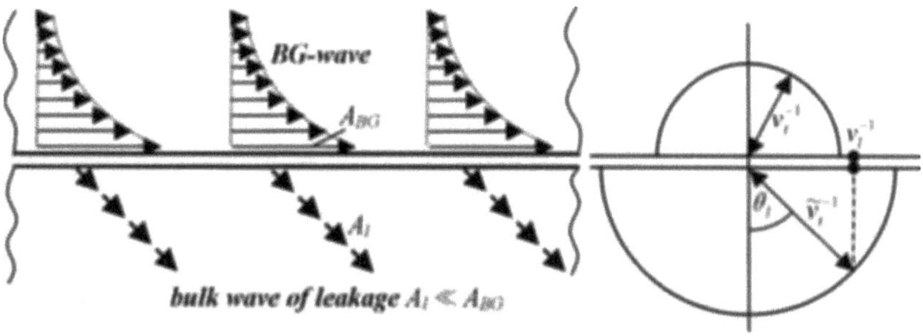

Fig. 10.4. A leaky wave eigenmode in the bicrystalline gap structure.

Thus, the considered transformation of the second Bleustein-Gulyaev wave creates the so-called leaky eigenmode:

$$u_l(\mathbf{r},t) = u_l(y)\exp[iK_l(x - V_l t)], \tag{10.23}$$

$$u_l(y) = \begin{cases} A_s \exp(-m^2 q k_l y), & y \geq 0, \\ A_l \exp[-ipk_l(y+h)], & y \leq -h, \end{cases} \tag{10.24}$$

where

$$q = \left(1 + \frac{\mu}{\mu_0}\frac{\mu_0 + \tilde{\mu}\tanh k_l h}{\tilde{\mu} + \mu_0 \tanh k_l h}\right)^{-1}, \tag{10.25}$$

p is again given by (10.20), and the relation for the amplitudes is

$$\frac{A_l}{A_s} = -ie^{-k_l h}\frac{m_{15}\tilde{m}_{15}}{\tilde{c}_{44}\mu}\frac{\mu_0 + (\mu - \mu_0)q}{(\tilde{\mu} + \mu_0)p}. \tag{10.26}$$

As it must be for a leaky mode, both the speed V_l and the wavenumber K_l prove to be complex:

$$V_l = v_l - iv_l', \tag{10.27}$$

$$K_l = k_l + ik_l' = \omega/(v_l - iv_l') \approx \frac{\omega}{v_l}\left(1 + i\frac{v_l'}{v_l}\right), \tag{10.28}$$

where

$$v_l = v_t(1 - m^4 q^2/2), \quad v_l' = v_t \frac{m^4 \tilde{m}^2 \mu \tilde{\mu} q^3}{p(\tilde{\mu}\cosh k_l h + \mu_0 \sinh k_l h)^2}. \tag{10.29}$$

The latter expression at $h = 0$ coincides with that presented in [191]. The limit $k_l h \to \infty$ in (10.23)-(10.29) again returns us to the Bleustein-Gulyaev solution (10.14), (10.15).

10.3 Resonance excitation of Bleustein-Gulyaev waves

(a) Resonant reflection

A slow decay of the surface component of the leaky wave (10.23), (10.24) which occurs along the surface $y = 0$ proportionally to $\exp(-k_l' x)$, clearly has nothing to do with energy dissipation. Such decrease of the wave amplitude is completely natural having in mind the energy leakage from the upper subspace provided by the bulk component with the amplitude A_l (10.24) propagating in the lower medium from the gap to the depth under the angle

$$\theta_l = \arcsin(\tilde{v}_t/v_l) \approx \arcsin(\tilde{v}_t/v_t), \tag{10.30}$$

see Fig. 10.2. In fact, this energy leakage together with the decay of the surface wave amplitude along the boundary $y = 0$ can be compensated if to direct the appropriate bulk wave incident in the lower medium from the depth to the gap under the angle $\theta \approx \theta_l$ so that the corresponding reflected wave would be close in direction of propagation with the wave of leakage. In view of (10.26), one can also expect that such a pumping wave of the amplitude A_i will excite in the upper medium the surface wave of much higher amplitude: $|A_s/A_i| \sim e^{k_l h}/m\tilde{m}$.

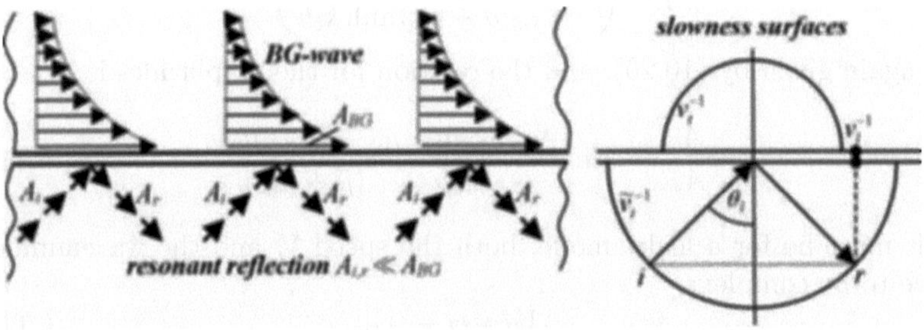

Fig. 10.5. A resonance reflection in the bicrystalline gap structure.

Thus we have come to the idea of a resonant reflection (Fig. 10.5) first introduced for a purely elastic half-space in [192] (see also [193-195]). Indeed, the direct calculation brings us to the wave picture which is in a good agreement with the above expectations. The resulting wave field of displacements is described by the following superposition:

$$u(\mathbf{r}, t) = u(y) \exp[ik_x(x - vt)], \tag{10.31}$$

$$u(y) = \begin{cases} A_s \exp(-m^2 q k_x y), & y \geq 0, \\ A_i \exp[ipk_x(y+h)] + A_r \exp[-ipk_x(y+h)], & y \leq -h, \end{cases} \tag{10.32}$$

where the tangential projection k_x of the wave vector of the pumping incident wave is supposed to be chosen close to the real part k_l of the wave number

K_l of the leaky eigenmode (10.28). The relations between the amplitudes A_s, A_i and A_r have quite a resonant character. The coefficient of excitation $S = A_s/A_i$,

$$S = \frac{\eta m (k_l k_l')^{\frac{1}{2}}}{k_x - k_l + ik_l'}, \tag{10.33}$$

$$\eta = -2(pq\tilde{c}_{44}/c_{44})^{\frac{1}{2}}, \tag{10.34}$$

as a function of k_x, forms a high and narrow peak with a maximum at $k_x = k_l$ (Fig. 10.6a) determined by

$$|S|_{\max} = -i\frac{p}{q}\frac{\tilde{c}_{44}(\tilde{\mu} + \mu_0)}{m_{15}\tilde{m}_{15}} e^{k_l h}. \tag{10.35}$$

The reflection coefficient $R = A_r/A_i$,

$$R = \frac{k_x - k_l - ik_l'}{k_x - k_l + ik'} = \exp(i\varphi), \tag{10.36}$$

is characterized by the phase shift φ,

$$\varphi = 2\mathrm{arc\,cot}\left(\frac{k_l - k_x}{k_l'}\right), \tag{10.37}$$

which changes its value by π in the very narrow region $-k_l' < k_x - k_l < k_l'$ (Fig. 10.6 b), manifesting an extremely high sensitivity to the incidence angle θ,

$$\kappa \equiv \left(\frac{d\varphi}{d\theta}\right)_{k_x = k_l} = \frac{2pv_t^2}{\tilde{v}_t v_l'} \gg 1. \tag{10.38}$$

The latter property promises a pronounced nonspecular reflection of the appropriate acoustic beam propagating in the lower medium under a resonance angle $\theta \approx \theta_l$ (10.30). In this case the general theory [196] predicts a lateral shift of the reflected beam along the surface on

$$\Delta x \sim \kappa/k_l. \tag{10.39}$$

If the beam width w is high enough and the resonance is not too narrow so that the beam divergence $\Delta k \sim w^{-1}$ remains inside the resonance range, i.e. $wk_l' \geq 1$, the estimate (10.39) in view of (10.38) and (10.28) gives

$$\Delta x \sim w. \tag{10.40}$$

The large shift (10.40) of the reflected beam must disappear at quite a small change $\delta\theta \sim k_l'/k_l$ of the incidence angle. The phenomenon of this kind is sometimes called a Schoch effect [197]. In addition to a lateral shift, it also manifests itself by a change of the width of the reflected beam and by a serious redistribution of the intensity along its cross-section [198, 199].

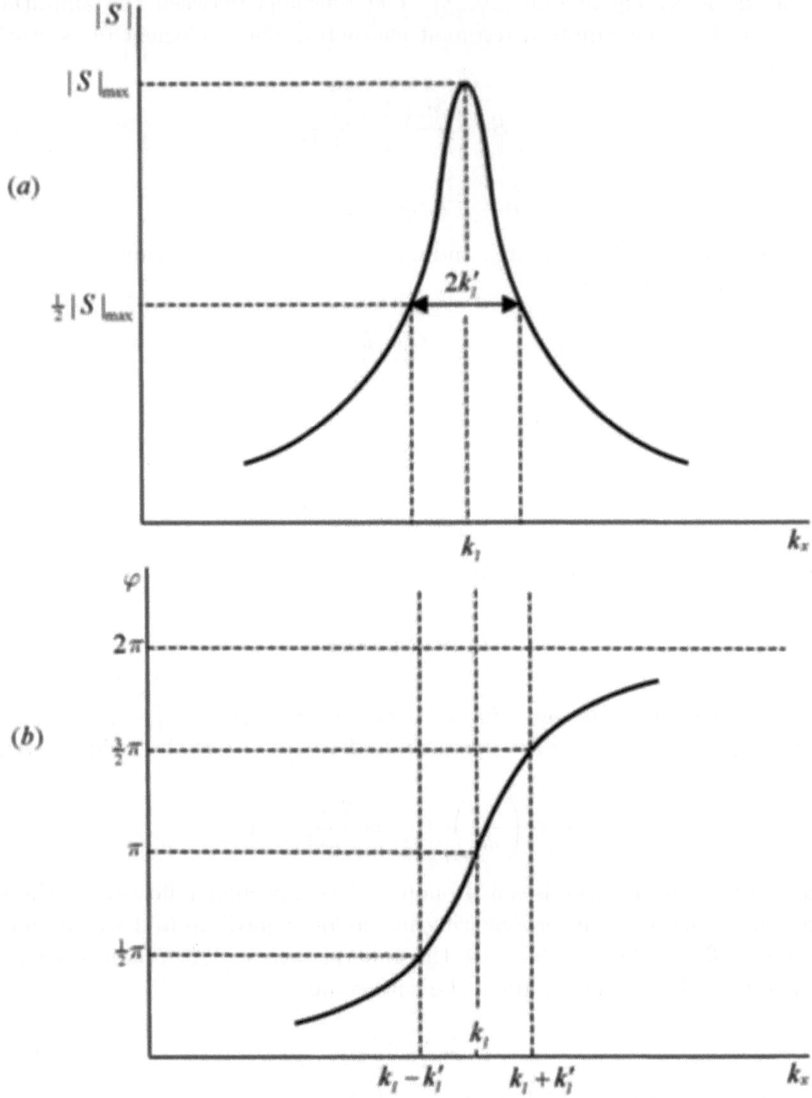

Fig. 10.6. The peak of the excitation coefficient (a) and the behaviour of the phase of the reflection coefficient (b) in the vicinity of the resonance.

(b) Some analysis of results

The above considered acoustical features are supposed to be created by a linearized magnetostrictional coupling under the bias magnetic field. It appears to be of interest to discuss the sensitivity of different acoustic characteristics to a change of this field H_0. In accordance with (10.2) and (10.13) the coefficients of magnetomechanical coupling m^2 and $\tilde{m}^2$ are proportional to H_0^2:

$$m^2 = (H_0/\bar{H})^2, \quad \tilde{m}^2 = (H_0/\bar{\tilde{H}})^2, \tag{10.41}$$

where

$$\bar{H} = (c_{44}\mu)^{\frac{1}{2}}/q_{44}, \quad \bar{\tilde{H}} = (\tilde{c}_{44}\tilde{\mu})^{\frac{1}{2}}/\tilde{q}_{44}. \tag{10.42}$$

Naturally, the larger are the magnetostrictional moduli q_{44} and $\tilde{q}_{44}$, the less are the characteristic fields $\bar{H}$ and $\bar{\tilde{H}}$, and the higher are the coefficients m^2 and $\tilde{m}^2$.

Let us estimate the magnetic influence on different acoustical parameters. For instance, the speeds of all considered surface waves of the Bleustein-Gulyaev type, v_{BG} (10.15), $\tilde{v}_{BG}$ (10.17), v_{St} (10.22) and v_l (10.29), at any field H_0 remain very close to corresponding bulk wave speeds v_t or $\tilde{v}_t$, and their regulation by means of the bias field does not look promising. On the other hand, the depths of localization of all those surface modes are determined by a dependence of the type

$$u(y) \propto \exp(-\alpha H_0^2 y), \tag{10.43}$$

being quite sensitive to the magnetic field H_0.

In contrast, the localization of the surface partial wave propagating in the upper medium (see eq. (10.19)) with the speed v_{St} (10.22) is practically independent of H_0. However its amplitude A_{St} (10.21) is proportional to H_0^2, i.e. it is an operated parameter. The same may be said about the amplitude A_l (10.26) of the bulk wave component of the leaky eigenmode (10.23), (10.24).

But the most sensitive parameter to the magnetic field H_0 proves to be an imaginary part of the leaky wave speed $v_l' \propto H_0^6$. Accordingly, the decay of the surface wave component of the leaky eigenmode along the boundary $y = 0$ is strongly dependent on H_0:

$$u(y=0) \propto \exp(-\beta H_0^6 x). \tag{10.44}$$

By the same reason the widths of the resonance zones of the functions (10.33) and (10.36), (10.37), $\Delta k \sim k_l' \propto H_0^6$, are also characteristics, which can be experimentally operated. For instance, in view of (10.38), (10.39) this feature must be observable on the magnetic dependence of the Schoch effect.

In fact, for a practical observation of the considered resonance reflection there are quite strict limitations for a choice of all physical parameters. For instance, in order to have the excitation coefficient $|S|_{\max} \sim 10$, one should have (at $h = 0$) the level of the magnetomechanical couplings m^2, $\tilde{m}^2 \sim 10^{-1}$, i.e. $H_0^2 \sim 10^{-1}\bar{H}\tilde{\bar{H}}$. In turn, it leads to the angular width of the resonance $\delta\theta \sim K_l'/K_l \sim V_l'/V_l \sim m^4\tilde{m}^2 \sim 10^{-3}$. For such a narrow peak the magnetic field H_0 can be used as a tuning parameter. Let us fix in (10.33) $k_x = k_l(H_0)$, i.e. suppose that the incidence angle (10.30) is fixed, $\theta_l =$ const, and corresponds to a maximum of $|S|$ at some level of H_0. Now consider the tuning characteristics of the resonance at changing $H_0 \to H_0 + \Delta H$:

$$|S|^2 = \frac{|S|^2_{\max}}{1 + \gamma^2 |S|^2_{\max} (\Delta H/H_0)^2}, \qquad (10.45)$$

where

$$\gamma = \frac{q_{44}\tilde{\mu}}{\tilde{q}_{44}\mu}. \qquad (10.46)$$

Thus this tuning function $|S|^2 = f(\Delta H)$ is also rather narrow: its "semi-width" δH, defined by the condition $f(\delta H) = \frac{1}{2}f(0)$, is equal

$$\delta H = H_0/\gamma |S|_{\max}. \qquad (10.47)$$

The choice of $|S|_{\max} \sim 10$ corresponds to the spectral width of the resonance $\Delta k_x \sim k_l' \sim 10^{-3}k_l \approx 10^{-3}\omega/v_t$. In order to observe this resonance on a beam, the width of the latter $\Delta k \sim w^{-1}$ should be small enough, being at least comparable with k_l'. Choosing $w \sim 1$cm we come to the estimate $k_l' \gtrsim 1\text{cm}^{-1}$, which determines the limitation for the frequency:

$$\omega \sim 10^3 v_t k_l' \gtrsim 10^8 s^{-1}. \qquad (10.48)$$

Note in conclusion, that we could not seriously consider coefficients of excitation of magnitude much higher than $|S|_{\max} \sim 10$. Indeed, for $|S|_{\max} \sim 10^2$ we would have too narrow peak ($\delta\theta \sim 10^{-6}$) and too high a frequency threshold ($\omega \gtrsim 10^{11}s^{-1}$).

10.4 Waveguide SH acoustic modes in a piezoelectric plate

(a) Statement of the problem

Waveguide acoustic SH eigenmodes in piezoelectric plates and layer-substrate structures have multiple applications in different acousto-electronic devices. So, efficiency of their excitation appears to be a problem of practical importance (see, e.g., the extensive review article [200] with ~400 references).

At a fixed wavenumber k a spectrum of waveguide eigenwaves in a plate contains infinite number of modes differing by tracing speeds v_n ($n = 1, 2, \ldots$). If one face of the piezoelectric plate is metallized and the other one is adjoined to a vacuum half-space, each eigenwave in the plate must be accompanied by a surface wave of electropotential in the vacuum. However in a layer-gap-substrate structure the initial eigenwave related to v_n can be transformed into a leaky mode due to an admixture in the substrate of a bulk wave outgoing from the interface (Fig. 10.7*a*), if v_n belongs to the "supersonic" region for the substrate. For an ordinary weak piezoelectric effect the amplitude of the bulk component and the resulting leakage of the energy from the initial eigenwave must be small. Such a leaky eigenwave is characterized by a complex tracing speed $v_l^{(n)} - iv_l^{(n)\prime}$, where $|v_l^{(n)} - v_n|, v_l^{(n)\prime} \ll v_n$.

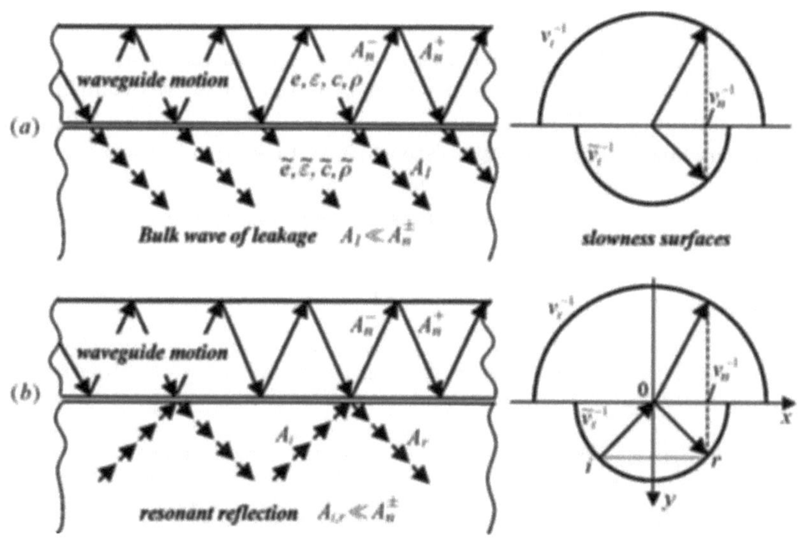

Fig. 10.7. A leaky wave mode (a) and a resonant reflection (b) in the layer-substrate structure.

As was shown in the previous part of this chapter and in papers [68, 69, 192-195] related to a theory of excitation of Rayleigh- and Bleustein-Gulyaev-type surface waves, reflections with incidence angles associated with tracing speeds v close to the real part of the leaky wave velocity have a resonance character. In our case at such reflection the amplitudes of the incident and reflected waves in the substrate must be much less than the amplitude of the waveguide mode in the layer Fig. 10.7*b*). This is a basic idea of the presented consideration.

We consider a layer-gap-substrate structure (Fig. 10.7) with the layer and the substrate made of transverse isotropic piezoelectric materials with the parallel principal axes. The layer faces are mechanically free of tractions, with the external face being electrically closed (metallized) and the internal face electrically open. The thickness of the layer is d and the thickness of the gap is supposed to be infinitely small (at least much less than the wavelength). The following material constants will be involved into our analysis for the layer and the substrate, respectively: the elastic moduli $c_{44} = c_{55} \equiv c$ and $\tilde{c}_{44} = \tilde{c}_{55} \equiv \tilde{c}$, the piezoelectric coefficients $e_{15} = e_{24} \equiv e$ and $\tilde{e}_{15} = \tilde{e}_{24} \equiv \tilde{e}$, and the electric permittivities $\varepsilon_1 = \varepsilon_2 \equiv \varepsilon$ and $\tilde{\varepsilon}_1 = \tilde{\varepsilon}_2 \equiv \tilde{\varepsilon}$. The sagittal plane (the wave vectors plane) is chosen to be orthogonal to the principal axes. In such media the SH bulk waves have the speeds $v_t = \sqrt{\bar{c}/\rho}$ and $\tilde{v}_t = \sqrt{\bar{\tilde{c}}/\tilde{\rho}}$, where $\bar{c} = c + e^2/\varepsilon$ and $\bar{\tilde{c}} = \tilde{c} + \tilde{e}^2/\tilde{\varepsilon}$.

(b) Basic equations and boundary conditions

For the considered SH wave system the displacement fields in the layer ($\mathbf{u}$) and in the substrate ($\tilde{\mathbf{u}}$) are orthogonal to the sagittal plane xy and can be completely described by scalar fields $u_z = u(x, y, t)$ and $\tilde{u}_z = \tilde{u}(x, y, t)$. Being parallel to the interface, these displacements cannot interfere even through an infinitely narrow gap. However, due to the piezoelectric coupling, the acoustic waves in the layer and in the substrate prove to be coupled. Together with the accompanied quasi-static electric fields characterized by the electropotentials $\varphi(x, y, t)$ and $\tilde{\varphi}(x, y, t)$ the studied acoustic field has the form

$$\begin{pmatrix} u(x,y,t) \\ \varphi(x,y,t) \end{pmatrix} = \begin{pmatrix} u(y) \\ \varphi(y) \end{pmatrix} e^{ik(x-vt)}, \quad \begin{pmatrix} \tilde{u}(x,y,t) \\ \tilde{\varphi}(x,y,t) \end{pmatrix} = \begin{pmatrix} \tilde{u}(y) \\ \tilde{\varphi}(y) \end{pmatrix} e^{ik(x-vt)}. \tag{10.49}$$

One can show that $\varphi(y) = (e/\varepsilon)u(y) + \varphi_c(y)$ and $\tilde{\varphi}(y) = (\tilde{e}/\tilde{\varepsilon})\tilde{u}(y) + \tilde{\varphi}_c(y)$, where

$$\begin{aligned} u(y) = A^- e^{-ipky} + A^+ e^{ipky}, \quad \tilde{u}(y) = \tilde{A}_i e^{-i\tilde{p}ky} + \tilde{A}_r e^{i\tilde{p}ky}, \\ \varphi_c(y) = B^- e^{-ky} + B^+ e^{ky}, \quad \tilde{\varphi}_c(y) = \tilde{B}^- e^{-ky}. \end{aligned} \tag{10.50}$$

Substituting (10.49), (10.50) into the standard equations of motion for piezoelectric bodies (see Ch. 1) one obtains

$$p = \sqrt{v^2/v_t^2 - 1}, \quad \tilde{p} = \sqrt{v^2/\tilde{v}_t^2 - 1}. \tag{10.51}$$

Having in mind that in a reflection problem the amplitude of the incident wave ($\tilde{A}_i$) is supposed to be known, there are six unknown amplitudes in (10.50): $A^\pm, B^\pm, \tilde{A}_r$ and $\tilde{B}^-$. They can be found from boundary conditions: all three surfaces are mechanically free, $\sigma_{yz}(-d) = \sigma_{yz}(0) = \tilde{\sigma}_{yz}(0) = 0$, the external surface is metallized, $\varphi(-d) = 0$, and the interface is electrically open, $\varphi(0) = \tilde{\varphi}(0), D_y(0) = \tilde{D}_y(0)$. For brevity we shall not present here either an explicit form of this system, or the details of its analysis. Results will be obtained only for narrow regions $|v - v_n| \ll v_n$ around the eigenspeeds

$$v_n = v_t\sqrt{1 + (\pi n/kd)^2}, \quad n = 1, 2, \ldots \tag{10.52}$$

(c) Results of analysis

As was expected, in the vicinity of v_n the reflection has a resonance character. The coefficients of reflection ($R_n = \tilde{A}_r/\tilde{A}_i$) and of excitation of the SH waveguide n−mode ($K_n = A_n^+/\tilde{A}_i$) have the form

$$R_n = -\frac{v - v_l^{(n)} - iv_l^{(n)\prime}}{v - v_l^{(n)} + iv_l^{(n)\prime}} = e^{i\Phi}, \quad K_n = \frac{v_n''}{v - v_l^{(n)} + iv_l^{(n)\prime}}, \tag{10.53}$$

where $v_l^{(n)} \equiv v_n - \Delta v_n$ and the speed parameters Δv_n, v_n'' and $v_l^{(n)\prime}$ are given by

$$\Delta v_n = \frac{v_t^2}{v_n}\frac{\kappa^2}{kd}\frac{\varepsilon \sinh kd + 2\tilde{\varepsilon}[\cosh kd - (-1)^n]}{\varepsilon \cosh kd + \tilde{\varepsilon}\sinh kd}, \tag{10.54}$$

$$v''_n = -\frac{v_t^2}{v_n}\frac{\kappa^2}{kd}\frac{\varepsilon(\tilde{e}/e)[\cosh kd - (-1)^n]}{\varepsilon \cosh kd + \tilde{\varepsilon}\sinh kd}, \tag{10.55}$$

$$v_l^{(n)\prime} = \frac{v_t^2}{v_n}\frac{\kappa^2\tilde{\kappa}^2}{kd}\frac{\varepsilon\tilde{\varepsilon}}{p_n}\frac{\sinh^2 kd + 2[1 - (-1)^n]}{(\varepsilon \cosh kd + \tilde{\varepsilon}\sinh kd)^2}. \tag{10.56}$$

In (10.54)-(10.56) $\kappa^2 = e^2/\varepsilon\bar{c}$ and $\tilde{\kappa}^2 = \tilde{e}^2/\tilde{\varepsilon}\tilde{\bar{c}}$ are coefficients of electromechanical coupling and $\tilde{p}_n = \tilde{p}(v_n)$. The coefficients κ^2 and $\tilde{\kappa}^2$ are supposed to be small parameters. Therefore, by (10.56), $v_l^{(n)\prime} \ll \Delta v_n, v_n''$. With these inequalities and eqs. (10.53), the reflection coefficient R_n must sharply change its phase Φ at small variations of the tracing speed v close to $v_l^{(n)} \approx v_n$, and the excitation coefficient $K_n \propto \tilde{\kappa}^{-2}$ must be large in the narrow resonant zone $\left|v - v_l^{(n)}\right| \sim v_l^{(n)\prime}$. Let us estimate max$|K_n|$ given by (10.53) at $v = v_l^{(n)}$:

$$\max|K_n| = \left|\frac{v''_n}{v_l^{(n)\prime}}\right| = \frac{\tilde{p}_n\tilde{e}/e}{\tilde{\kappa}^2\tilde{\varepsilon}}\frac{[\cosh kd - (-1)^n](\varepsilon \cosh kd + \tilde{\varepsilon}\sinh kd)}{\sinh^2 kd + 2[1 - (-1)^n]}. \tag{10.57}$$

In particular, for symmetric modes described by even $n = 2m$ ($m = 1, 2, \ldots$):

$$\max|K_{2m}| = \frac{\tilde{p}_{2m}\tilde{e}/e\tilde{\varepsilon}}{2\tilde{\kappa}^2}\left[\varepsilon + 2\tilde{\varepsilon}\tanh(kd/2) + \varepsilon \tanh^2(kd/2)\right]. \tag{10.58}$$

And for antisymmetric modes $n = 2\mathrm{m} + 1$:

$$\max|K_{2m+1}| = \frac{\tilde{p}_{2m+1}\tilde{e}/e\tilde{\varepsilon}}{2\tilde{\kappa}^2}\frac{\varepsilon \cosh kd + \tilde{\varepsilon}\sinh kd}{\cosh^2(kd/2) - \tanh^2(kd/2)}. \tag{10.59}$$

For $\tilde{e}/e \sim 1$, $\tilde{c}/c \sim 1$ and $\tilde{\kappa}^2 \sim 10^{-1}$ one can see that the excitation coefficient in its maximum (10.57)-(10.59) is universally large at any n and *kd*. That is why the considered method appears to be very promising in a practical sense.

(d) Some concluding remarks

Thus, as we have seen, it is possible to excite in the layer any of waveguide SH eigenwaves with large amplitudes by means of reflection of a weak pumping bulk wave in the substrate with the appropriate choice of incidence angles. The shape of peaks of excitation $|K_n(v)|$, arising close to the speeds v_n of the corresponding waveguide modes, is characterized by the height $\max|K_n| \propto \tilde{\kappa}^{-2}$ and by the width $\left|v - v_l^{(n)}\right| \propto \tilde{\kappa}^2$. Of course, there is no problem here with the energy conservation law, because we solved a steady-state problem. It is evident that for a non-stationary approach to the problem the time of establishing the steady-state regime would be the more, the larger is $\max|K_n|$.

On the other hand, the paradoxical result, that the weaker is piezoelectric coupling, the bigger is $\max|K_n|$, does not automatically provide a physical increase in efficiency of excitation. Indeed, as was mentioned, with decreasing $\tilde{\kappa}^2$ the peaks $|K_n(v)|$ simultaneously become more and more narrow. As a result, the efficiency of the excitation will be eliminated when the width of the peak becomes much less than the natural divergence of the acoustic beam, which should replace in any application the above-considered incident plane wave. This shows that in practice one should optimize parameters in order to get maximum efficiency of excitation. Still, as our analysis shows, such optimization is quite possible and the considered method can be efficient.

References

1. Curie P (1908) Oeuvres de Pierre Curie. Societe Francaise de physique, Paris
2. Voigt W (1910) Lehrbuch der Kristallphysik. B.G. Teubner, Leipzig
3. Tiersten HF (1968) Linear Piezoelectric Plate Vibrations. Plenum Press, New York
4. Auld BA (1990) Acoustic Fields and Waves in Solids, 2nd Ed., Vol. I. Krieger, Malabar Florida
5. Dieulesaint E, Royer D (1980) Elastic Waves in Solids, Application to Signal Processing.J. Willey, New York
6. Nowacki W (1975) Dynamic Problems of Thermoelasticity. Noordhoff, Leiden and P.W.N., Warsaw
7. Nelson DF (1979) Electric, Optic and Acoustic Interactions in Dielectrics. J. Willey, New York
8. Ristic VM (1983) Principles of Acoustic Devices. J. Willey, New York
9. Eringen AC, Maugin GA (1989) Electrodynamics of Continua. Foundations and Solid Media, Vol. I. Springer, New York
10. Truesdell C, Toupin R A (1960) The Classical Field Theories. In: Fluege S. (ed) Encyclopedia of Physics, Vol. III/1. Springer, Berlin
11. Hutter K, van de Ven AAF (1978) Field Matter Interactions in Thermoelastic Solids. Lecture Notes in Physics,#88 Springer, Berlin
12. Eringen AC (1980) Mechanics of Continua. Krieger, New York
13. Maugin GA (1985) Nonlinear Electromechanical Effects and Applications. Series in Theoretical and Applied Mechanics. Hsieh RKT (ed) World Scientific, Singapore
14. Maugin GA (1988) Continuum Mechanics of Electromagnetic Solids. North-Holland, Amsterdam
15. Nowacki W (1983) Efecty Electro-Magnetyczne w Stalych Cialach Odksztalcalnych. P.W.N., Warszawa
16. Rymarz Cz (1993) Mechanika Osrodkow Ciaglych. P.W.N., Warszawa
17. Chung MY, Ting TCT (1995) J Appl Mech 62: 423-428
18. Chung MY, Ting TCT (1996) Int J Solids Struct 33: 3343-3361
19. Kirchner HOK, Alshits VI (1996) Phil Mag A 74: 861-885
20. Dunn ML, Wienecke HA (1996) Int. J. Solids Struct. 33: 4571-4581
21. Haojiang D, Chenbuo, Liangjian (1997) Int J Solids Struct 34: 3041-3057
22. Haojiang D, Yuwei C, Fenglin G (1999) Int J Solids Struct 36: 2613-2631

23. Qin Q-H (1999) Arch Appl Mech 69:133-144
24. Lu Pin, Tan MJ, Liew KM (2000) Int. J. Solids Struct. 37: 1065-1078
25. Mikata Y (2000) Int J Engng Sci 38: 605-641
26. Pan E, Yuan FG (2000) Int J Engng Sci 38: 1939-1960
27. Nowacki JP, Alshits VI, Radowicz A (2000) Int J Appl Electromagnetics & Mechanics 12: 177-202
28. Nowacki JP, Alshits VI, Radowicz A (2002) J Tech Phys 43: 133-153
29. Nowacki JP, Alshits VI, Radowicz A (2002) Int J Engng Sci 40: 2057-2076
30. Nowacki JP, Alshits VI, Radowicz A (2001/2002) Int J Appl Electromagnetics & Mechanics 14: 429-433
31. Merten L (1964) Phys Kondens Mater 2: 53-65
32. Merten L (1964) Phys Kondens Mater 2: 66-79
33. Merten L (1964) Z. Naturforsch 192: 788-
34. Merten L (1964) Z. Naturforsch 192: 1161-
35. Kosevich AM, Pastur LA, Feldman EP (1968) Sov Phys-Crystallogr 12: 797-801
36. Saada G (1971) Phys Stat Sol(b) 44: 717-731
37. Faivre G, Saada G (1972) Phys Stat Sol (b) 52: 127-140
38. Eshelby JD, Read WT, Shockley W (1953) Acta Metall. 1: 251-....
39. Smirnova IS (1974) Sov Phys–Solid State 15: 1543-1546
40. Huston AR, Walker LR (1979) J Appl Phys 50: 6247
41. Stroh AN (1958) Phil Mag 3: 625-646
42. Stroh AN (1962) J Math Phys 41: 77-103
43. Malen K (1971) Phys Stat Sol (b) 44: 661-672
44. Teodosiu C (1982) Elastic Models of Crystal Defects. Springer, Berlin
45. Barnett DM, Lothe J (1973) Phys Norv 7: 13-
46. Bacon D, Barnett DM, Scattergood RO (1979) Prog Mater Sci 23: 51-262
47. Indenbom VL, Orlov SS (1967) Sov Phys-JETP 6: 274-
48. Lothe J (1967) Phys Norv 2: 153-
49. Lothe J (1970) In: Simmons JA, De Wit R, Bullough R (eds) Fundamental Aspects of Dislocation Theory, Vol. 317, Spec Publ. Nat. Bur. Stand., Washington, p. 11
50. Indenbom VL, Lothe J (eds) (1992) Elastic Strain Fields and Dislocation Mobility. North-Holland, Amsterdam
51. Barnett DM, Lothe J (1974) J Phys F 4: 1618-
52. Barnett DM (1986) In: Subramyan K, Imam MA (eds) Structure and Deformation of Boundaries. Metallurgical Society, London, pp. 31-50
53. Wu K-C, Chiu Y-T (1995) Int J Solids Structure 32: 543-552
54. Alshits VI, Kirchner HOK (1995) Phil Mag A72: 1431-1470
55. Nabarro FRN, Kostlan EJ (1978) J Appl Phys 49: 5445-
56. Moss WC, Hoover WG (1978) J Appl Phys 49: 5449-
57. Stagni L, Lizzio R (1986) Int J Engng Sci 24: 471-
58. Gutkin MYu, Romanov AE (1991) Phys Stat Sol A 125: 107-
59. Gutkin MYu, Romanov AE (1992) Phys Stat Sol A 129: 363-
60. Jergan E (1970) Diploma Thesis, University of Trondheim, Norwegian Institute of Technology, Trondheim
61. Barnett DM, Lothe J (1975) Phys Stat Sol (b) 67: 105-111
62. Alshits VI, Darinskii AN, Lothe J (1992) Wave Motion, 16: 265-283
63. Lothe J, Barnett DM (1976) J Appl Phys 47: 1799-1807
64. Lothe J, Barnett DM (1977) Physica Norvegica 8: 239

65. Alshits VI, Darinskii AN, Lothe J, Lyubimov VN (1994) Wave Motion 19: 113-123
66. Bleustein L (1968) Appl Phys Lett 13: 412-413
67. Gulyaev YuV (1969) JETP Letters 9: 63-65
68. Nowacki JP, Alshits VI, Lyubimov VN (2000) Int J Appl Electromagnetics & Mechanics 11: 223-232
69. Nowacki JP, Alshits VI, Lyubimov VN, Radowicz A (2004) Int. J. Appl. Electromagnetics and Mechanics, 19: 685-688
70. Minagava S (1984) Phys Stat Sol (b) 124: 565-571
71. Minagawa S, Shintani K (1985) Phil Mag A 51: 277-285
72. Minagawa S, Shintani K (1987) Phil Mag A 56: 343-352
73. Minagawa S, Hsieh RKT, Nowacki JP, Shintani K (1988) Int J Engng Sci 26: 1169-1175
74. Cosserat E et F (1909) Theorie des corps deformables. A. Hermann et Fils, Paris
75. Aero EL, Kuvshinskii EV (1961) Sov Phys- Solid State 2: 1272-1281
76. Kuvshinskii EV, Aero EL (1964) Sov Phys- Solid State 5: 1892-1897
77. Aero EL, Kuvshinskii EV (1965) Sov Phys- Solid State 6: 2141-2148
78. Eringen AC (1965) Mechanics of micromorphic materials. In: Proc. 11th Int. Cong. Appl. Mech. Springer, Berlin, pp. 131-138
79. Eringen AC (1966) J Math Mech 15: 909-924
80. Green AE, Rivlin RS (1964) Arch Rat Mech Ann 17: 113-147
81. Mindlin RD (1964) Arch Rat Mech Ann 16: 51-78
82. Mindlin RD (1965) Int J Solids Struct 1: 265-271
83. Neuber H (1965) On the general solution of linear-elastic problems in isotropic and anisotropic Cosserat continua. In: Proc. 11th Int. Cong. Appl. Mech. Springer, Berlin, pp.153-158
84. Sandru N (1966) Int J Engng Sci 4: 81-94
85. Schaefer H (1967) ZAMN 47: 485-498
86. Cowin SC (1970) Int J Solids Struct 6: 389-398
87. Cowin SC (1970) Zeit f Ang Math u Physik 21: 494
88. Nowacki W (1972) Theory of micropolar elasticity. Springer, Vienna New York
89. Nowacki W (1973) Bull Acad Polon Sci, Ser Sci Tech 21: 295-
90. Nowacki JP, Nowacki W (1977) Bull Acad Polon Sci, Ser Sci Tech 25: 151-159
91. Nowacki JP (1977) Bull Acad Polon Sci, Ser Sci Tech 25: 207-216
92. Nowacki JP (1977) Bull Acad Polon Sci, Ser Sci Tech 25: ???
93. Nowacki JP (1985) J Tech Phys 26: 315-323
94. Nowacki JP, Glockner PG (1980) Int J Solids Structures 16: 337-345
95. Nowacki JP, Shindo Y (1992) Variational approach to micropolar magnetoelasticity. In: Enokizono M, Nowacki JP (eds) Electromagnetic Phenomena and Computational Techniques. Elsevier, Amsterdam, pp. 155-165
96. De Wit R (1970) Linear theory of static disclinations. In: Simmons JA, De Wit R, Bullough R (eds) Fundamental Aspects of Dislocation Theory, Vol. 317, Spec Publ Nat Bur Stand, Washington
97. Antony KH (1970) Arch For Rat Mech Anal 39: 43-
98. Claus WD, Jr, Eringen AC (1969) Three dislocation concepts and micromorphic mechanics. In: Developments in Mechanics, Vol. 6, Proc. Of the 12th Midwestern Mechanics Conf.
99. Nowacki JP (1973) Bull Acad Polon Sci, Ser Sci Tech 21: 585-592

100. Nowacki JP (1974) Bull Acad Polon Sci, Ser Sci Tech 22: 379-387
101. Nowacki JP (1977) Arch Mechanics 29: 531-545
102. Nowacki JP (1978) Bull Acad Polon Sci, Ser Sci Tech 26: 93-100
103. Mindlin RD (1972) J Elasticity 2: 217-282
104. Mindlin RD (1968) Int J Solids Struct 4: 637-642
105. Toupin RA (1956) J Rat Mech Anal 5: 849-915
106. Suhubi ES (1969) Int J Engng Sci 7: 993-997
107. Ludwig W (1967) Recent developments in lattice theory. Springer, Berlin
108. Mead CA (1962) Phys Rev 128: 2088-
109. Pine AS (1970) Phys. Rev. B 2: 2049-.
110. Tosi MP (1964) Solid State Physics, V. 16, Academic Press, New York, p. 92
111. Germer LH, Macrae AU, Hartman CD (1961) J. Appl. Phys. 32: 2432-
112. Nowacki JP, Hsieh RKT (1986) Int J Engng Sci 24: 1655-1666
113. Nowacki JP (1987) Dislocations in dielectrics with polarization gradient. In: Yamamoto Y, Miya K (eds) Electromechanical Interactions in Deformable Solids and Structures. North Holland, Amsterdam, pp. 389-393
114. Nowacki JP (1988) J Techn Phys 29: 99-103
115. Nowacki JP (1989) Micromechanics of materials with piezoelectric effect, dislocations in dielectrics. In: Miya K (ed) Applied Electromagnetics in Materials. Pergamon, Oxford, pp. 295-304
116. Nowacki JP (1990) Dislocations dynamics in elastic dielectrics. Force acting on dislocation. In: Hsieh RKT (ed) Mechanical Modellings of New Electromagnetic Materials. Elsevier, Amsterdam, pp. 289-295
117. Chowdhury KL, Glockner PG (1977) Int J Solids Structures 13: 1173-1182
118. Chowdhury KL, Epstein M, Glockner PG (1978) On the thermodynamics of nonlinear elastic dielectrics, Report No 119, Department of Mechanical Engineering, University of Calgary
119. Nowacki JP, Glockner PG (1983) Constitutive equations of the thermoelastic dielectrics. In: Brulin O, Hsieh RKT (eds) New Problems in Mechanics of Continua. University of Waterloo Press, pp. 17-27
120. Nowacki JP, Trimarco C (1989) Note on thermal inclusion in elastic dielectric material. Atti Sem Mat Fis, Univ Modena 38: 371-378
121. Nowacki JP (2004) Archive Mech 56: 33-57
122. Mindlin RD, Toupin RA (1971) Int J Solids Structures 7: 1219-
123. Nowacki JP, Glockner PG (1979) Int J Solids Structures 15: 183-191
124. Nowacki JP, Glockner PG (1981) Int J Engng Sci 19: 603-613
125. Majorkowska-Knap FK (1981) Bull Ac Pol: Tech 29: 387-
126. Nowacki JP (1982) J Thermal Stresses 5: 183-194
127. Nowacki JP (1984) Some dynamical problems of elastic dielectrics with polarization gradient.In: Maugin GA (ed) The Mechanical Behavior of Electromagnetic Solid Continua. North Holland, Amsterdam, p. 223
128. Nowacki JP, Trimarco C (1988) A dynamic problem for a piezoelectric medium. In: Parker DF, Maugin GA (eds) Recent developments in surface acoustic waves. Springer, pp. 193-200
129. Nowacki JP (1989) Electromechanical waves in elastic dielectrics. In: McCarthy MF, Hayes MA (eds) Elastic Waves Propagation. North Holland, Amsterdam, pp. 607-613
130. Nye JF (1957) Physical Properties of Crystals. Clarendon, Oxford

131. Sirotin YuI, Shaskolskaya MP (1975) Fundamentals of Crystal Physics (in Russian). Nauka, Moscow; translation into English: (1982) MIR Publishers, Moscow
132. Kröner E (1956) Kontinuumstheorie der Verzetzungen und Eigenspannungen. Springer, Berlin
133. Nowacki W (1965) Proc of Vibr Probl 6: 3
134. Berlincourt DA, Curran DR, Jaffe H (1964) Piezoelectric and piezomagnetic materials and their function in transducers. In: Mason WP (ed) Physical Acoustics, Vol. 1A, Ch. 3. Academic, New York
135. Birss R (1966) Symmetry and Magnetism, Chs 4 and 5. North Holland, Amsterdam
136. Auld BA (1971) Magnetostatic and magnetoelastic wave propagation in solids. In: Wolfe R (ed) Applied Solid State Science, Vol 2. Academic, New York, pp. 2-103
137. Burgers JM (1939) Proc K Ned Akad Wet 42: 293
138. Nabarro FRN (1952) Advances in Physics, 1: 269
139. Indenbom VL (1992) Dislocations and internal stresses, Chap.1 in the book [50], pp. 1-174
140. De Wit R (1960) Solid St Phys 10: 249-292
141. Kosevich AM (1979) Crystal dislocations and the theory of elasticity. In: Nabarro FRN (ed) Dislocations in Solids, Vol 1. North Holland, Amsterdam, pp. 33-141
142. Read WT, Jr. (1953) Dislocations in Crystals. McGraw-Hill, New York
143. Cottrell AH (1953) Dislocations and Plastic Flow in Crystals. Oxford University Press, Oxford
144. Friedel J (1956) Les Dislocations. Gauthiers-Villars, Paris
145. Nabarro FRN (1967) Theory of Crystal Dislocations. Oxford University Press, Oxford
146. Hirth J, Lothe J (1982) Theory of Dislocations, 2nd ed. Wiley, New York
147. Synge J.L. (1957) The Hypercircle in Mathematical Physics: A Method for the Approximate Solution of Boundary-Value Problems. Cambridge University Press, Cambridge
148. Barnett DM (1972) Phys Stat Sol (b) 49: 741-...
149. Indenbom VL, Orlov SS (1968) Proc. Kharkov Conf. on Dislocation Dynamics. Physical-Technical Institute of Low Temperatures, Academy of Sciences of the USSR, Kharkov, p.406
150. Lothe J (1992) Dislocations in anisotropic media, Chap 4 in the book [50], pp. 269-328
151. Mura T (1963) Phil Mag 8: 843-857
152. Brown LM (1967) Phil Mag 15: 363
153. Willis JR (1970) Phil Mag 21: 931-949
154. Steeds JW, Willis JR (1979) in: Nabarro FRN (ed) Dislocations in Solids, Vol. 1. North Holland, Amsterdam, pp. 145-165
155. Eshelby JD (1956) Solid St Phys 3: 79-144
156. Gavazza SD, Barnett DM (1975) Scripta Metall 9: 1263
157. Peach M, Koehler JS (1950) Phys Rev 80: 434
158. Barnett DM, Swanger LA (1971) Phys Stat Sol (b) 48: 419
159. Indenbom VL, Alshits VI (1974) Phys Stat Sol (b) 63: K125
160. Barnett DM, Lothe J (1973) Phys Norv 7: 13

161. Asaro RJ, Hirth JP, Barnett DM, Lothe J (1973) Phys Stat Sol (b) 60: 26
162. Minagawa S, Oshiro T (1987) Int J Engng Sci 25: 405-
163. Emaletdinov AK (1977) Sov Phys - Physics of Metals 44: 984-989
164. Basinski LJ (1960) Phys Austr 13: 354
165. Klyavin OV, Nikiforov AV (1973) Izv. AN SSSR, ser. fiz. 37: 2409
166. Govorkov VG, Chistyakov IG, Sizova NL, Gorina II, Petuchov BV, Akchurin MSh (1977) Sov Phys - Doklady 234: 1067-1069
167. Sizova NL, Gorina II, Chistyakov IG (1980) Mol Cryst Liq Cryst 59: 63-70
168. Sizova NL, Natsik VD, Gorina II, Regel VR (1983) Phys Stat Sol (a) 76: 447-451
169. Urusovskaya AA (1968) Sov Phys - Uspekhi 96: 39-
170. Witworth RW (1975) Adv Phys 24: 203-
171. Stepanov AV (1933) Zs Phys 81: 560-
172. Tyapunina HA, Belozerova EP (1988) Sov Phys - Uspekhi 156: 683-
173. Mindlin RD (1973) Int J Solids Struct 9: 233-235
174. Mindlin RD (1974) Int J Solids Struct 10: 1307-1314
175. Chowdhury KL, Glockner PG (1977) Int J Engng Sci 15: 481-493
176. Chowdhury KL, Glockner PG (1979) Bull Acad Polon Sci, Ser Sci Tech 27: 5/6
177. Schwarz J (1969) Int J Solids Struct 5: 1209-121?
178. Askar A, Lee PCY, Carmark AS (1971) Int J Solids Struct 7: 523-5??
179. Gou PF (1971) Int J Solids Struct 7: 1467-147?
180. Ingebrigtsen KA (1969) J Appl Phys 40: 2681
181. Ting TCT (1996) Anisotropic elasticity. Oxford University Press, Oxford
182. Kröner E (1980) Physics of Defects. In: Balian R *et al* (eds) Les Houches session XXXV, Course 3. North Holland, Amsterdam
183. Gulyaev YuV, Kuzavko YuA, Oleinik IN, Shavrov VG (1984) Sov Phys - JETP 60: 674-676
184. Kosevich YuA (1985) Sov Phys - Solid State 27: 193-195
185. Kaganov MI, Kosevich YuA (1986) Poverkhn, Fiz Khim Mekh (USSR) 5: 148-154
186. Maugin GA (1988) Shear horizontal surface waves on solids. In: Parker DF, Maugin GA (eds) Recent Developments in Surface Acoustic Waves. Springer, Berlin
187. Balakirev MK, Gorchakov AV (1986) Poverkhnost' 5: 80
188. Bakhturin MP (1989) Sov Phys - Acoust 35: 990
189. Chernozatonskii LA, Ermoshin SN, Gramotnev DK (1991) Phys Lett A 154: 465
190. Alshits VI, Darinskii AN, Shuvalov AL (1993) Crystallography Reports 38: 22-43
191. Alshits VI, Lyubimov VN (1995) Phys Solid State 37: 1014-1020
192. Alshits VI, Lothe J (1981) Wave Motion 3: 297-310
193. Alshits VI, Gorkunova AS, Lyubimov VN, Nowacki JP (1999) Crystallography Reports, 44: 592-602
194. Alshits VI, Gorkunova AS, Lyubimov VN, Kotowski RK (1999) Crystallography Reports, 44: 799-805
195. Alshits VI, Gorkunova AS, Lyubimov VN, Gierulski W, Radowicz A, Kotowski RK (1999) Methods of resonant excitation of surface waves in crystals. In: Maruszewski BT, Muschik W, Radowicz A (eds) Proc. of the International Symposium on Trends in Continuum Physics (TRECOP '88). World Scientific, Singapore, pp. 28-34

196. Brekhovskikh LM (1973) Waves in Layered Media [in Russian]. Nauka, Moscow
197. Schoch A (1950) Ergb Exact Naturwiss 23: 127
198. Bertoni HL, Tamir T (1973) J Appl Phys 44: 157
199. Alshits VI, Darinskii AN, Kotowski RK, Shuvalov AL (1988) Sov Phys - Crystallogr 33: 318-325
200. Chimenti DE (1997) Appl Mech Rev 50: 247-284

MIX
Papier aus verantwortungsvollen Quellen
Paper from responsible sources
FSC® C105338

If you have any concerns about our products,
you can contact us on
ProductSafety@springernature.com

In case Publisher is established outside the EU,
the EU authorized representative is:
Springer Nature Customer Service Center GmbH
Europaplatz 3, 69115 Heidelberg, Germany

Printed by Libri Plureos GmbH
in Hamburg, Germany